Time to Quarry: The Archaeology of Stone Procurement in Northwestern New South Wales, Australia

Trudy Doelman

BAR International Series 1801
2008

Published in 2016 by
BAR Publishing, Oxford

BAR International Series 1801

Time to Quarry: The Archaeology of Stone Procurement in Northwestern New South Wales, Australia

ISBN 978 1 40730 0288 1

BAR Publishing is the trading name of British Archaeological Reports (Oxford) Ltd. British Archaeological Reports was first incorporated in 1974 to publish the BAR Series, International and British. In 1992 Hadrian Books Ltd became part of the BAR group. This volume was originally published by Archaeopress in conjunction with British Archaeological Reports (Oxford) Ltd / Hadrian Books Ltd, the Series principal publisher, in 2008. This present volume is published by BAR Publishing, 2016.

Printed in England

BAR
PUBLISHING

BAR titles are available from:

BAR Publishing
122 Banbury Rd, Oxford, OX2 7BP, UK
EMAIL info@barpublishing.com
PHONE +44 (0)1865 310431
FAX +44 (0)1865 316916
www.barpublishing.com

Preface

This research was carried out in fulfilment of a Ph.D. at the School of Archaeology, La Trobe University, Melbourne, Australia.

2002 Time to Quarry: The Archaeology of Stone Procurement in Northwestern New South Wales. Ph.D. thesis, School of Archaeology, La Trobe University, Melbourne.

A portion of this study is published elsewhere:

2005 Doelman, T. Quarrying time: the influence of tool manufacture, use and discard on the formation and composition of a quarry assemblage. *Lithic Technology* 30 (1): 13-28.

2005 Doelman, T. Patterns among the masses: an approach to the analysis of a quarry assemblage. *Archaeology in Oceania* 40:50-61.

2001 Doelman, T., J. Webb and M. Domanski. Source to discard: patterns of lithic raw material procurement and use in Sturt National Park, northwestern New South Wales. *Archaeology in Oceania* 36:15-33.

Abstract

The quarry has been considered a cornerstone in understanding lithic production systems. However, the methodological problems associated with the investigation of a quarry assemblage often leads to inadequate recording. The lack of detailed quarry research in Australia focusing on non-axe quarries has meant that they are poorly understood and for this reason a plethora of potentially valuable research regarding the role of the quarry in the organisation of lithic technology is virtually absent. There is a real need to develop quarry studies in Australia and worldwide. It is hoped that this study aides in the expansion of quarry research by providing a sound methodological and analytical approach to the study of quarry assemblages.

A detailed technological and spatial analysis of quarries and occupations sites was used to determine the organisational strategies used to acquire and reduce the stone resources available in the arid zone margin of New South Wales, Australia, and identify the reasons why these particular strategies were employed during the late Holocene. Comparisons are made between quarried and non-quarried stone to identify their 'role' in the organisation of lithic technology.

The prehistoric acquisition of stone from the quarries was goal driven and reflects different problem-solving approaches to the organisation of lithic technology. While planning, standardisation and an investment of time and labour are integral parts of quarrying behaviour, these do not occur in the procurement of stone from non-quarried stone resources. The tools manufactured from this latter stone source represent a situational response to an immediate demand.

The theoretical framework incorporates aspects of non-site distributional archaeology. The individual artefact is the basic methodological and theoretical building block from which greater scales of variation in the distribution and composition of the archaeological record can be examined. This examination uses the concept of 'risk' as the heuristic device with which to explore the costs and benefits of employing different technological strategies. Hence the form of an artefact, its position in space and its time in the cultural system, are the key components of this study. By using a combination of these approaches it is possible to identify not only the many factors that contribute to the formation and distribution of stone resources but also the ways Aboriginal people organised their stone technology during the late Holocene.

LITHIC PROCUREMENT, NON-SITE DISTRIBUTIONAL ARCHAEOLOGY, SCALES OF TIME AND SPACE, TECHNOLOGICAL ORGANISATION, ARID AUSTRALIA

Acknowledgments

I am grateful to so many people for their assistance and support during this research. This project was undertaken with the permission and backing of the Tibooburra Aboriginal Lands Council and the Wangkumara Tribal Lands Council. Special thanks to the Tibooburra Aboriginal Land Council for all assistance. I am indebted to the Ebsworth family for all their good humour, support and encouragement. Many sincere thanks to the people of Tibooburra for their warmth and friendship during my numerous visits to the area. A special thanks to the staff at the Tibooburra NSW Parks and Wildlife headquarters for all their logistical support to the project. A vote of thanks to all the numerous volunteer fieldworkers for their long hours – this project would not have happened without you. Dr John Webb and Mark Domanski La Trobe University helped to establish the geological background and thanks to Dr Patricia Fanning for the geomorphological surveying of Quarry 35.

This research was funded by the Humanities Grants program at La Trobe University and supported by ARC collaborative grants awarded to Associate Professor Simon Holdaway, Dr Dan Witter and Dr Patricia Fanning. A special thanks to Simon for access to the Stud 1 and 2 data. Dr Dan Witter was a fount of information and endlessly supportive. I am also indebted to those that read and commented on this research; Dr Nicola Stern, Associate Professor Simon Holdaway and Dr Phillip Edwards. I am especially grateful to Dr Peter White who read the final draft.

I would like to thank my extended family here in Australia and back in New Zealand. I am very fortunate to have all the love and encouragement from my husband, Jamie and our beautiful daughter Lily.

Table of Contents

List of Figures

List of Tables

Chapter 1

Introduction

The Mt Wood Ranges located in the arid zone of northwest New South Wales are rich in stone resources. Two sources of stone, massive silcrete outcrops and pavements of sub-spherical cobbles of gibber dominate this region. This study identifies the types of technological strategies used by the late-Holocene Aboriginal inhabitants to manage their stone technology. This aim was achieved by characterising the flaking practices used to reduce the stone at two quarried outcrops and three lithic scatters, representing areas where the two types of stone resources are available. The results are used to examine the factors that influenced the distribution, formation and composition of these assemblages and ultimately establish the 'role' that different stone sources played in the organisation of lithic technology. This study is contextualised within a multi-scaled framework which encapsulates different time processes, from the manufacture individual artefact to the creation of archaeological pattern within the wider landscape.

The study of stone resources has long been acknowledged as central to understanding all aspects of stone-production systems yet there is a paucity of detailed studies focusing on quarries in Australia (Hiscock and Mitchell 1993:72-73). The lack of quarry research is somewhat surprising since stone forms the most durable element and the bulk of the archaeological record in Australia. There is no underestimating the problems of accurately recording and analysing a quarry assemblage. Most quarries typically have an imposing volume of material forming a "shattered, overlapping, sometimes shallow, nondiagnostic, undatable, unattractive, redundant...record" (Ericson 1984:2). The challenge of creating a sound sampling methodology and interpreting the results to identify meaningful patterns regarding the organisation and technology of the prehistoric inhabitants of an area is, needless to say, a daunting scenario. Herein are the specific goals of this study. I hope that this approach contributes to the development of appropriate field and analytical methodologies and encourages further quarry research in Australia and worldwide.

The database for this study includes the in-depth technological and spatial analysis of two quarries (Quarry 35 and Quarry 27) and three lithic scatters located within the Mt Wood Ranges. The data obtained from two of the lithic scatters (Stud 1 and Stud 2) form part of a larger database collected by the Western New South Wales Archaeological Project (WNSWAP) directed by Dr. Simon Holdaway (Holdaway *et al.* 2000a; 2004). The analysis of the lithic scatters, known as landscape sampling locations, has a two-fold objective; they represent the reduction of non-quarried cobbles obtained from nearby pavements of gibber and highlight the form that quarried material was discarded away from the source.

While this study does not trace the entire distribution of stone away from the quarries across the landscape, the detailed investigation of a relatively small area can provide interesting insights into the *ways* stone was procured, reduced and discarded in a lithic resource rich environment and establish the reasons *why* stone resources were managed in such a manner.

This study incorporates aspects of non-site distributional archaeology in which the individual artefact forms the empirical unit of data collection, analysis and interpretation. Each artefact, from the 76 653 artefacts which form the database for this study, has a unique position in space (in three-dimensional coordinates), an associated geomophological context, linked to both metrical and descriptive attributes of each artefact in a Geographic Information System (GIS). By undertaking such a detailed spatial and technological investigation the scale of observation and analysis can be expanded from the individual artefact to encompass the entire study area. For example, not only are the spatial and technological relationships between artefacts within an assemblage compared, but also how the landscape features, such as a quarried outcrop, influenced the character and distribution of the assemblage. In this way many factors that operate at different scales of space and time to form the archaeological record can be identified. Only once these factors are identified can the concept of 'risk' can be applied as a heuristic device to explore the 'role' of the stone resources in the organisation of lithic technology.

The Theory and the Method

The interpretative and analytical framework was drawn from studies of site formation processes, non-site distributional archaeology and an approach, derived from evolutionary ecology, referred to as technological organisation.

By using Schiffer's (1972, 1976) behavioural chain analysis as a heuristic device the "flow" of artefacts from two different stone sources moving from the systemic context to the archaeological context can be determined. The life history of an artefact passes through many stages from its procurement, manufacture and use to its discard. During the systemic life of an artefact may be transported, retouched, resharpened and recycled before being discarded throughout the landscape to become part of the archaeological record. Gould (1977) recognised that artefacts made from quarried and non-quarried stone sources have different life histories. While it is impossible to trace the life history of every artefact, cumulative patterns in the characteristics and distribution of artefacts acquired from the different stone sources are examined and compared in the study area.

Spatial variation in the archaeological record can be observed at ever increasing scales of observation and resolution (Steffen *et al.* 1998:135). At each scale of observation, different research questions direct the analysis and its interpretation. Within a non-site distributional approach the single empirical unit of analysis is the individual artefact which has a unique position in space and a unique set of characteristics (e.g. Ebert 1992). Hence, the individual artefact represents the finest scale of resolution (Steffen *et al.* 1998). By removing the 'site' boundary the challenge has been to find pattern and meaning in the archaeological record. Using the single artefact as the unit of analysis frees interpretation by allowing variation in the distribution of artefacts to be observed and examined at any scale. Using hierarchical relationships, and expanding the 'view' from the individual artefact, variation in the distribution of artefacts can be observed within locations, between locations or across a landscape. The advantage of using a Geographic Information System (GIS) is in allowing any scale of spatial variation to be compared and interpreted by varying the 'view' of the landscape.

Studies of technological organisation aim to identify the many social and environmental variables that influence the behavioural strategies used by people to adapt to an environment (Carr 1994; Nelson 1991; Shott 1986). The principals of natural selection and adaptation, founded in evolutionary ecology, govern an organisation of technology approach to lithic analysis (Carr 1994:2). Technology is viewed as an adaptive response to the social and physical environment and reflects how people in the past undertook different problem-solving strategies to survive (Binford 1977, 1978a, 1978b, 1979; Torrence 1989a, 1989b; Nelson 1991). Studying the distribution, characteristics and composition of stone artefact assemblages can identify these strategies and enables the researcher to distinguish the reasons why stone was acquired, reduced, transported, used and discarded in particular ways.

The methodological approach used in evolutionary ecology to identify the process of adaptation is the costs/benefit analysis. In this approach a currency that equates to reproductive success, such as energy or time, is used as a way of comparing the costs and benefits of undertaking a particular behaviour (e.g. Smith 1983; Bettinger 1980). This method creates optimisation models which compare the optimal or ideal strategy with the observed strategy to identify evolutionary processes. In archaeology, the problems associated with measuring time or energy from the archaeological record has led some researchers to advocate risk-avoidance strategies as being an appropriate currency to measure how technology is adapted to suit particular goals (e.g Torrence 1989b; Elston 1990; Bamforth and Bleed 1997). Risk equates to the probability that the costs of an activity outweigh the benefits of that activity (Elston 1990:154). Technology can reduce the risk of failure to meet subsistence needs and increases the chance of survival (Torrence 1989b; Bamforth and Bleed 1997:115). The severity of risk influences the types of strategies people used to obtain stone, reduce it and design of tools. For example, if the risk of failing to obtain adequate resources is high, then more time and labour would be invested in the manufacture of tools that were designed to be reliable and maintainable which enabled greater long-term benefits (Torrence 1989b).

Information about the use of risk-avoidance strategies is sought for this study by undertaking a detailed investigation of how stone was reduced and discarded at two quarries and three lithic scatters. Exploring the costs and benefits of particular strategies is used as a heuristic device to compare how stone from different types of stone resources was procured and why it was procured in such a manner. In turn, the results are used to make inferences about the 'role' of the stone sources in the organisation of technology.

The Development of Quarry Studies in Australia

If, as many quarry researchers have argued (e.g. Ericson 1984; Torrence 1984, 1986; Arnold 1985, 1987a, 1987b; Reher 1991; Root 1992), the quarry is the key to understanding manufacturing technology, and the organisation of production and stone distribution, then the first step in a lithic analysis should be to identify the factors that influenced the formation and composition of a quarry assemblage. Similarly, it is vital to the study of technological organisation to define the factors that contributed to types of technological strategies employed. Past research has recognised a variety of factors, such as the quality and abundance of stone sources, the degree of mobility and the settlement pattern, that have influenced technological organisation (Binford 1973, 1977, 1978a, 1978b, 1979, 1980; Bleed 1986; Parry and Kelly 1987; Andrefsky 1994a, 1994b; Bamforth 1986; Torrence 1989b; Elston 1990; Nelson 1991; Bamforth and Bleed 1997). Hence, it is important to determine which factors influenced the strategies used to manage their stone resources.

Previous studies in Australia highlight that raw material was a primary influencing factor that dictated the distribution and composition of the archaeological remains (eg. Hiscock 1986b, 1988, 1996; McNiven 1994; Gould 1977, 1980). This study aims to move beyond single-factor determinations of the archaeological record and argues for a multiple, interrelated causality operating at different scales of space and time that ultimately influenced the form and distribution of the artefacts in the study area. Hence, how short and long-term time processes, discussed by Bailey (1981) and Fletcher (1992), have influenced the formation, distribution and composition of the archaeological record within the study area is also explored within this study.

This study further explores the many factors identified in the literature that may influence the ways stone was discarded in the landscape. At a regional scale is important to understand the geological context, the types of raw materials available and their different qualities, and where and how they outcrop (e.g. Bamforth 1992;

Andrefsky 1994a, 1994). These factors influenced past decisions about which outcrop to quarry. At each quarry various features may affect where people worked and discarded their cores and flaking debris through time. The most obvious feature is the outcrop. On or near the outcrop material testing and initial core reduction tends to take place (e.g Root 1992). As distance increases from the outcrop cores undergo greater reduction until they are spent and discarded (e.g. Byrne 1980; Hiscock 1986b, 1988). Furthermore, the topography, landform surface and the location of water may also influence the distribution of the cores and flaking debris at the quarries (e.g. Torrence 1986; Witter 1992).

The acquisition of stone from the quarries is often goal driven, which involved both planning and the anticipation of risk. Hence, planning, standardisation and an investment of time and energy are typically integral parts of quarrying behaviour. At quarries specific products, designed for particular purposes, were manufactured and removed. These products have a different set of design variables, involving different amount of times and labour invested, resulting from different problem-solving approaches to lithic organization (Nelson 1991). For example, manufacturing blades allowed a technological strategy that was more efficient, transportable, maintainable and anticipated future needs thereby reducing the 'risk' of not having the appropriate tool at the required time. By using the concept of 'risk' as a heuristic device and contrasting the reduction strategies employed at Quarry 35, Quarry 27 with the reduction of non-quarried stone at the landscape sampling locations the different 'roles' played by these resources in organisation of lithic technology.

Underlying Themes

Several underlying methodological themes were also developed within this study which considers a number of important issues in Australian archaeology and elsewhere in the world. These issues focus on developing the research potential of surface archaeology, often haunted by the problems of time and assemblage conflation and the methodological and theoretical problems of studying a quarry assemblage.

Although surface scatters form the bulk of the archaeological remains within Australia they have been largely ignored by archaeologists until recently (Robins 1997:24). The problems of time conflation, overprinting, post-depositional disturbance and the difficulties of establishing appropriate sampling methodologies have hindered the development of surface archaeological studies in Australia. However, there is a real need to tackle the surface remains and investigate new methods of gaining meaningful interpretations from the archaeological record (Holdaway *et al.* 1998). This study contributes to methodological and analytical approaches to promote the research potential of the surface remains in Australia.

The development of an appropriate sampling methodology for the analysis of the core reduction and the spatial distribution of different stages of reduction within a quarry assemblage are further goals of this study. The spatial differentiation of artefacts within a quarry has received little attention within Australia (Hiscock and Mitchell 1993:78-79) and yet it is vital to the accurate interpretation of core reduction strategies. A number of studies have focused on axe quarries in Australia (e.g. McBryde 1978, 1979, 1984a, 1984b; Baker 1987; Wilson 1994) but few detailed studies have concentrated on other types of quarries. Similarly, quarry assemblages have been used to examine the wider context of raw material use and distribution (e.g. Bryne 1980; Hiscock 1986b; 1988), yet an underlying understanding of the reduction strategies employed at the quarries has been lacking. Within this study an appropriate and well justified sampling methodology is established that allows wider questions of technological organisation to be answered. The very detailed investigation of the composition and distribution of two quarry assemblages undertaken for this study has allowed the factors that contributed to their formation to be identified. It is hoped that quarry studies will be further developed in Australia to fulfil a greater role within lithic studies.

The term 'quarry' is troublesome. Like the 'site', the quarry has been identified in the archaeological record along purely subjective lines and has been used in many different ways by archaeologists (Hiscock and Mitchell 1993:19-23). What is a quarry? The term and its applications are explored within the realms of this research leading to a redefinition of a quarry in Chapter 9.

Summary

This chapter highlights the main aims, theoretical components and methodological issues which underpin this research. The goal of this study is to identify the factors that influenced the formation and composition of a quarry assemblage by contrasting how the types of reduction strategies implemented, the landscape features and time processes influenced the discard of artefacts. Comparisons between Quarry 27 (Chapters 5 and 6) and Quarry 35 (Chapter 7) exemplifies how similar and yet different two quarries can be. These results are then compared to the use of other available stone sources (the gibber pavement) at the landscape sampling units (Chapter 8) to establish the 'role' that these stone sources had in the organisation of technology.

The following chapter provides the detailed theoretical framework which guides this study.

Chapter 2

Developing the Theoretical Framework

2.1 INTRODUCTION

The theoretical framework adopted in this study draws on three different aspects of archaeological method and theory. These aspects include the formation of assemblages, the study of stone artefact analysis referred to as technological organisation and non-site approaches to the analysis and interpretation of artefact distributions. This chapter firstly presents a brief overview of quarry studies, introduces each of the analytical and interpretative frameworks and then integrates these into the theoretical foundation of the research.

This study aims to identify whether any similarities or differences in the way stone from different raw material sources were acquired, reduced, used and discarded in the study area. The results are used to determine which factors interact and influence the formation and composition of an assemblage. It is essential to understand the process of assemblage formation and its composition before identifying the role of quarried and non-quarried stone sources in the organisation of lithic technology.

2.2 PREVIOUS QUARRY STUDIES AND THE DEVELOPMENT OF QUARRY RELATED THEORY

Root (1992:14-22) defined three distinctive periods of quarry research over the last 100 years that follow a transition from purely descriptive accounts of quarries to placing the quarry within a wider settlement system. From 1900 to the 1960s quarry studies aimed at defining the range of artefact forms present, the product, the activities undertaken at the site and the ownership of the quarry (Ericson 1984:2; Root 1992:14). Most of these early quarry studies focused on the technological analysis of the core and flaking debris at a quarry (Ataman *et al.* 1992:2). From the 1970s, under the guidance of Binford's (1977, 1980) functionalist arguments, attempts were made to find the role of the quarry within the settlement system by focusing on theories surrounding economic and social organisation. Common themes addressed by subsequent quarry studies centred on craft specialisation (McCoy and Gould 1977; Torrence 1986; Arnold 1987a, 1987b; Root 1992) exchange systems (Ammerman and Andrefsky 1982; Singer and Ericson 1977; Ericson 1981, 1982) and the technological strategies used to procure stone (*cf* Johnson 1984a, 1984b; Vehik 1985b; Ellis and Lothrop 1989; *cf* Johnson and Morrow 1987; Torrence 1989a; Gramly 1980, 1984; Reher 1991).

An early advocate for the importance of quarry studies was Jonathon Ericson. Ericson's (1984:1-2) previous work at the Bodie Hills quarry in eastern California (Singer and Ericson 1977) and other quarries in the region (Ericson 1981, 1982) led him to refine the concept of a "lithic production system." The lithic production system incorporated all the activities and places involved in the procurement, manufacture and use of stone from a quarry. The quarry was central to the understanding of a lithic production system and pivotal in understanding the relationship between technology, production and exchange. Ericson (1984:5-8) proposed a series of research directions that sought to understand the structure of the "lithic production system." These directions include, understanding the lithic resources available, defining the procurement and reduction strategies and identifying the social and economic relationship between producers and consumers. The labour invested in procurement, the costs of stone transportation and the social organisation of the group involved in the lithic production system were also seen as valid lines of inquiry. The foremost quarry studies of the time were presented in the same volume edited by Ericson and Purdy (1984) which focused on many of the research directions defined by Ericson (1984). Most of the studies identified the stone resources available (Findlow and Bolognese 1984; Sappington 1984; Torrence 1984; Gibson 1984), discussed procurement strategies (Gramly 1984; Torrence 1984; Sappington 1984) or reduction strategies (Singer 1984; Leach 1984), the rate of production (Torrence 1984; Luedtke 1984) and the social organisation of the group (Spence *et al.* 1984).

The advent of quarrying studies in the 1980s is highlighted by the number of edited volumes published (*cf* Ericson and Purdy 1984; Vehik 1984a; Butler and May 1984; Sieveking and Newcomer 1987). In addition, an issue of *World Archaeology* (1984) was devoted to the study of quarries from around the world (e.g. Jones 1984; Fladmark 1984; McBryde 1984b).

During the 1980s, Torrence (1984, 1986) developed a number of analytical methods designed to measure the organisation of quarry production, seen in the degree that labour was organised during the procurement of stone. Her work focused on the analysis and comparison of two obsidian quarries on the island of Melos in the Aegean Sea. Whether the quarries were exploited by specialists or not was tested by a variety of methods. The methods were designed to assess the degree of standardisation in core reduction, the number of knapping errors and the amount of spatial segregation of activities, all seen as measures of efficiency. That specialists rarely used the quarries was highlighted by the absence of these characteristics. Procurement of obsidian was based on the household needs of fishermen and not a major commercial exchange system (Torrence 1986:201-216).

Arnold (1987a, 1987b), following the work already undertaken by Torrence (1984, 1986), studied the microblade industry in the Channel Islands off southern California. Her research also focused on the development of craft specialisation. Using a technological analysis she estimated the volume of production, error rates, technological efficiency and standardisation in reduction. She concluded that specialised production of microblades was linked to an increase in social complexity over time.

Root (1992) expanded on the work undertaken by Arnold (1987a, 1987b) and Torrence (1984, 1986). The study aimed to define the organisation and technology of stone tool production at the Knife River flint quarries, western North Dakota. The changing level of stone working 'efficiency' over time was used to compare the role of the specialist versus procurement for 'household' use. He argued that exchange maintained social relations, such as marriage, between bands of hunter-gatherers and prompted the need to employ specialists in production during the Late Paleoindian and Late Archaic periods. Interestingly, these periods are separated by ~3000 years indicating the changes in the organisation of production did not always become more complex through time.

In 1990 a symposium on raw material economics was held jointly by the Department of Anthropology at University of Kansas and the Institute of Archaeology of the Jagiellonian University in Krakow. The papers presented in the volume edited by Montet-White and Holen (1991) indicate that most of the research on lithic procurement focuses on assemblage variability within and between occupation sites rather than on quarries in Central Europe and the Levant. However, while raw material sources and quarries are described or discussed in the volume (e.g. Kozlowski 1991; Dibble 1991; Freeman 1991; Reher 1991; Soffer 1991; Bar-Yosef 1991) no detailed analysis of a quarry assemblage was presented. Similarly, another symposium on eastern Paleoindian archaeology at the Society for American Archaeology (edited by Ellis and Lothrop 1989) focuses on lithic resource use but fails to incorporate any detailed studies on quarrying. Many of the papers discuss stone procurement but mainly focus on finding wide-scale patterns of organisation (e.g. Ellis 1989; Meltzer 1989; Lothrop 1989; Curran and Grimes 1989). If, as Ericson (1984) argues, the quarry is central to understanding patterns of raw material use then there seems to be a real need to undertake more explicit quarry studies.

Considerable research has focused on quarries throughout the world. In contrast, quarry studies have remained under-developed in Australia. In an attempt to consolidate what was known about quarries in Australia a report was undertaken by Peter Hiscock and Scott Mitchell (1990) for the Australian Heritage Commission. The report presents an overview of the role quarries play in the development of archaeological theory worldwide by dividing the major areas of research into four groups; studies of manufacturing technology, the organisation of production, the organisation of stone distribution and settlement pattern studies (Hiscock and Mitchell 1993:78-80). The study of manufacturing technology describes how stone was reduced at a quarry (e.g. Leach 1984). The organisation of production examines how the labour involved in acquiring material from a quarry was organised either through specialist craftsmen or simple, unorganised egalitarian procurement (e.g. Torrence 1986; Arnold 1987a). The organisation of stone distribution focuses on how material was distributed across a landscape through complex exchange networks or different mobility patterns. A number of studies in Australia have focused on the distribution of material removed from a quarry (e.g. McBryde 1978, 1984a, 1984b, 1986; McBryde and Harrison 1981; McBryde and Watchman 1981; Hiscock 1986; Bryne 1980). Arguments surrounding logistical and settlement patterns centre on how stone was obtained (e.g. Binford 1979; Fladmark 1984; Gould and Saggers 1985; Torrence 1983; Reher 1991) and the degree of mobility a group maintained while occupying a region (Binford 1977, 1979, 1980; Bamforth 1986, 1990, 1991; Shott 1986; Blanton 1984). Hiscock and Mitchell (1993:72-80) conclude that most of the quarry studies in Australia have been purely descriptive and often have failed to deal with any wider theoretical applications defined in their four groupings.

A number of ethnographic studies of quarrying behaviour were undertaken by archaeologists in Australia (Binford and O'Connell 1984; Jones and White 1988; Tindale 1965; Gould 1977) to understand how material was procured. These studies provide interesting insights into how material was worked at a quarry. However, the use of ethnographic analogies to interpret the archaeological record should be used with care. Warnings have been given about the use of ethnography in archaeology (e.g. Gould and Watson 1982; Murray and Walker 1988; Wobst 1978). Indeed the danger is always to stretch the archaeological imagination beyond the bounds of the data to create a 'good story.' This scenario is often the case with many quarry studies where the field methodology and data obtained do not support the resulting theoretical implications made by the researcher.

Previous studies that have focused on quarrying have been criticised due to the researcher's failure to combine the data on technological analysis obtained from quarries with more theoretically driven questions surrounding the role of the quarry in wider economic systems (Ataman *et al.* 1992:4). A relatively new approach that incorporates the study of quarries as a key element is the study of technological organisation. This approach forms the foundations for this study and builds on previous theoretical ideas described above. An outline of the approach is presented below but firstly how assemblages form is discussed.

2.3 THE FORMATION OF AN ASSEMBLAGE

In the 1970s Schiffer (1972, 1976) sought to develop an approach to the interpretation of archaeological data that explained past human behaviour by taking into account the many cultural and non-cultural processes that contribute to the formation of the archaeological record. Schiffer (1972, 1996:644) hoped to define the various factors that influence the formation of assemblages to gain both a predictive and explanatory model to interpret variation in human behaviour. Finding the relationship between people and the artefacts they discarded became the goal of behavioural archaeology. The life history of an artefact was used as a framework for organising behavioural studies which concentrated on processes of formation which ultimately create the archaeological record.

The life history of each material object refers to the transition of the item from the systemic context into the archaeological context. The systemic context is the condition (form and function) of the element in the behavioural system passing through the culture from initial procurement to the final place of discard (Schiffer 1972:157). Before being discarded each artefact undergoes a series of changes or transformations within the cultural system, where different influences (cultural and non-cultural) interact. Continuous and similar interactions produce patterns in the distribution of artefacts which can be related to different 'stages' in the cultural system. The system provides a useful heuristic device, which allows the collective life histories of material objects with similar characteristics to be observed within the archaeological record (e.g. Schick 1986, 1987a).

Schiffer (1972:158) further distinguishes material objects as being durable, such as tools, or consumable, such as food. A durable element passes through five main 'stages': procurement, manufacture, use, maintenance and discard. This transition is reflected in the refuse disposal pattern as either a primary refuse disposal where the element is used and discarded in the same context, or a secondary refuse where the element is used and transported before being discarded.

Gould (1977) expanded on Schiffer's model by dividing stone resources into two distinctive groups, quarried and non-quarried stone. This distinction can be related to Binford's (1973, 1977, 1979) notion of curated (from quarried stone) versus expedient, situational tools (from non-quarried stone), later developed by Andrefsky (1994a) into 'formal' and 'informal' tool types. Andrefsky (1994a) used the 'formality' of tool types to determine the degree of effort in the production. A 'formal' tool type was characterised as requiring a high cost of manufacturing time and having a long use-life within the cultural system. In contrast, an 'informal' tool was used expediently with relatively little time invested into manufacturing and had only a short use-life. Working in Australia Gould (1977:167) predicted that more formal tool types (or high maintenance tools), such as adzes and scrapers, would be derived from quarried material whereas informal tool types ('flake knives') would result from an opportunistic collection of stone from the immediate surrounds. A formal tool (made at a quarry) would be designed, transported and maintained while an informal tool (from non-quarried stone) would be made, used and discarded in the same context. As Gould (1977:167) suggests non-quarried stone (e.g. gibber cobbles) should be absent from habitation sites unless surface sources are present in the immediate vicinity. In this study both quarried and non-quarried sources of raw material are available locally. Comparing the ways these sources were acquired and used as tools provides a test of Gould's contrasting models of tool production.

One of the primary contributions *Behavioural Archaeology* (Schiffer 1976) made to the development of archaeological theory was in the realms of taphonomy, causing many archaeologists to revise their view of the archaeological record (Stern 1980). However, several problems with the underlying theoretical premise have been observed. Schiffer's (1972) behavioural chain analysis reduces the formation of the archaeological record to its simplest form – the event. This approach does not cater for a succession of many overlapping events occurring at one place, account for all sources of artefact variability or acknowledge the impossibility of defining all the non-cultural formation processes acting on an assemblage (Binford 1981:204; Stern 1995:50; Wandsnider 1996:372). The challenge for archaeologists is still to link the patterning of material culture with human behaviour.

For this study, the use of Schiffer's "site formation processes" is helpful in viewing the process of artefact accumulation and patterning across the landscape and hence can be used to organise and contrast data sets from different raw material sources. Each artefact has a "life history" but it is the patterns of discard that the archaeologist uses to make inferences about past life ways. One of the first goals for this study is to identify the stages through which stone from different raw material sources passed to characterise the composition of the assemblage at the quarries and landscape sampling units. It can be argued, based on Gould's (1977) observations, that there are fundamental differences in the way stone from gibber cobbles and quarried outcrops were used. By contrasting the 'life history' of stone from quarry and gibber sources different patterns in the procurement, manufacture, use and final placement of discarded artefacts can be assessed (Figure 2.1).

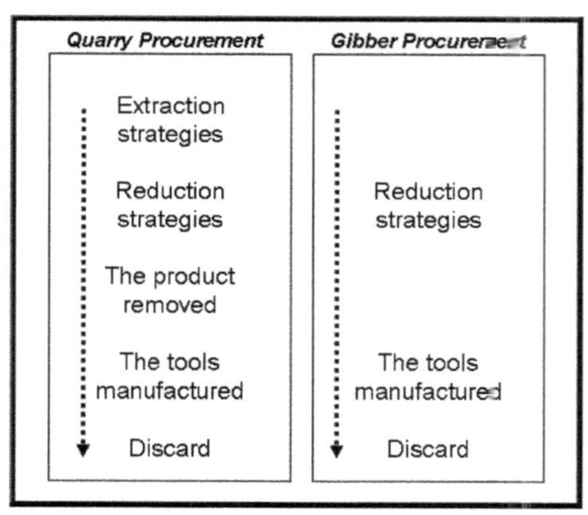

Figure 2.1: Stages in the Life History of Raw Material Sources (based on Gould 1977:167)

2.4 STUDIES OF TECHNOLOGICAL ORGANISATION

2.4.1 The Development of the Study of Technological Organisation

In the 1960s arguments surrounding the reasons for assemblage variation focused on the Bordes-Binford debate (*cf* Dibble and Rolland 1992:3-4). Bordes (1968) viewed variability as a result of a series of discrete cultural traditions, defined by tool typologies and used to infer changes in culture over time. In contrast, Binford (e.g. Binford and Binford 1966:245) advocated a functional hypothesis which linked particular tool types with discrete activities and not any single cultural tradition.

The functionalist approach was advanced by Binford's (1978a, 1978b, 1979, 1980) and Yellen's (1977) ethnographic observations of how mobility strategies influenced tool manufacture and use in different contexts. These observations of present-day hunter-gatherers sought reasons to explain patterning in artefact variability and to test the influence of tool function as being a major contributor towards the composition of each assemblage, thus defining the 'site function'. As a result of these ethnographic studies the formation of assemblage was seen as being far more complex than originally thought leading Binford (1977, 1978a, 1978b, 1979, 1980) to propose several theoretical models. These models aimed to explain how assemblages were created in terms of the types of organisational strategies employed. Through these studies an understanding of the complex factors which contribute to artefact variability and to assemblage composition were acknowledged and ultimately used to develop an organisation-of-technology approach (Carr 1994; Nelson 1991).

The theory of evolutionary ecology, an approach developed from studies of animal behaviour, ecology and evolution, has provided the backdrop to most studies of technological organisation (Jochim 1989; Carr 1994; Torrence 1989a, 1989b; Jeske 1992). This approach was derived from Darwinian evolutionary theory and uses the mechanism of natural selection as a way of studying adaptation to an environment (Carr 1994:2). Consequently, studies of technological organisation aim to establish the link between the form of the discarded artefacts, the types of technological strategies employed and way these were used to solve subsistence problems and successfully live in a particular environment (Torrence 1983, 1989a; Nelson 1991:57; Carr 1994:1).

Often adaptations in evolutionary ecology are measured in terms of a currency, such as time or energy, which can gauge the costs and benefits of employing a particular behavioural strategy and its reproductive success (Jochim 1983:157-167; Elston 1990). In the biological sciences energy has often been used as the currency and this currency has been adopted by many archaeologists (e.g. Winterhalder 1981; Morrow and Jefferies 1989; Lurie 1989). Using optimisation models a series of predictions can be made about the best way to capture energy, ideally the costs of procuring resources should not outweigh the benefits, and these can be tested against archaeological data. It is the difference between what is optimal and what is observed in the archaeological record, reflected in the technology, which provides an understanding of the underlying evolutionary processes that influence human behaviour (Bamforth and Bleed 1997:118; Nelson 1991:61). Hence, the researcher's goal is not to solely identify the 'optimal solution' and 'observed solution' but also to understand why particular behavioural strategies were employed.

The problem is how to measure 'time' or 'energy' in the archaeological record. Each activity, in any environment, requires a meaningful value to compare the costs and benefits of that activity with other activities (Elston 1990). Furthermore, the underlying premise of archaeological studies using theory derived from evolutionary ecology is that technology will optimise efficiency because natural selection has created the appropriate adaptation to a particular set of constraints (Johnson and Morrow 1987). However, there is no theoretical foundation supporting the notion that behaviours reconstructed from the archaeological record are always optimal or adaptive (Pavilides 1999:139).

The use of energy as a currency in archaeological studies of evolutionary ecology was questioned by Torrence (1989a) and Elston (1990). They considered that measuring the energy invested by humans is too difficult as many different, complex social factors influence behaviour. Furthermore, Torrence (1989a:3) questioned whether energy is ever sufficiently scarce enough to place a selective pressure on human behaviour. In response, both Torrence and Elston advocate the concept of 'risk' to measure the costs and benefits of human action. Risk relates to the probability that the costs of an activity outweigh its benefits (Elston 1990:154). The composition and distribution of the assemblage thus reflects the behavioural strategies used to manage the resources; the costs and benefits of obtaining those resources, is dependent on the time invested in procurement ('venture risk') and the 'contingency risk' of not having the required resource at the appropriate time (Elston 1990:154). As Torrence (1989a:4) states "...tool-using, procurement, production, and maintenance are better understood as being one of the primary means used by humans to reduce risk." Hence, there is a strong relationship between the problem-solving strategies used to reduce the available stone, the environment and the organisation structure of a group. The combination of these factors leads to either a successful adaptation or the demise of the individual or group. For example, Hiscock (1994) has argued that the advent of the Australian Small Tool Tradition enabled the 'risk' of colonising new and often difficult landscapes to be reduced.

Pavilides (1999:173) has argued that 'risk' can not be measured in the archaeological record but can only be used as a heuristic device to assess "how stone technology and tool design might alter cost/benefit equations". For her work in West New Britain she determined the potential 'risks' of living in a tropical rainforest environment and related these to the changes in settlement organisation and subsistence that occurred over 30 000 years of occupation. The problem-solving strategies people employed to successfully adapt to their environment were reflected in the way they organised their technology.

Other currencies, such as the availability of raw material resources have also been used by researchers to make arguments surrounding the organisation of lithic technology (Andrefsky 1994a, 1994b; Bamforth 1990, 1991, 1992; Hayden 1989). The reasoning behind these arguments is that people in the past responded to the quality and availability of raw material sources. If a region has few, high quality stone resources then material acquisition was planned and cores were carefully reduced to conserve material. In contrast, a lithic rich region allowed a wasteful reduction of stone. The use of raw material as a currency was criticised, as technology does not simply reflect the geological setting but the actions of people (Torrence 1989a:3, 1989b:64; Elston 1990). Furthermore, the link between theory and raw material availability is tenuous as technology has changed through time despite a consistency in the availability of raw material types (Johnson 1987a:7; Jochin 1989:107). Similarly, any link between raw material and reduction success, a vital component of evolutionary ecology, is difficult to establish using archaeological data (Holdaway and Stern 2004:76).

There are many difficulties with the use of adaptive models in archaeology. So many factors contribute to the types of technological strategies people employed that it is often impossible to assess which factor, such as raw material availability, contributed the most to long-term adaptations. Furthermore, each social and environmental setting requires a different "blend of strategies" to solve the associated problems (Torrence 1994:129). How to measure and compare these interacting factors and how these factors relate to evolutionary processes is a challenging problem. Hence, studies of technological organisation should aid in theory-building, moving beyond writing a prehistory describing how people behaved and explaining why they behaved in such a fashion (Torrence 1994:125; Simek 1994:121).

2.4.2 The Goal of Technological Organisation Studies

The goal of an organisation-of-technology approach is to examine past human behaviour by linking the patterning and variability of artefacts with the various technological strategies used by hunter-gatherers to successfully adapt to their particular environment (Carr 1994:35; Larsen 1994). Ultimately it is a way of investigating the interplay between past human behaviour, the environment and the causes of human adaptation and cultural change (Nelson 1991:57; Shott 1989:221). The key to examining technological organisation is to firstly understand the environmental aspects which may have influenced artefact variability. Defining these aspects enables the underlying behavioural strategies employed by a group to successfully adapt and survive within a region to be inferred. Technology is then viewed as a way of long-term 'problem solving' prompted by responses to the physical and social context (Torrence 1989b; Nelson 1991; Larsen 1994:57; Elston 1990). As a construct the approach provides a framework for assessing artefact variability both at the scale of the artefact and across the landscape to understand how patterning occurs across time and space (Kelly 1988).

Any number of questions about past human behaviour can be examined within an organisation-of-technology framework (Carr 1994:2). However, one of the major focuses of technological organisation has been on mobility patterns (e.g. Amick 1987; Andrefsky 1991; Lothrop 1989; Bamforth 1986; Shott 1986; Elston 1990). Other types of information have also been sought using this approach such as social strategies (e.g. Sassaman 1994), socio-political interaction and labour specialisation (e.g. Shott 1989:221) and the use of raw material sources (e.g. Bamforth 1991, 1992; Ingbar 1994:54).

Ultimately an organisation-of-technology approach can be seen as a way of trying to gain an understanding of the different factors which have influenced artefact variability and distribution. While different researchers have viewed different factors as having a greater impact on assemblage composition it is more likely that a combination of factors contribute to artefact patterning. It is these factors that will be identified within the realms of this study as they underlie the types of technological strategies employed.

2.4.3 The Influence of Technological Organisation on the Composition of an Assemblage

Below a brief overview identifies the many factors that influence the procurement and use of stone resources. The following discussion emphasises how a combination of different factors forms an assemblage and influences the composition of an assemblage.

The Influence of the Raw Material Quality and Availability

The availability, abundance and quality of raw material sources have been seen by many archaeologists as the prime cause of variation in the composition of an assemblage (e.g. Hiscock 1986b, 1988; Andrefsky 1994a, 1994b; Bamforth 1990, 1991, 1992; Dibble and Rolland 1992). However, debate has centred on the degree of influence raw material availability and quality had on the use and distribution of artefacts. It has been argued that the availability and quality of stone governs the amount of tool maintenance and resharpening (Bamforth 1986, 1991, 1992; Hayden 1989; Kelly 1988; Andrefsky 1994a,

1994b; Hiscock 1986b). Consequently, where the geology limits the availability of high quality stone, its reduction must be carefully controlled to maximise its use. Many researchers (e.g. Ozbun 1991; Bamforth 1991, 1992; Ingbar 1994:54; Andrefsky 1994a, 1994b) have argued that the distribution of lithic raw materials should be a fundamental component of all lithic studies as it is an important variable influencing the types of technological strategies employed. Others have questioned whether raw material availability solely dictates the implementation of different technological strategies as technology can change over time while the geological setting remains constant (e.g. Binford 1977, 1979, 1980; Kelly 1988; Goodyear 1989; Shott 1986; Edmonds 1987; Torrence 1989a). There is no doubt that raw material quality and availability does influence technology but other aspects of settlement organisation and subsistence also impact on the strategies people undertook to manage their raw material resources. The availability and abundance of high quality stone may simply influence the 'basic cost' of procuring, extracting and processing (Elston 1990:155-160).

The type of raw material, in particular the physical properties, influences the products of stone tool manufacture (Cotterell and Kamminga 1979; Domanski and Webb 2000). The type and quality of the material allows more control, ease of manufacture and retouch, and enables the greater maximisation of material often leading to the association of particular raw materials to different tool types (Dortch 1984; Elston 1990:156). For example, Goodyear (1989) found that high quality, cryptocrystalline material was ideal for portable, curated tools. This type of raw material could be easily shaped into a variety of forms, with minimal waste, and was used to form sharp tool edges that could be resharpened. Hence, using high quality stone for tools reduces the 'contingency risks' by increasing the utility (the amount of time a tool can be used) and reduces the 'costs' of obtaining more stone (Elston 1990:156-157).

The study area is rich in lithic resources. A large abundance of high quality raw material sources is available as outcrops in the Mt Wood Ranges, while the gibber pavement forms an almost continuous source of stone. If raw material was the main factor influencing the types of technological strategies employed then little evidence of material conservation and the controlled reduction of cores for the manufacture of particular tool types should be evident in the assemblages. Furthermore, following Andrefsky's (1994a, 1994b) arguments there should be little difference in the use of quarried and non-quarried stone sources in the study area.

Procurement Strategies

Lithic procurement is simply the way people acquired stone from any geological context for the manufacture of stone tools. The process of lithic procurement are strategic "problem-solving" approaches based on a knowledge of sources, the priorities of the social system, the proximity to stone sources and food resources, the territory size, the distribution of raw material, its value and the costs and energy required to acquire stone (Torrence 1983, 1986, 1989b; Nelson 1991; Seeman 1994; Leonard *et al.* 1989; Curran and Grimes 1989). The costs, such as the time expended in procuring a stone resource, should be outweighed by the benefits of having that resource (Edmonds 1987:157-158). Possessing stone was motivated by the need to obtain "subsistence security", that is, obtaining stone was seen as vital to the procurement of food resources in some contexts (Edmonds 1987:158). While different resources may have a higher or lower value within the system, if the value of a particular source is high then the need to acquire the resource influences the movement of the group to obtain the material. Binford (1979) termed this strategy as "direct procurement". In contrast, an "embedded procurement" strategy is a secondary objective in the collection of a primary subsistence resource such as food. Binford (1979) argued that the acquisition of stone by mobile hunter-gatherers was "embedded" in other subsistence activities and required no 'cost' in terms of the energy expended. Procurement was only 'direct' "when things went wrong" (Binford 1979:259).

The needs of the hunter/gatherers and availability of stone in the region may simply dictate whether procurement was embedded or direct. A contrasting observation to Binford (1979; Binford and Stone 1985) was made from data collected in the western desert of Australia (Gould 1977, 1980; Gould and Saggers 1985). A "direct procurement" strategy was used to acquire stone from quarries, in order to manufacture complex tools, such as tula adzes, while the procurement of non-quarried material was embedded in other residential activities and used for the manufacture of instant tools (Gould 1977). Furthermore, Gould (1980:124) observed that obtaining stone was the "most time-consuming and laborious part of the entire toolmaking process". A similar observation was made for the Spanish Diggings quarries in Wyoming where an intensive amount of labour was invested in core reduction (Reher 1991). Reher (1991:278) argued that obtaining quarried stone involved considerable 'costs' in production and supported the use of economising or cost-benefit framework in the analysis of lithic procurement.

Lithic procurement strategies can provide insights into the ways people organised and managed their resources within any given environment. Based on evolutionary-ecology models, Torrence (1983, 1989b) has argued for a time-stress management for stone procurement that is dependent on scheduling retooling events around settlement-subsistence strategies. If forced to collect raw material seasonally, then groups are more mobile, implement high levels of tool curation and invoke resharpening strategies. Often the limited amount of time available to acquire stone prompts a direct, specialised procurement and tool production. The expenditure of time and energy in "gearing-up" suggests a high degree of planning and organisation (Frison 1978; Binford

1979). The benefit of investing time into the procurement of stone outweighs the risk of not having the stone (Elston 1990:154). In comparison, if an abundant and even distribution of resources occurs then the risk in obtaining any particular resource is low requiring only a minimal time and energy to procure material, allowing an expedient use of stone and a more generalised procurement strategy.

The type of settlement-subsistence system used to acquire resources also influences the procurement strategy. The forager-collector continuum relates to extremes in the organisational structure of hunter-gatherer groups (Binford 1980). Foragers are typically highly mobile, with a central residential base from which resources are locally acquired. The even spread of resources across the landscape allows groups to "map onto" resources, thereby exploiting and exhausting the environment around the camp and forcing the group to move to another location. In contrast, for collectors the procurement of resources is seasonally dependent and uneven in distribution requiring a specialised task group to seek and collect resources before returning to the residential base. The dispersed nature of the resources and the subsequent movement of the task groups typically created different "activity points" in the landscape.

In response to Binford's (1979, 1980) settlement-subsistence models, Gould and Saggers (1985) offered an alternative explanation to understanding the variability in artefact assemblages of the Central and Western Deserts of Australia. This explanation focused on the utilitarian properties of raw material, that influence the making and using of stone tools, in order to demonstrate the relationship between the artefact context and the stone source. Procurement was therefore seen as a "techno-mechanical" result rather than an outcome of a settlement-subsistence strategy as advocated by Binford (1980).

Procurement strategies are of prime importance to studies revolving around the analysis of quarries (Reher 1991:252). Many factors, such as whether stone was acquired through a direct or embedded strategy, the cost of obtaining stone versus the risk of not having stone, and the availability of stone resources, interact to influence the strategies people employed to procure stone. The type of technological strategies employed defines the role of the quarry in the organisation of lithic production. For this study, how material was procured from the two quarries investigated and the non-quarried stone sources is examined and compared to determine their 'role' in the organisation of technology.

Technological Strategies

Binford (1973) originally defined two contrasting technological strategies that affected artefact form and assemblage composition as expediency and curation. Although Nelson (1991:62-66) added an "opportunistic technological strategy" to the two already established I would question how easily this strategy can be distinguished from expediency in the archaeological record, particularly in the context of this study. Nelson's (1991:62) definition of expediency "anticipates [plans] the presence of sufficient material and time" through stockpiling or caching material at a given place. There was no need to stockpile or cache material in the study area since the gibber pavement formed an almost continuous source of stone and required a 'minimised technological effort' to exploit. The gibber pavement acts as a stockpile. The gibber pavement could also be exploited using an "opportunistic technological strategy" that was unplanned and satisfied an immediate demand for a tool. The artefacts discarded from each strategy would be identical. To avoid the confusion of using these three terms in the study area only curation and expediency were used following the definitions made by Binford (1973, 1979) and others, and not those used by Nelson (1991).

An expedient technological strategy implies that little effort was made to control the reduction of the cores needed to manufacture simple, unmodified flakes that were used causally (Binford 1977:34). The cores and tools are generally abandoned in the same context, usually near the source of the stone. Expedient tools have no standardised shape, were often made opportunistically as a unplanned response to an immediate demand (Andrefsky 1998:213; Binford 1979; Hayden *et al.* 1996) and had a short use-life (defined by Schiffer (1987:48-49) as the service life of an artefact). Examples of expedient tools may include utilised flakes with edge damage. Raw material sources can also be used expediently if there is a large abundance of material available nearby that could be liberally reduced (Hayden *et al.* 1996; Nash 1996:86; Marks 1988). The identification of the expedient use of material, in this context, is the presence of a high number of cores and tools discarded near the source of the stone. The cores should show little evidence of reduction, remain large in size, have only few flakes removed and still have a large amount of cortex on their surfaces. To sum, implementing an expedient technological strategy usually occurs in areas where a high abundance of stone resources is available allowing tools to be manufactured, used and discarded as an unplanned need arose (Parry and Kelly 1987).

Binford (1973:242-244) introduced the concept of a 'curated technology' to define the transportation, preservation and efficiency of tool use within a mobile hunter/gatherer population. Bamforth (1986:39) further classified curation as being the planned "production of implements prior to use, design of implements for multiple uses, transportation of implements from location to location, maintenance and recycling of tools." Curation can be influenced by any number of interrelated factors which ultimately affect the assemblage formation and distribution of the artefacts. The amount of curation employed by a group is determined by the social and economic concerns, the degree of mobility, time management strategies, stone availability and quality, function, the need to transport resources and how extensively tools were used (Nelson 1991; Kuhn 1991; Elston 1990; Bleed 1986, 1997; Bamforth 1986; Torrence 1983). A curated core reduction strategy reduces the risk of not having the appropriate tool at the appropriate time (Elston 1990:156) and requires more effort and time invested in manufacturing and advanced preparation, seen in the controlled and standardised core reduction, which anticipated future needs (Torrence 1983:11-13, 1989b; Bamforth 1986). The manufacture of blades is an example of a curated core reduction strategy.

The notion of curation has become many different things to many different archaeologists and has recently become a widely debated topic as to the cause and effects of curation strategies (*cf* Shott 1996; Odell 1996; Nash 1996). Researchers have defined tools, raw material, assemblages and technology as being curated depending on the definition employed (Nash 1996:81-82; Binford 1979). In a recent review of curation by Shott (1996) a redefintion of the concept was undertaken based on what curation was not. To Shott (1996:264-265) curation was not multifunctionality, maintenance, hafting, repair, complexity, in anticipation of use, transportability, use-life, recycling or efficiency. Curation was simply defined as being the "degree of use" implying a continuum of tool use. An expedient tool, such as an unmodified flake with usewear, can then be equated to low curation (Shott 1996:267-268). In contrast, adze flakes (tulas) in the arid zone of Australia are considered the most curated tool type (Gould 1977; Gould and Saggers 1985; Hiscock and Veth 1991). Hence curation and expediency are not mutually exclusive but simply different planning options (Nelson 1991:65).

In Nash's (1996) and Odell's (1996) critiques of the term curation they conclude that the term should either not be used or, if used, the term should be clearly defined. One of the problems with a curated technological strategy was how to identify it within the archaeological record and how it can be separated from scarcity-induced economising activities prompted by a low availability of stone (Odell 1996). After carefully examining the concept of curation, Odell (1996:74-75) recommended that the term should be restricted to research questions surrounding the mobility and settlement of a hunter/gatherer group and not aspects of tool conservation. Curation strategies can be explored by defining whether careful planning was involved in the reduction of cores, the design of tools with multiple uses and identifying the form material was transported from location to location.

Groups of people employ different technological strategies depending on the type of settlement-subsistence system. These strategies were divided into a specialised manufacturing strategy or a generalised manufacturing strategy (Binford 1979, 1980; Torrence 1983). A settlement-subsistence system based on collecting resources typically has a specialised strategy requiring more sophisticated tool types that are highly specialised in function, time consuming to manufacture and heavily curated (Torrence 1983; Bamforth 1986). The tool types are standardised in design in order to maintain maximum efficiency in food procurement (Bettinger 1991:69). In comparison, a typical foraging group has a generalised strategy, seen through a relatively unsophisticated, limited set of tools needed to fulfil the requirements of food procurement. However, these strategies should not be viewed as dichotomies and were often 'complementary strategies' used in combination within one system (Edmonds 1987:156).

The design variables deemed important when manufacturing a tool, such as reliability, maintainability, versatility and flexibility, influence the type of technological strategy employed (Bleed 1986:739-742, 1997; Shott 1986; Nelson 1991:66-73). The difficulty in using these terms lies in being able to identify their presence in the archaeological record (Odell 1996). Nevertheless, each of these terms should be defined as they have potential value in describing assemblage variation. Each design variable is manipulated to suit the technological strategy (curation or expediency) required and is responsive to resource constraints, such as stone availability, or social determinates, such as the type of mobility pattern employed (Figure 2.2).

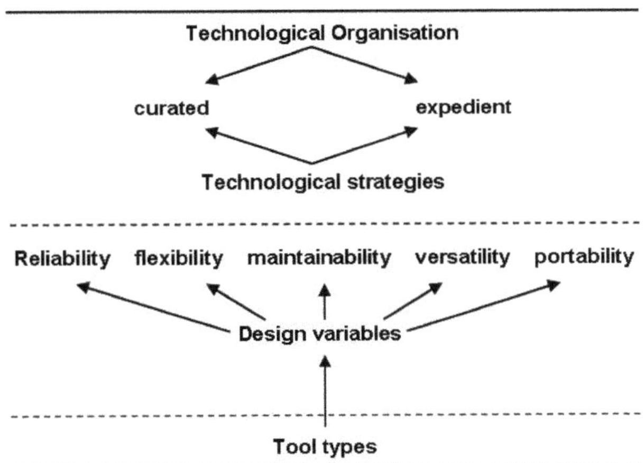

Figure 2.2: Defining Technological Organisation (adapted from Nelson 1991)

Reliability requires a tool to be dependable when needed and is often achieved by an 'over design' and maintenance before use (Nelson 1991:67; Myers 1989). The cost of a *reliable* tool is in the time invested in obtaining suitable material, manufacture and maintenance but the resulting tools are stronger and better suited to the task (Torrence 1983). *Reliability* can be seen as a standardisation of replacement parts, such as in blade reduction or the use of hafted implements, which allow tools to be dependable (Arnold 1987a, 1987b). Hence, *reliable* tools reduce the risk of not successfully completing a task (Torrence 1989b; Nelson 1991:66-67; Bleed 1986).

Maintainability requires a tool to work easily under a variety of conditions (Bleed 1986; Bamforth 1986; Shott

1986:40; Nelson 1991:70). A maintainable design can be either versatile or flexible (Nelson 1991:71). *Versatility* requires materials to be used in a variety of purposes whereas *flexibility* allows materials to be resharpened for a variety of needs (Shott 1986:19; Nelson 1991:70).

Another factor considered important for mobile hunter/gatherers is the portability of their tools (Nelson 1991; Parry and Kelly 1987; Shott 1986; Torrence 1983). The amount of gear transported was limited by what could be efficiently carried from location to location. Tools need to be lightweight, maintain a long use-life, easily fixed, employed for a variety of tasks and ideally made from high quality material. The interrelationships between these variables are reflected in the distribution and characteristics of the artefact types and tool forms.

Fundamental differences may occur in the types of technological strategies that were used to reduce the stone discarded at the quarries and landscape sampling units. The artefacts at the landscape sampling units reflect the strategies used to manufacture tools from non-quarried stone sources (gibber cobbles) and the form that material was transported from the quarried outcrops and discarded. By comparing the products discarded from these two sources it may be possible to establish the design variables selected and identify the technological strategies employed.

Distance from the Stone Source and the Discard of Artefacts

Discard occurs throughout every stage of the manufacturing process from initial core preparation to the final depletion and abandonment of a tool. The place of discard may be dependent on the context, such as at a source of stone or the depletion and abandonment of a tool at an occupation location. In turn, patterns of discard can also be influenced by the social and technological strategies employed by a group.

The raw material type, tool use, distance of transport and degree of mobility influence the discard patterns. These factors play a role in the movement and distribution of lithic material. Schiffer (1972) noted that secondary refuse involved transport of the raw material from the quarry to the location of use whereas primary refuse was discarded where it was manufactured and used. As each tool was used, broken, rejuvenated or laterally recycled the artefacts become smaller and the use-life decreases until the artefact is discarded. The longer the tool was in the cultural system the more likely that it was transported a considerable distance from the stone source.

The quarry is the beginning of core reduction where large densities of artefacts were discarded and other artefacts are removed and discarded over an extensive area. The 'fall-off' rate was observed by Renfrew (1977:71-90) who modelled a quantitative drop in the number of artefacts away from a quarry as the distance increased from the source over 100 000 km^2. A more recent study suggests that even at small distances from the source, less than 1 km, an economising behaviour was observed (Marks *et al.* 1991). Within a 1 km radius of the quarry preferences in procurement and conservation of particular raw material types were observed for material acquired directly from the source. However, it must be remembered that items were often transported over considerable distances from different sources before being discarded (Bird 1985). Artefacts that have undergone this process are present in any assemblage and influence the composition of an assemblage. Distributional studies in Australia show that the greatest fall-off rate, observed in the frequency of raw material types and changes in artefact size, form, tool types and core rationing, occurs within a 1 to 5 km radius of the quarry (O'Connell 1977; Bryne 1980; Hiscock 1986b, 1988). By focusing on a small area within the Mt Wood Ranges where the different types of stone sources were available patterns in their use and distribution may be examined.

For this study a distance-decay profile is measured for the Stud Creek catchment. The distance from the quarry to the three landscape sampling locations is small, less than 2 km. By comparing the stone obtained from quarries with non-quarried stone obtained immediately adjacent to the landscape sampling locations it may be possible to determine the form quarried stone was brought to these locations and how that form changed even at such a small distance from the source.

The Degree of Mobility

The dichotomy between foragers and collectors was established by Binford (1980) to describe how the degree of mobility of a hunter-gatherer group has been related to the type of settlement organisation. While foragers are seen as producing little spatial differentiation in assemblage composition across the landscape, due to the focus of activities concentrating on the residential base camp, collectors can produce a variety of activity based locations (Binford 1980). Foragers range out daily from their camps to gather resources. The foraging radius and distance between camps is dependent on the availability of resources and the number of people in the group (Binford 1980:9). As a result, a more spatially dispersed artefact distribution occurs for collectors based on the movement of the specialised task groups to procure resources than for foragers. Aboriginal hunter-gatherers in Australia were considered an example of foragers (Binford 1980:9-10).

The degree of mobility influences the costs of obtaining high quality material (Elston 1990:158). Low levels of mobility in a lithic poor area means that places were occupied for a longer duration increasing the use of low quality, local stone sources and increasing the "contingency risks" of not having the appropriate tool eventually prompting the need to directly procure stone (Elston 1990:160). A reliance on the use of low-quality sources increases the amount of stone processed through increased wastage. In contrast, high mobility allows procurement to be embedded in the seasonal round, restricted the amount of material transported and required specialised "gearing-up" strategies to be undertaken (Elston 1990:158).

Shott (1986:25) argued that the diversity of tool types in an assemblage reflects the degree of mobility. The underlying premise of this argument was that as mobility increased the tool diversity decreased. The longer a group occupied a location the greater the likelihood that tools with longer use-lives would be discarded. That the duration of occupation influenced the composition of an assemblage was also observed in the Mousterian assemblages of west-central Italy Kuhn (1991:97) and the Middle Palaeolithic assemblages of Western Europe (Dibble and Rolland 1992). The same argument can be used as a measure of the time invested in reducing stone at a quarry. The cores and flaking debris discarded at a quarry reflect the types of technological strategies employed. If a more 'curated' core reduction strategy was undertaken then the assemblage should indicate a controlled and standardised reduction. Furthermore, the diversity of 'formal' tools at a quarry, suggesting a number of retooling events and repeated visits, should be higher than at a stone source that was used more expediently.

Manufacturing techniques may be examined through the aspects of intensity (Dibble and Rolland 1992:10-12). Intensity can be observed in two ways; intensity of occupation and intensity in the utilisation of lithic raw materials. The degree of intensity can be related to the availability of stone sources and the degree of mobility (Dibble and Rolland 1992:18). If the duration of an occupation is long at any particular place then tools were resharpened or recycled. A high number of retouched tools within an assemblage indicates a more intensive use of lithic resources (Dibble and Rolland 1992:13).

2.4.4 The Use of a Technological Organisation Approach for this Study

Studies of technological organisation involve the analysis and interpretation of the different products of tool making, seen as sets of activities that were integrated into the organisation of prehistoric peoples lives (Thomas 1975). As Nelson (1991:87-88) states in examining strategies of technological organisation, archaeologists must assess the factors, such as environmental conditions, resource selection and activity scheduling, that affect how resources are acquired, how mobility and settlement occurred and what items were transported. These factors can be investigated by examining the different technological strategies used to acquire and reduce stone obtained from quarried and non-quarried sources in the Mt Wood Ranges. This approach acknowledges how different factors of artefact variability contribute to the creation of assemblage compositions and searches for 'cause-giving' explanations to patterning within the archaeological record (Dunnell 1992:219).

Any number of factors can influence the composition and distribution of artefacts across the landscape. An organisation-of-technology approach allows a search for the underlying behavioural strategies which are seen as adaptive responses to these factors. By firstly searching for pattern in the accumulation of the discarded artefacts and linking these with the quality and availability of raw material sources, the ways in which these sources were used and why they were used in such a manner can ultimately be inferred. Several factors must be taken into consideration when assessing lithic production: raw material abundance, raw material quality, raw material form, manufacturing techniques and the required function of the stone tool. The varying combinations and degrees of influence for each of these factors is reflected in the types of procurement strategies, and technological strategies used to organise the settlement-subsistence system.

The advantage of using an organisation-of-technology approach is that it allows the underlying environmental variables, such as the influence of raw material sources, to be identified. As a result, a fundamental requirement to an organisation of technology approach is to determine the environmental context before any inferences about behavioural strategies can be made. Patterns of artefact distribution can then be linked with the settlement-subsistence and risk-reduction strategies used by the past inhabitants to acquire and use the stone resources within the study area and establish their 'role' in the organisation of technology.

Within the study area variability in stone use between individual raw material sources, quarries and non-quarried, are compared in detail with the products discarded at three locations within the study area. It is argued that this discard reflects the long-term strategies made by the past inhabitants regarding the use of stone resources.

Several questions which direct the research for this study stem from an organisation of technology approach:

- How was stone procured from the different raw material sources?
- Were different technological strategies used to acquire and work the stone that was obtained from quarried and non-quarried sources?
- Does the availability and quality of the raw material sources impact on the technological strategies employed?
- What other factors contribute to the procurement, reduction and use of material obtained from quarried and non-quarried stone sources?
- What is the 'role' of the quarry in the organisation of technology?

2.5 DISTRIBUTIONAL ARCHAEOLOGY

2.5.1 The History of Past Archaeological Approaches to the Landscape

The development of spatial archaeology was founded in early European studies of the Austro-German school of "anthropo-geographiers" in the late nineteenth century (Clarke 1977:2). Studies focused on mapping the relationships between ancient settlement distribution and environmental factors. Within Britain a similar approach was advocated by Fox (1922) based on the premise that landscape and human geography were implicitly related. As a result, he published a series of archaeological and environmental distribution maps which covered vast regions over different time periods.

In America the development of spatial archaeology was embedded in anthropology, which focused on social organisation rather than geographical tradition. Until the advent of Gordon Willey's (1953) work in the Viru Valley of Peru few studies systematically focused on the landscape. Prompted by Julian Steward's (1937, 1938) pioneering studies in the American Southwest, Willey sought to explain the social organisational aspects through the study of settlement patterns within the region (Wandsnider 1998:92). The study of settlement patterns focused on geographical processes such as, political spheres of interaction, chronological developments and economic relationships within a regional level of sedentary societies. The unit of spatial examination was the "site" which was used to define the settlement system. Other studies, namely those in Mesopotamia (*cf* Adams 1965) and Mexico (*cf* Sanders *et al* 1979) focused on a rank-size distribution of sites using site size and inter-site distance to determine settlement hierarchies. These approaches were developed and became established in archaeology during the 1970s. However, the researchers themselves acknowledged the difficulty of moving from survey information to settlement inference (Wandsnider 1998:92).

Within the bounds of anthropological social theory, direct ethnographic analogies were made between present day observations of hunting and gathering groups and the archaeological record. This led to the belief that since human behaviour is patterned, it has structure in the archaeological record allowing organisational inferences to be made from the archaeological record. As a result intrasite studies initially focused on the interpretation of "living floors" and "tool kits" based on ethnographic accounts, implying a short time span, single occupation event using theory derived from ecological studies (e.g. Hietala 1984; Whallon 1973). The approach was challenged by behavioural archaeology (Schiffer 1976), incorporating the ideas of formation processes, and the results of actualistic studies, (ethnoarchaeology and experimental archaeology) which were used to determine how the record was formed. Consequently, the notions of single events and tool kits were debunked (Whallon 1978; Wandsnider 1996:323).

The primary problem most researchers face when undertaking spatial archaeology is the definition of the site; this was especially the case in hunting and gathering archaeology. It was deemed impossible to define site, size, function or chronology, thereby defining any interrelationships between sites was extremely difficult. The term 'site' was considered a flawed unit of analysis as they only reflected "density peaks" in a continuous surface of occupation which may or may not be contemporary (Isaac and Harris 1975; Thomas 1975; Wandsnider 1998:92; Dunnell 1992). In response non-site archaeology was developed based on the acknowledgement of previous problems within spatial archaeology.

2.5.2 Non-Site Approaches to Distributional Archaeology

Distributions of archaeological materials can be studied using either the traditional "site" concept as a basic unit of analysis or individual artefacts as the unit of analysis (Dunnell 1992; Dunnell and Dancey 1983; Thomas 1975; Foley 1981a, 1981b; Ebert 1992; Nelson 1991; Kelly 1988). The more traditional view of the "site" in hunter-gatherer archaeology defined functions to individual locations such as residences, camps, quarries and limited activity sites where a fixed itinerary of activities were undertaken (Nelson 1991:84-85). These activities were interpreted as being observable ethnographic moments in time. Direct links between ethnographic studies and the archaeological record were made which supported these assumptions (Binford 1979; Yellen 1977). It is these components of a 'site-based approach' that a number of researchers found fundamentally flawed.

Non-site archaeology has its foundations in the 1970s through the work of David Thomas (1975), Glynn Isaac (Isaac and Harris 1975) and Robert Foley (1981a, 1981b). Each took into account the changing nature of the environment and the problems of post-depositional influences to conclude that the site approach was essentially inadequate when viewing the archaeological record within a regional context. The 'site' approach limited the use of a landscape to isolated locations and ignored the wider, continuous behavioural interactions across a region thereby restricting any interpretation of pattern. Furthermore, the 'site' as an analytical unit is fraught with methodological problems because it is subjective, observer biased (Dunnell 1992:36). Their criticisms ultimately led to the development of landscape archaeology and distributional archaeology that aimed at viewing the totality of landscape use. These approaches are based on the "ontological, epistemological and theoretical premises which state that human behaviour does not generate spatially or temporally discrete archaeological residues but is continuous and overlapping within a geomorphological context" (Zvelebil *et al.* 1992:194).

In a non-site approach the artefact is the basic unit of analysis (Dunnell 1992:34). By using the artefact as the unit of analysis, and incorporating both geomorphology and actualistic studies combined with the development of new concepts in method and theory, a better understanding of the behavioural and ecological factors that influenced an artefact's location is gained (Thomas 1975; Dunnell and Dancey 1983; Dunnell 1992). Hence, the physical and cultural worlds are interrelated and can not be understood individually (Ebert 1992:73).

Ebert (1992) advocated a non-site approach, which he called distributional archaeology, while trying to examine patterns of debris and utilised flakes that were the products of an expedient technology in Southwestern Wyoming. Ebert (1992:72-73) found fault with past non-site studies that used simplistic environmental correlations and function to assess artefact variability; in response he advocated the search for technological systematics or trends in the stages of a lithic production system repeated across the landscape. He argued that these trends can be observable in patterns of procurement, manufacturing, curation, tool use and discard.
By using the individual artefact as the basic empirical unit of association the relationships between the artefact, landscape and technological organisation can be investigated. It is generally agreed that the distribution of objects in the archaeological record is a consequence of abandonment, loss or discard (e.g. Shott 1989). Hence, the aim of distributional studies is to understand how technological organisation influences the distribution and discard of tools and manufacturing debris within a landscape (Ebert 1992). In turn, distributional studies also aim to examine how these systems of adaptation depend upon the interrelationship between the cultural and physical world (Ebert 1992:69). This enables the study of the relationships between technological organisation, distribution and cultural formation processes to occur. Hence, distributional studies define the consequences of technological strategies employed (Nelson 1991:77-78).

Few non-site approaches have been used in Australia. To date only four projects have been published: the work of Holdaway *et al.* (e.g. 1998; 2000b, 2004) in Western New South Wales; Robins' (1997) work at the Currawinya Lakes, Southwest Queensland, and; McNiven's (1992) midden survey of the Coolooba region of southeastern Queensland. In Australia, a non-site approach is valuable to regional studies focusing on artefact distributions, particularly in the arid zone where artefact visibility is high.

2.5.3 The Use of Non-site, Distributional Archaeology for this Study

A criticism of non-site distributional archaeology is that human behaviour has been averaged, derived from repeated responses to environmental constraints, to create a synchronous interpretation of the past (Stern 1995). This approach reduces human behaviour to a mechanistic response to the environment. For this reason, there is a need to search for the many 'dimensions of variability' (Hodder 1987:8) and identify the many 'structuring processes' (Wandsnider 1996:373) that influence the formation and composition of an assemblage. Accordingly, a quarry is a term used in a 'site' based approach. In a non-site approach the label of quarry can be removed and each artefact within the location can be examined as to the factors which influenced its discard and position in the landscape. Consequently, the factors that influenced the discard of artefacts can be pulled apart and used to redefine what is a 'quarry' within a 'siteless' approach.

Most non-site distributional studies have examined wide areas of the landscape (e.g Ebert 1992; Robins 1997; Bamforth 1991; Shott 1986). It was thought that only at such a large scale could patterns of artefact distribution and organisation be found (Binford 1992:52-57). If as Gramly (1984:21) has argued "...the logical place to begin an investigation of a stone-tool-using culture is at the lithic sources" then before a large scale investigation can be undertaken the stone sources must first be understood. This view was reiterated by Bamforth (1992:131) who advocated an understanding of the stone resource base and how material was procured and reduced at the quarry before attempting to make statements about the overall organisation of lithic technology within the region. By examining the procurement, reduction and use of stone sources, both quarried and non-quarried, available within the Mt Wood Ranges it is possible to address larger organisational questions.

It is important to note that although the collection of data appears to be "site" based, the field methodology and analytical techniques used to collect and interpret the data has its foundations in non-site distributional archaeology. The landscape sampling locations and quarries merely represent spatial aggregates of discarded artefacts, formed over time, influenced by post-depositional processes, and are highly visible within the landscape. However, each artefact at each of the locations investigated has a unique position in space and a unique set of characteristics that reflect the reduction strategies employed and its place of discard. Spatial patterns in the location of the different stages of reduction can be examined, used to identify the various factors which influence the distribution of artefacts and to make inferences about the reasons why these distributions occur in such a manner.

For this study the distribution of individual artefacts at two worked outcrops and three landscape sampling units is used to document the different strategies employed to reduce stone derived from quarried and non-quarried stone sources. Analysis was undertaken by comparing the quarry assemblages to find common factors, such as the environmental or social variables, that have influenced the distribution of the archaeological record at each quarry. These factors were created by long-term processes (Bailey 1983:180), through the repeated use of a location, forming a 'tendency' for artefacts to be grouped in a particular manner. It is the patterns of variation created from the distribution of artefacts obtained from each of the sources which provide the basis for making inferences about technological organisation. These inferences can be used to make predictions about the role of the stone sources in the organisation of technology.

2.5.4 Surface Archaeology and the Non-site

Surface archaeology provides one way of documenting the distribution of archaeological remains across a landscape. A series of problems defined by Zvelebil (1992:196-197) when dealing with surface remains include the lack of chronological resolution, the effects of palimpsests, defining a regional scale, bias due to taphonomic processes, paleoenvironmental reconstruction and modern land use. However, if an understanding of the potential biases that influence the formation of the archaeological record can be gained then surface material has considerable research potential (Lewarch and O'Brien 1981:297; Tainter 1998).

The primary disadvantage of surface archaeology is its dependence on the geomorphological environment. Surface archaeology is more suited to an arid or semi-arid environment where large areas are exposed. Even within this environment artefact exposure is biased based on the amount of erosion, human impact or landform, often producing a limited view of the archaeological record. Recording this view in terms of total occupation impact will always be imprecise. However, if the aim is to interpret patterns across space, as products of technological organisation, it is ideal. Foley (1981a) supports this view by arguing that surface evidence is a product of behaviour, discard, accumulation and post-depositional processes. The distribution of archaeological remains will reflect the interplay between these events giving the record structure and therefore explanatory potential. A cautionary note was made by Stern (1994:101-102) who argued that potentially many different behaviours created the archaeological record in any one temporally bounded unit, therefore the archaeological record should not be viewed as being an average representation of behaviour. However, by undertaking a detailed study of a small area the many factors, the dimensions of assemblage variability, that contribute to formation and composition of the archaeological record can be teased apart. Hence, there is a need to understand why variability, not just what variability appears in the archaeological record as this reflects the organisation of lithic technology.

Only in the last three decades in Australia have archaeologists focused on surface remains. Earlier O'Connell (1977) showed that the examination of the relationship between raw material characteristics, availability and assemblage variability could be useful in studying surface scatters normally thought to have limited information. Since then several studies have focused on surface remains, such as Byrne (1980), Bird (1985), McNiven (1992) Hiscock (1988), Webb (1993), Robins (1997) and Holdaway *et al.* (e.g. 1998, 2004). The lack of previous studies was caused in part by the problems of post-depositional processes, the difficulty in dealing with amalgamations of time and overprinting, pure methodological problems of coping with vast amount of material and lastly the research questions in Australian archaeology which have tended to focused on defining chronology not space (Robins 1997). The history of the archaeology at the Willandra Lakes highlights the difficulties of analysing surface remains, with a 40 000 year span of occupation during which a vast number of artefacts were discarded (Johnson and Clark 1998). As a result hearths dated to 35 000 BP are in close spatial association with recent hearths (Johnson and Clark 1998:113). However, surface remains form the bulk of the archaeological record and the quest for cultural resource management has prompted the need to deal with these remains in Environmental Impact Statements (EIS). The present cultural, archaeological, political and economic climate has thus created a greater emphasis on tackling the difficult questions of surface archaeology (Robins 1997:24).

It is a relatively rare set of circumstances which degrades a surface vertically and remains relatively intact horizontally however this has occurred at Stud Creek where modern-day human intervention has interplayed with environmental conditions to cause extensive erosion and degradation (discussed further in Chapter 3). Although the condition of the land surface has influenced the degree of artefact exposure, the surface remains offer considerable potential for studying the acquisition, reduction and distribution of the raw material sources at the landscape sampling units.

2.5.5 How can Distributional Studies deal with Time?

How time has influenced the distribution of the archaeological record forms an important role in the analysis and interpretation of the archaeological remains within this study. As Bailey (1983:167) points out the archaeological record is an "amalgamation and intersection of many different processes operating over different time spans and defined by different time boundaries". Hence, both short-term processes and long-term processes create pattern in the archaeological record. The difficulty lies in trying to relate short and long-term behavioural processes to the archaeological record. Certainly the very formation of the archaeological record dictates that its interpretation must take into account the structure of the archaeological record produced by short and long-term time processes (Fletcher 1992:36; Bailey 1983:180-182).

Stoltman (1978:705) argues that temporal models can be constructed for any unit of space. In this sense, short-term processes can be examined by advocating a distributional, non-site approach where the archaeological record is reduced to the single artefact. The single artefact reflects several moments in time from procurement to final discard, while aggregations of artefacts across space and through time produce patterns of discard behaviour. These patterns of discard behaviour would be overlooked if time were narrowed to only one perspective, as long-term processes are not visible at the ethnoarchaeological scale (Binford 1981; Bailey 1981; Frankel 1988, 1995). Long-term patterns can also be examined using a non-site distributional approach which searches for patterns indicating a tendency for particular 'behaviours' to occur within a given space. However, these tendencies should not be viewed as an "average" but as a palimpsest of all past depositional and post-depositional activities that have created the archaeological record (Stern 1994:101, 1995).

With any degradation of the landscape surface archaeological remains which are not removed are conflated resulting in a loss of stratigraphic integrity. Furthermore, the longer or more repeated an area was occupied the greater the over-printing of many behavioural events which have occurred. Overprinting has been seen as a problem in dealing with surface remains as many different 'occupations' create the archaeological record (Robins 1997:23). However, overprinting can enhance patterns of technological organisation within the same geographic context, creating a spatial redundancy in its distribution, thereby producing long-term structure in the archaeological record that was formed by many depositional events (Foley 1981a, 1981b). The danger that lies in approaching the archaeological record this way is to explain all human behaviour in terms of environmentally determined responses, rather it is better to examine the structure of the archaeological record and explore the dimensions of its variability (e.g. Gamble 1986; Stern 1995:49).

2.6 FRAMEWORK FOR THIS STUDY

By using a non-site distributional approach the "notion of site" has been removed leaving a somewhat daunting blanket of archaeological remains that are seemingly vast and continuous across a landscape (Dunnell 1992). How can the array of artefacts become meaningful? Here I am presenting one attempt to gain pattern out of the apparent chaos by identifying the technological strategies employed to procure, reduce and use material obtained from the available stone sources within a small, defined area of the Mt Wood Ranges. In doing this, the multitude of factors that cause variability, such as the landscape features or types of technological strategies employed, which affect the formation and composition of an assemblage are teased apart and used to determine the role the different stone resources played in the organisation of technology.

For this research four scales of temporal and spatial variation are examined (Figure 2.3). The scales of interpretation span from the individual artefact, reflecting a discard point in space and time, to the distribution of artefacts acquired from different stone sources and discarded through time at the three landscape sampling locations and two quarries investigated. Each scale represents an increasing amalgamation of depositional events that occurred across space and through time. The cumulative effect of the discard and distribution of many artefacts creates pockets of variation in space where "hotspots" of overlapping artefact concentrations occur that may represent similar technological strategies, for example stone procurement at a quarry. Various factors, such as the landscape features, might influence the location of these "hotspots" and the composition and distribution of their assemblages.

As Stoltman (1978:705) points out geographic limits and arbitrary, culturally or naturally defined boundaries of time units are vital to constructing temporal models in archaeology. The study area falls within the Mt Wood Ranges and encompasses all the sources of lithic raw materials available in the region. The quantity and quality of the available raw material types are an unchanging variable within any given geological area such as the Mt Wood Ranges. If the time of occupation was narrowly constrained or bracketed (discussed in Chapter 3) how was technology organised within a given time and geological context? If occupation is limited to an envelope of time then questions related to technological organisation of the different raw material sources can be asked. In Chapter 3 the geological limits and the time-depth for the data used in this study are defined using the results of radiocarbon dating and typological indicators. Needless to say, while time may be bracketed and similar patterns of discard may have occurred within this time period these patterns should not be viewed as an 'average' of all human behaviour. Rather, it is important to identify the factors that created the pattern, not just the pattern, as these provide insights into human behaviour.

When interpreting the distribution of artefacts varying scales of space can be viewed. For this research the 'view' moves from the small scale, the single artefact to the largest scale - the study area (Figure 2.3). As Wandsnider (1998:88) notes a hierarchical relationship between units of observation exists from which interpretation can be gained at ever increasing spatial scales ranging from the individual artefacts to the wider landscape. Two quarries were analysed in detail to access the factors which influenced their formation, assemblage composition and the distribution of the artefacts within each quarry assemblage. To achieve this goal an understanding of the reduction process must be gained; that is, how cores were reduced and what was removed from each quarry. Hence, decisions were made as to which attributes on the artefacts discarded at the quarries best represented the types of reduction strategies employed (discussed in Chapter 4). The results from the technological and spatial analysis of the two quarries were used to compare the core reduction processes undertaken at the quarries and establish whether similar factors contributed to the formation, distribution and composition of the quarry assemblages. The reduction and use of material derived from the quarries were further compared to the reduction and use of material derived from non-quarried stone, gibber cobbles, found within the

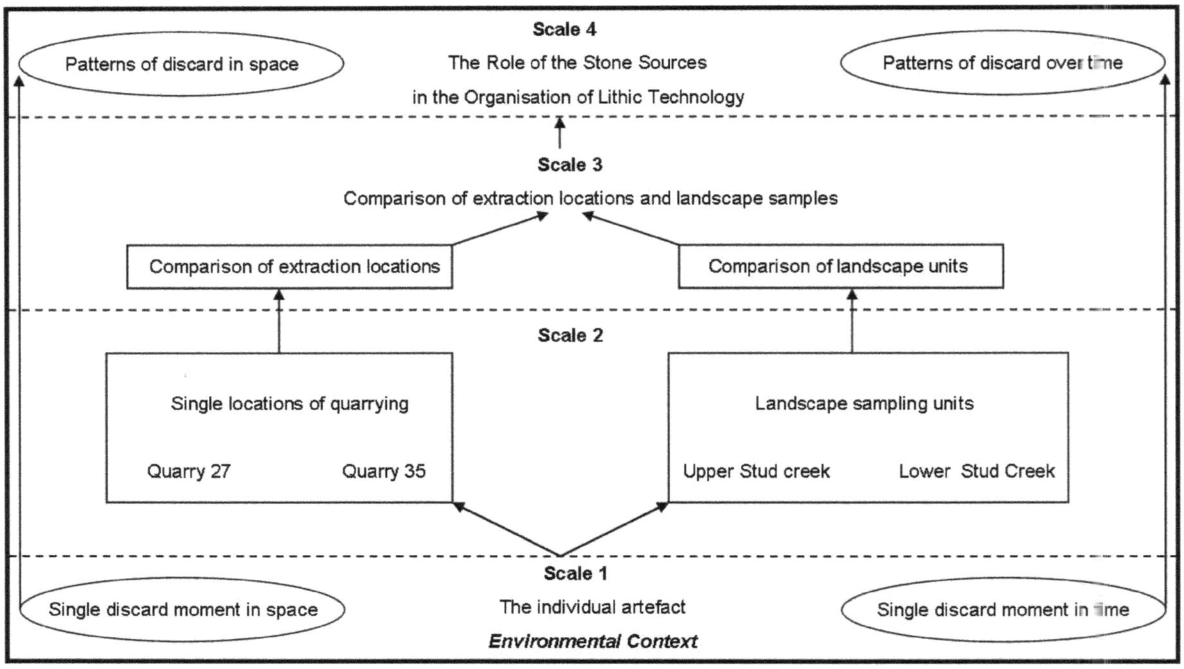

Figure 2.3: Framework for the Analysis of Spatial and Temporal Variation

landscape sampling units. This comparison enables the 'role' of each type of stone resource to be identified using an organisation of technology approach.

This research attempts to discover what variation in the distribution of the artefacts occurs in the study area and the reasons why variation may occur. The results are then used to determine how stone from the different stone sources was procured, why it was procured in such a manner, the types of technological strategies employed, and the 'role' they played in the organisation of technology.

By amalgamating formal units of measurement (the description of artefact attributes, and an artefact's position in space) with time, then these three dimensions of the archaeological record can be used to understand strategies people undertook to organise their lithic technology within the study area. For this study time is simply viewed as the life span of an artefact within the cultural system, the time invested in stone procurement and manufacture and the time spent at a particular place in the landscape (discussed in Chapter 9). Each of these time-scales was influenced by the procurement and reduction strategies employed. Hence, how time has influenced the formation and composition of the assemblages examined is yet another way of comparing the use of the different lithic raw material sources.

An important factor that may have contributed to the types of technological strategies employed within the study area is the concept of 'risk' (Torrence 1989b; Elston 1990; Bamforth and Bleed 1997). The anticipation of risk, the failure to obtain food resources, may have motivated the ways in which stone resources were used and deserves to be further explored within the realms of this study. The way stone was procured, reduced, fashioned into tools and transported reflects the problem-solving strategies people in the past undertook to live in their environment. Furthermore, the tool design variables, such as reliability, flexibility and maintainability, applied to the reduction of stone and the manufacture of tool types from different stone sources is important in defining the technological strategies employed.

2.7 SUMMARY

There is no simple way of dealing with the prehistoric past as the archaeological record is the result of complex interrelationships between time, space, environmental context and human behaviour. This study attempts to deal with these interrelationships by trying to construct both a theoretical and methodological approach that aims at examining how stone resources were used in the past. By using a non-site, distributional approach to technological organisation it is hoped that pattern can be found from the seemingly vast, overlapping and continuous array of archaeological remains that occurs in the study area.

The theoretical and methodological approach employed (discussed further in Chapter 4) needs to account for the different scales of spatial variation in the distribution of artefacts. Research involving surface archaeology and quarry studies must acknowledge that spatial variation occurs and find new ways to approach the archaeological record. If spatial variation within an assemblage is not taken into account then the results will always be biased and based on the sampling strategy that was used. Hence for this study, the characteristics of the individual artefacts and their location, given the appropriate framework, will eventually help to answer the research questions asked about the organisation of technology in the study area. This approach also combines environmental factors and geomorphological studies (discussed in the following chapter) to assess their influence on distribution, formation and composition of the assemblages in the study area.

Chapter 3

The Corner Country – A Rocky Place

3.1 INTRODUCTION

An understanding of the landscape context and the available resources is an essential part of any study seeking to understand the factors influencing the distribution of archaeological debris. Recognition and accommodation of the complex interplay between natural and cultural variables is required to determine the factors influencing the use of different resources in the landscape. This interplay underlies the need to place any archaeological record within an environmental context. In the following chapter, an assessment of the region highlights the broad environmental features, the types, abundance and quality of the stone sources and the distribution of lithic resources across the landscape.

3.2 LOCATION OF THE RESEARCH AREA

The research area is located within Sturt National Park, near Tibooburra in Western New South Wales, 350 kilometres north of Broken Hill (Figure 3.1, 3.2). The park is bounded by the Queensland border to the north and the South Australian border to the west and is known locally as 'the corner country'. This region is known locally as the corner country. Sturt National Park was formed in 1972 when the National Parks and Wildlife Service (NSW) identified five pastoral leases (Mt Wood, Olive Downs, Mount King, Binerah Downs and Fort Grey) as being unsuitable for agriculture and proceeded to manage the area as a national park (NSW National Park and Wildlife Service 1996). The area now encompasses over 300 000 hectares making it the largest reserve in the western region of New South Wales.

The study area extends approximately 15 km north-south and 30 km east-west, transecting two major landforms, plains and ranges, and two large water systems, Thomson Creek and Twelve Mile Creek (Figure 3.3). The Mt Wood Range forms the backbone of the study area, while the Tibooburra Range bounds the western margin. The majority of the fieldwork has focused on the Stud Creek catchment, a drainage system of the Mt Wood Range (Figure 3.2).

3.3 REGIONAL ENVIRONMENT

3.3.1 Geological Setting

Sturt National Park lies within the Great Australian Basin covering an area of 1.7 million square kilometres (km^2). The Great Australian Basin combines five smaller Jurassic-Cretaceous sedimentary basins; Carpentaria, Eromanga, Surat, Clarence and Moreton (Cramise and Hawke 1985). The Tibooburra Township and Sturt National Park are situated within the Eromanga Basin. A section through the Eromanga Basin gives an average elevation of 150 metres (m) with variations in height from 80 m Above Sea Level (ASL) at the Bulloo Embayment to 330 m ASL along the Grey Range (Witter 1992:114).

The present landscape reflects previous faulting and folding events along the Mt Poole, New Bendigo and Warrata faults, forming north-south anticlinal uplift and exposing Cambrian, Silurian and Cretaceous rocks overlaid with Tertiary sediments (Rose *et al.* 1967). In the places where uplift has occurred along the ranges, silicified Tertiary sediments (silcrete) have formed a resistant cap rock (duricrust) over steep-sided scarp faces (Langford-Smith and Dury 1965). The cappings occur in elevated areas, usually along the scarp edge, with an average thickness ranging from approximately three to ten metres giving a total scarp profile of 60 m (Hawke and Burke 1985). The scarps form an almost continuous bluff along the Mt Wood Ranges creating distinctive landform profiles of flat-lying platforms, weathered mesas and cuestas (Figure 3.1). Later erosion, within the ranges, along creeks and drainage channels have formed valley catchments with smaller mesas of silcrete surrounded by an undulating landscape of Cretaceous rolling-downs sediments, with pockets of residual and Quaternary colluvial sands and gravels (Cramise and Hawke 1985). Stud Creek and the Arcoola Stream catchments were formed through this process of erosion (Figure 3.3).

The Tibooburra Range (Figure 3.3) has an older geological makeup, formed from the core of resistant Pre-Cambrian slates and Silurian granites, and has undergone block uplift. A contact metamorphic zone, composed mainly of hornfels, surrounds the plutonic core (Witter 1992:123-124).

Figure 3.1: The Mt Wood Range, Sturt National Park

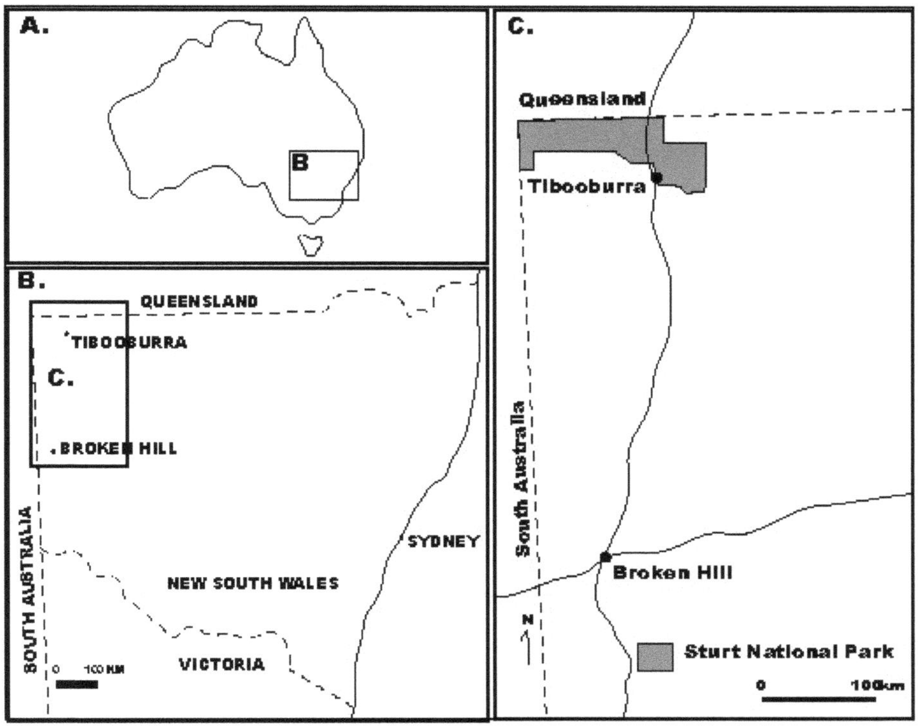

Figure 3.2: Location of the Study Area

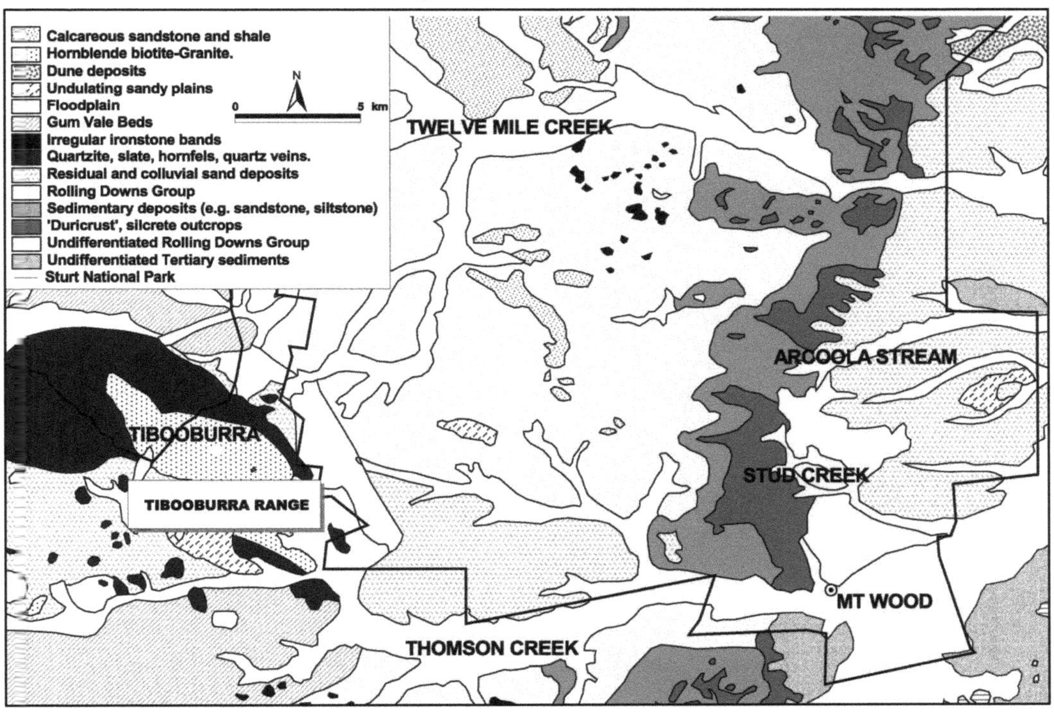

Figure 3.3: Landscape Features in the Study Area

3.3.2 Local Landscape Setting

Witter (1992:120-125) described Sturt National Park in terms of four distinct landscape units, riparian waterways, ranges, plains and dunes, based on the NSW Soil Conservation Service Land Systems Series Sheets of Milparinka SH54-7 and Urisino SH54-8 (Table 3.1). The study area incorporates two riparian units, two ranges and several of the surrounding plains (Figures 3.3, 3.4). These units and the two main upland catchment areas found in the study area will be described in order to understand the local landscape setting.

Landform Units	***Name***
Range	Tibooburra Range Grey Range, Barrier Range Mt Wood Range
Plains	Bulloo Plain 12 Mile Plain Whitabrinnah Plain
Riparian Waterways	Riparian Waterways 12 Mile Creek Thomsons Creek
Upland Catchments	Stud Creek Arcoola Stream
Dune Systems	Pindera Dunes Millers Dunes Bulloo Dunes

Table 3.1: The Landforms Found in the Eromanga Basin

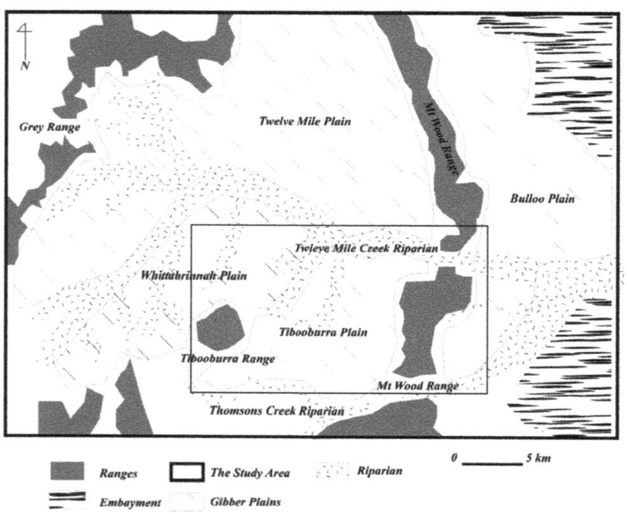

Figure 3.4: The Location of the Landforms in the Eromanga Basin

Riparian Landscape Unit

The study area has two major waterways, Twelve Mile Creek and Thomsons Creek (Figure 3.3). Both of these systems intersect the Mt Wood Ranges to form steep-sided gorges. Where the creeks narrow and deepen a series of braided channels and ephemeral waterholes occur. Flooding has redeposited alluvial sands along the margins of these creeks creating extensive plains and limiting artefact visibility. These waterways form the main drainage systems from the Tibooburra Range to the Bulloo Embayment and tend to flood when rainfall is high. Within these gorges massive exposures of silcrete outcrop are abundant especially in the Twelve Mile Gorge. The availability of water, combined with richer soils, supports a dense and diverse vegetation. Along the banks of the main water systems large stands of coolabah (*Eucalyptus microtheca*), red gum (*Eucalyptus camaldulensis*) and gidgee (*Acacia canbagei*) are found.

Plains

Between the Grey Range and the Bulloo Embayment stretch intermittently the Twelve Mile, Bulloo, Tibooburra and Whittabrinnah Plains (Figure 3.4). These plains are paved with gibber cobbles. The plains form the largest landscape unit in the region. They are typically flat, open areas with sparse vegetation consisting only of low shrubs due to the lack of available water, however, during the wetter periods they become open grasslands (Figure 3.1)

Ranges Landscape Unit

The most prominent landforms in the region are the Mt Wood Ranges (Figure 3.1) or "jump-ups" (mesas) and the granite hills of the Tibooburra Range (Figures 3.1, 3.4). The geology of these ranges has been discussed previously, however, it is important to note that the majority of the flakeable lithic raw materials are found within these ranges. While the vegetation is sparse in the Tibooburra Range, the Mt Wood Ranges have pockets of low *Acacia* woodland along the scarp slopes and watercourses.

Upland Catchments

The two main catchment areas within the study area are Stud Creek and the Arcoola Stream (Figure 3.3). The catchments were formed from the erosion of the eastern slope of the Mt Wood Ranges as it gently dips towards the Bulloo Embayment. The catchments are large (30 km^2) with interconnecting tributaries collecting the downhill runoff before joining to form the main stream channel (Figure 3.3).

3.3.3 Geomorphological Landscape History

The current landscape reflects 90 years of pastoral land management. Agricultural practices have seen the clearing of vegetation and the introduction of sheep (Gerritson 1981). The present land surface has been degraded, through human intervention, causing soil destabilisation and erosion by wind and water action. Consequently, soils have eroded from the upland catchments and been re-deposited along the stream margins to form large alluvial plains (Fanning 1999). The destabilisation of the landscape is also shown in the stream movements meandering across the valley floors, cutting deeper channels and infilling older streambeds.

Until 1972, the Stud Creek catchment was used as a stud

paddock resulting in high density grazing and creating a large-scale degradation of the catchment. In contrast, the neighbouring Arcoola Catchment has more ground cover, trees and soil, is more geomorphologically stable and archaeological debris is almost invisible (Figure 3.3). Only one high density, lithic scatter was found in the Arcoola catchment. In this sense, the high artefact density seen at Stud Creek catchment is a consequence of erosion.

Intensive geomorphological study at both a large scale (macro) and small-scale (micro) level within the Stud Creek catchment has revealed the continuing process of land reformation (Fanning and Holdaway 2001a). Erosion and redeposition is dependent on relief, vegetation and rainfall. Within the upland catchments massive soil loss from the ranges, ranging from 30.6 tonne per hectare per year, ($t\ ha^{-1}\ year^{-1}$) to over 200 $t\ ha^{-1}year^{-1}$, has occurred due to overgrazing in the late nineteenth century (Fanning 1994:173-187). After 3000 BP and before European settlement soil loss was 50 times less per year (Fanning 1999). These results imply a relatively stable environment for the time of Aboriginal occupation.

The lagging of valley surface floors adjacent to waterways has combined with both wind and water erosion to remove the topsoil and expose artefacts. The artefacts were deposited on a slightly organic, brown, loamy material interpreted as a vegetative floodplain surface (Fanning 1999, Fanning *et al.* 2000b:16-17). Stockton (1977:10) also noted the high concentration of artefacts near waterways during the Moomba pipeline survey through Northwestern New South Wales and Southwest Queensland. Witter (1992:158) describes these concentrations as 'base camps' or 'occupation locations' due to the assemblage composition, the presence of ephemeral waterholes and the association of hearths. Elsewhere, on the range tops and gibber plains the density of artefacts, excluding the quarries, reduces substantially.

Only a set of interacting behavioural and environmental factors, such as grazing, rainfall, topography and soil composition, can create a surface where high artefact visibility is possible. Along Stud Creek the low-lying areas are subject to ongoing depositional and erosional processes that have disturbed the behavioural aspect of the archaeological record. This disturbance combined with a lack of stratigraphic control implies that the reoccupation of locations has created an overprinting of events without any clear behavioural connections. Overprinting has also occurred within the quarries, however, the uplands of the Mt Wood Ranges, where the quarries are located, are free from the major post-depositional processes that have influenced the lowland areas. At the quarries other factors such as the surface gradient, vegetation and animal disturbance have impacted on the artefact distribution.

3.3.4 Lithic Resources

Four types of rock: silcrete, quartzite, hornfels and quartz were used to make stone tools within the study area. Here a brief description of their geological characteristics, their formation, and their location within the Tibooburra region is given. The variety and abundance of stone resources in the region is epitomised by the Aboriginal name for the town, Tibooburra, meaning a rocky place.

Silcrete

Silcrete is the most common rock found in the region. Both geologically and archaeologically this rock type is difficult to classify (Bird 1985:186). In the past, there has often been confusion in the identification of silcrete and quartzite, coarse-grained silcrete has been identified as quartzite and fine-grained silcrete has been termed chert (Hughes *et al.* 1973:224; Sullivan and Simmons 1979:51). Whereas, quartzite has been recrystallised, cementing the quartz grains together, the quartz grains in silcrete remain independent of each other after silicification. As a result of their different formation processes the fracture path passes through the quartz grains in quartzite and around the quartz grains in silcrete. These two factors aided in the identification of silcrete and quartzite in the field.

Silcrete is defined as a brittle, intensely indurated rock composed of quartz clasts (ranging from sand to boulders), cemented by a matrix which may be well-crystallised quartz, cryptocrystalline quartz or amorphous (opaline) silica (Langford-Smith 1978:3). The actual texture of the silcrete reflects the parent rock, which may be a claystone, sandstone, siltstone, basal conglomerate or porcellinite (Watts 1978:41). Hence, the size of the quartz grains in the parent rock is an important factor which dictates the quality of silcrete. Silcrete is usually found in semi-arid environments and is usually formed by the cementation of quartz-rich fluvial sediments (Lapidus1990:472; Hill *et al.* 1994:11)

Sources of silcrete can be found as two main habits, massive exposures or gibber pavements within the Mt Wood Ranges (Watts 1974). The massive exposures occur as intermittent outcrops within the gullies and along the scarp ridges, while the gibber consists of an almost continuous pavement of weathered silcrete nodules that have been rounded to form a resistant, smooth cortex. The process of the gibber pavement formation is little understood. It is probable that as the silcrete weathers and erodes nodules of silcrete migrate from the outcrop at the scarp edge to form extensive gibber plains, while isolated pavements can occur on the catchment floors as a result of slope erosion and re-deposition (Tait 1998:74). On the floor of the Stud Creek catchment the gibber pavements are almost circular in shape and have a higher elevation then the surrounding landforms.

Quartz

Quartz is a mineral composed of silicon dioxide (silica) which has a hardness of 7 on the Mohs hardness scale and a vitreous lustre (Fuller 1995:46-48). Quartz can fracture conchoidally but usually has an irregular fracture making it difficult to control the reduction of cores (Witter 1992:43). A variety of quartz types (crystalline and non-crystalline varieties) can be found in the study area including rock crystal, chalcedony, milky quartz and agatised or petrified wood. The rock crystal quartz, a type of crystalline quartz, has a colourless, hexagonal crystal growth (Fuller 1995:46). The only known source of rock crystal in the region is the quartz veins found in the Tibooburra Range (Figure 3.2). Milky quartz, another type of crystalline quartz, is white, nearly opaque, and also found in veins outcropping within the Tibooburra Range or as gibber pebbles.

Non-crystalline quartz forms in the region include agatized wood and chalcedony. Occasional nodules of petrified wood, chalcedony and milky quartz have eroded from the scarp edges along the Mt Wood Range and have subsequently become incorporated into the gibber pavement. While quartz is found only rarely as gibber within the Mt Wood Ranges and surrounding plains, the gibber around the Tibooburra Range is predominantly quartz.

Quartzite

Quartzite is a metamorphic rock created through either the cementation of quartz bearing solutions from pure sandstones (orthoquartzites) or by the recrystallisation, under high temperature and pressure, of pure sandstone (metaquartzites) (Talent 1970:258). As a result of the formation process, quartzite has a finely mottled appearance with white, buff or pinkish grains of feldspar combined with quartz grains and a scattering of flaky minerals such as mica (Talent 1970:102). The majority of quartz particles are rounded and less than 2 millimeters (mm) in size, reflecting the parent sandstone. Quartzite can also be extremely fine-grained with a sugary appearance (Sullivan and Simmons 1979:50). In the study area metaquartzites can be found in the Tibooburra Range as nodules or as occasional cobbles found in the streambeds of the Mt Wood Ranges (Figure 3.3).

Hornfels

Hornfels is a black, fine-grained metamorphic rock formed by thermal metamorphism in a contact zone surrounding an igneous intrusion (Lapidus 1990:277). Under a hand lens the grains are equal in size and show no preferred orientation. Hornfels can only be found as massive outcrops around the granite intrusion of the Tibooburra Ranges (Figure 3.3).

3.3.5 Classification of Silcretes

Due to the complex geological formation of the silcrete and its associated physical characteristics, a need arose to tightly define and classify the silcrete within the study area. The classification system was based on the Sullivan and Simmons (1979) and Watts (1974, 1978) doctoral research on silcrete in the region. Analysis of geological samples from the study area was also undertaken by Tait (1998), John Webb and Mark Domanski from the Department of Earth Sciences, La Trobe University (Doelman *et al.* 2001). The silcretes were classified in order to distinguish between the different material types, the probable effects of material on flake form and tool form, and the possibility of tracing the material to the quarry location.
In 1979 Sullivan and Simmons recognised the difficulties of silcrete recognition and classification. They used Wopfner (1978) to distinguish between silcrete types based on the mineralogical composition of the matrix. The matrix type (i.e. the proportion and size of the clasts) influences the flaking ability of the stone. If the number of clasts is small and the matrix is composed of fine sand grains or even finer material, such as silts or clays, then the fracture is conchoidal and the cutting edge can be retouched (Sullivan and Simmons 1979:55).

Group	Lithology	Texture	Matrix	Type
Group I	sandstone	coarse	sorted	Coarse clast
Group II			unsorted	Medium clast
			sorted	Fine clast
Group III	siltstone	fine	sorted	Breccia
				Amorphous
				Banded
				Matrix-dominated

Table 3.2: The Classification of Silcrete Types

For this project, silcrete was subdivided into seven types. The types were classified as amorphous, matrix dominated, breccia, banded, fine clast, medium clast and coarse clast (Table 3.2). These could be divided into two textures: coarse and fine, two lithologies: sandstone and siltstone, and three groups based on Wopfner (1978). The groups were further divided by the "sorting" of the matrix. Group I silcrete was composed of large, similar sized or "sorted" quartz clasts. Group II was divided into medium and fine clast silcrete. Medium clast silcrete has a mixed clast size, while fine clast silcrete has smaller clasts of the same size. The highest quality silcrete is the Group III material which has an amorphous matrix without any visible clast structure under a hand lens. Group III silcrete was further subdivided into amorphous, banded (bands of varying colours present), breccia (recemented silcrete) and matrix dominated (an amorphous matrix with scatterings of quartz clasts). The groups were further divided into the colours, pink, white, grey, yellow and clear, to further differentiate the source of the material and identify the variety of colours in an outcrop (Appendix 1).

3.4 ENVIRONMENTAL CHARACTERISTICS OF THE REGION

3.4.1 Climate

The study area has a harsh continental climate with a low annual rainfall, low humidity and high evaporation creating a typical arid environment. During the summer, the temperature is high, with 50 Degrees Celsius (°C) being the highest recorded temperature (Cramsie and Hawke 1985). In contrast, the winter can be extremely cold due to the south-southwesterly prevailing winds (Hill *et al.* 1994:1). The rainfall tends to be poor with a mean annual rainfall of only 200 mm per annum and falls with no regular seasonal pattern (Hill *et al.* 1994:1). Sturt National Park only receives rain when the prevailing high-pressure system is destabilised or occasionally, if a tropical cyclone changes direction and moves inland (Reardon 1995:76-77). Consequently, prolonged droughts are common in the region, however, when rain does occur flooding can be severe (Gerritsen 1981:115). In contrast, the area also has a high density of watercourses indicating wetter periods in the last few thousand years (New South Wales National Parks and Wildlife Service 1995; Fanning *et al.* 2000a).

Past oscillations in the climate include the mid-Holocene amelioration from 7000 to 4000 Before Present (BP). The amelioration is associated with an increased rainfall during 6000 to 4500 BP (Ross *et al.* 1992:94; Veth 1993:9) and higher water levels from inland lakes (Ross 1992:82; Dobson 1974a, 1974b). This wetter period must have reactivated the waterways, shown as an extended depositional stage, in the Mt Wood Ranges (Fanning 1999).

3.4.2 Soils

Due to recent deflation and erosion of the landscape, the present soils are generally very poor with only skeletal soils present in the valley catchments. The presence of palaeosoils, observed along stream cutting, indicates that a much richer, fertile soil profile was present in the past (Holdaway *et al.* 1998; Fanning 1999). Many soil horizons are now lagged, concentrating stone fragments at the soil surface, due to erosion causing surface deflation and the upward sorting of stones by smectites and biological activity (Hill *et al.* 1994:12).

3.4.3 Fauna and Flora

Since European occupation, clearing and the introduction of exotic pastures and pests have reduced the variety of natural vegetation types (Denny 1994). The vegetation in the Eromanga Basin is heavily dependent on rainfall and soil composition (Cramsie and Hawke 1985). This dependence on rainfall implies that both seasonal and general climatic patterns influenced vegetation productivity.

Vegetation is typical of arid Australia reflecting both landform and geology. The indigenous vegetation is dominated by the *Eucalyptus* and *Acacia* genera. The plains are sparsely timbered with low shrubs such as mulga (*Acacia aneura*) and saltbush (*Atriplex sp.*) with Mitchell grasses (*Astrebla* sp.) dominating after rainfall. The low rainfall means that trees are small and slow growing, apart from the river red gums (*Eucalyptus camaldulensis*) and coolabah (*Eucalyptus microtheca*) along the watercourses. However, in a good season, there is a profusion of plant life throughout the landscape.

Along the scarp edges of the Mt Wood Ranges, mulga (*Acacia aneura*) and cabbage trees (*Acacia cana*) occur, together with low-lying shrubs and Mitchell Grass. The presence of dead gidgee (*Acacia canbagei*) along the hilltops suggests that the climate was periodically wetter. comm.). Mulga tends to occur on plateau surfaces where increased waterflow concentrates in low-lying areas (Irons and Quinlan 1988). Within the catchments, low shrubs and stands of mulga and gidgee are present along with tussock grasses. Charcoal from hearths in the Stud Creek catchment indicates that most of the burnt wood was gidgee (Reeves 1997:59).

The fauna is typical of an arid environment. Red kangaroos (*Macropus rufus),* emus (*Dromaius novaehollaniae*), euros (*M. robustus*), eastern grey kangaroos *(M. giganteus*) and western grey kangaroos (*M. fuliginosus*) are present in large numbers within Sturt National Park (NSW National Parks and Wildlife Service 1995). The number of individuals varies seasonally depending on rainfall (Reardom 1995). Conservation counts in Sturt National Park estimate 36 000 red kangaroos, 18 other native species of mammals, 166 bird species and 48 reptile species within the park boundaries (NSW National Parks and Wildlife Service 1995). Movement of fauna can be extensive to exploit the abundance of food after the rains.

The early explorers in the region noted a wider variety of animals, such as greater bibies (*Jalparous macrotis*), greater stick-nest rat (*Leporillus conditor*), jerboa (hopping mice) and the long-haired rat (*Rattus villosisimus*) (Denny 1994). Since European arrival vegetation cover has been reduced by half in the region creating a heavy soil and resulting in an estimated disappearance of 38% of all animal species (Denny 1994; Dickman 1994).

3.4.4 Water Resources

There is limited surface water within the park with only ephemeral water holes found in the gorges of the Mt Wood Ranges. The Bulloo Embayment receives the runoff from the Mt Wood, Grey and Barrier Ranges (Figure 3.4). In the Bulloo Embayment water may be present for weeks before percolating into the ground or evaporating (Cramise and Hawke 1985). The Twelve Mile Creek is the major watercourse in the area, mainly channelling water to the Bulloo Embayment. Along this water system, especially within the Mt Wood Gorge, there are several natural waterholes (Witter 1992). Only during periods of heavy rain do these channels flow, however, light rains can collect in waterholes lined with clay and these can hold water for several weeks. Soaks, formed by natural springs from ground water supplies, also occur. The presence or absence of water is both seasonally and climatically dependent resulting in a difficult landscape to occupy.

3.5 THE CHRONOLOGY OF SETTLEMENT

Previous archaeological research suggests that the surface scatters in Sturt National Park reflect a late-Holocene human habitation of the area (Lance 1992; Witter 1992). Witter (1992) identified a Late Holocene occupation in Sturt National Park by using typological and geomorphological indicators. Tool types have been regularly used to differentiate between the Australian Core Tool and Scraper and the Small Tool Traditions (White and O'Connell 1982). The Australian Small Tool Tradition is marked by the introduction of new tool types, such as tulas and backed blades, and is thought to occur no earlier then c. 4500 BP (White and O'Connell 1982:120), although recently Hiscock and Attenbrow (1998:55) argued that this date should be extended to 7500 BP. In the study area the strong presence of tool types from the Australian Small Tool Tradition indicates a Late Holocene occupation.

Results of the radiocarbon dating from 28 hearths in the Stud Creek catchment further support a late-Holocene occupation (Table 3.3; Holdaway *et al.* 2000a:13, 2004, 2005). Reeves (1997) argued for at least two discrete phases of occupation within the catchment based on the first 15 hearths dated in 1997. The first occupation was at around 1286± 21 BP and the later at 723± 28 BP giving a minimum gap of 290 uncalibrated years. Even with calibration, to 95.4% confidence levels, a gap of 90 years still exists. Within these two groups, clusters of dates pool to indicate four episodes of occupation. Reeves *et al.* (1998) simulated various depositional scenarios, based on Shott (1992:220-4), to statistically differentiate the two phases into four occupation episodes. While potentially four known occupation events occurred in the catchment it is probable that more untraceable overprinting events also took place. Further research suggests that the gap between the two phases of occupation stretches between 800 and 1100 years BP (Fanning and Holdaway 2001b; Figure 3.9)

A geomorphological envelope of 5290-4870 BP (95% confidence), based on radiocarbon dates from a grey clay unit underlying the surface on which the artefacts are located, indicate a maximum age of occupation (Reeves 1997:11-12; Fanning 1999; Fanning *et al.* 2000a, Fanning *et al.* 2001b). This dating envelope suggests that human settlement occurred after the reactivation of an intensive depositional cycle represented by multiple cut and fill events of the stream channel (Fanning 1999). Where artefact scatters overly the grey clay unit it is possible to suggest that occupation fell into approximately the last 5000 years.

Hearth ID	Laboratory ID	dC[13]	% Modern	Result (BP)	Weight (g)
H98-75	WK 6632	-23.1±0.2	97.3±0.6	220±55	3.2
H98-16	WK 6621	-23.6±0.2	95.4±0.6	380±50	10.4
H98-46	WK 6625	-23.2±0.2	94.5±1.4	450±120	2.15
H98-13	WK 5332	-23.3±0.2	94.3±0.6	470±50	15.15
H98-12	WK 5127	-23.0±0.2	93.1±0.6	580±60	5.82
H98-59	WK 6627	-24.1±0.2	92.5±1.4	630±130	1.51
H98-11	WK 5125	-22.6±0.2	92.2±0.5	660±50	15.06
H98-71	WK 6631	-23.4±0.2	92.1±0.6	660±50	3.38
H98-21	WK 5330	-23.6±0.2	91.8±0.5	690±50	10.10
H98-32	WK 6624	-24.0±0.2	91.4±0.6	720±55	3.15
H98-4	WK 6038	-23.0±0.2	93.1±4.9	790±50	7.00
H98-60	WK 6628	-23.3±0.2	90.7±0.8	790±75	2.10
H98-15	WK 5329	-22.0±0.2	90.3±0.5	820±50	15.04
H98-65	WK 6629	-23.3±0.2	86.4±1.4	1170±130	1.95
H98-28	WK 5124	-22.8±0.2	86.0±0.5	12[illegible]0±50	10.11
H98-22	WK 5122	-25.5±0.2	85.5±0.4	1260±40	15.08
H98-25	WK 5126	-23.2±0.2	85.5±0.6	1260±60	8.89
H98-19	WK 6622	-22.7±0.2	85.2±0.6	1280±60	4.85
H98-10	WK 6036	-23.2±0.2	85.2±0.4	1290±50	10.70
H98-20	WK 5331	-23.2±0.2	85.1±0.5	1300±50	15.10
H98-30	WK 6037	-23.4±0.2	84.9±0.5	13[illegible]0±50	5.90
H98-54	WK 6626	-23.0±0.2	84.8±1.5	1330±60	1.42
H98-23	WK 6623	-22.6±0.2	84.5±0.8	1350±150	2.01
H98-8	WK 6620	-23.6±0.2	84.1±0.7	1390±75	3.50
H98-27	WK 5123	-23.6±0.2	83.9±0.5	14[illegible]0±70	7.16
H98-2	WK 6039	-22.9±0.2	83.6±0.6	1440±60	5.70
H98-9	WK 6035	-22.5±0.2	83.4±0.5	1460±50	10.30
H98-66	WK 6630	-23.6±0.2	81.7±-0.5	1630±50	1.63

Table 3.3: Calibration Results from Hearths Located in the Stud Creek Catchment (Holdaway *et al.* 2000a:13)

3.6. ARCHAEOLOGICAL CONTEXT

3.6.1 Previous Archaeological Research

The Western New South Wales Archaeological Project (WNSWAP) was established by Simon Holdaway (University of Auckland) in collaboration with Dan Witter (NSW Parks and Wildlife) and a geomorphologist Patricia Fanning (Macquarie University). An Australian Research Council (ARC) Large Grant, Small Grants, and grants from the Aboriginal and Torres Strait Islanders Commission (ATSIC) and La Trobe University funded the project. Its objectives incorporate both archaeological and geomorphological aims in trying to determine prehistoric Aboriginal landuse within the region (Holdaway *et al.* 1998, Holdaway *et al.* 2000a). These aims include defining the extent of "overprinting" (that is the repeated reuse of different locations within the landscape), investigating post-depositional effects on artefacts (including deflation, water movement and visibility), defining a chronology of locational reuse and developing a predictive model of site reuse for heritage management (Holdaway *et al.* 2000a). The methodology includes the three dimensional surveying of individual artefacts and geomorphological features using an electronic theodolite and the regeneration of these distributions within a Geographic Information System (GIS) allowing a spatial integration of artefactual and environmental data. My study was undertaken in conjunction with WNSWAP and incorporates the

WNSWAP database from the two sampling locations, Stud 1 and Stud 2, in the Lower Stud Creek Catchment (Figure 3.5).

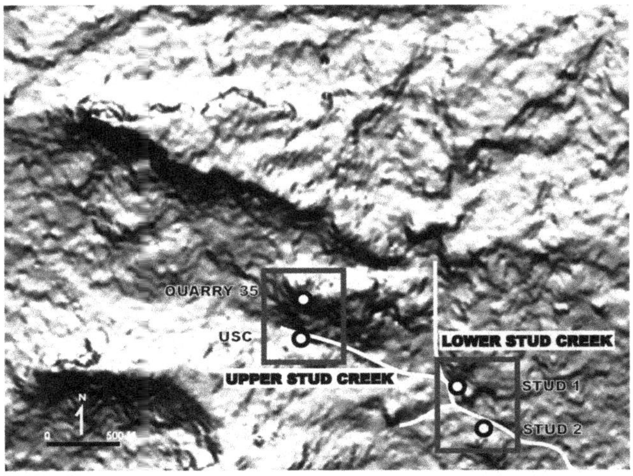

Figure 3.5: Digital Elevation Model of the Stud Creek Catchment Showing the Locations of the Study Areas

Five periods of fieldwork have been undertaken by WNSWAP since February 1995. This fieldwork included an initial general reconnaissance of the study area in February 1995, followed by a trial run of computer-aided analytical techniques undertaken at Mt Stuart quarry in June/July 1995 and the recording of an open settlement area known as TIB13 (Witter 1992; Holdaway *et al.* 1998). The surveying of TIB13 was completed in Easter 1996 with 10 thousand artefacts being recorded (Holdaway *et al.* 1998). The success of this project allowed WNSWAP to tackle the larger, more complex locations of Stud 1 (June/July 1996), Stud 2 (June/July 1997) and Stud 3 (June/July 1998). This fieldwork established an understanding of the archaeological and geomorphological context of the study area (Holdaway *et al.* 2000a, 2004).

Prior to WNSWAP the only research based archaeological work to have occurred within the region was Dan Witter's doctoral research which focused on changing utilisation of the landscape in a transect across eastern Australia (Witter 1992). Only three contract reports have been undertaken within the area. Lance (1992) produced a report for the Pipeline Authority, which analysed the lithics from an open site near Tibooburra. A general site recording survey was undertaken for the Moomba Pipeline survey (Stockton 1977). A survey of the proposed extension to the power grid incorporated a transect along Thomson's Creek (Dallas *et al.* 1995).

3.6.2 Characteristics of the Archaeological Record within the Study Area

To reiterate, although the region has an almost continuous distribution of artefacts the density of the debris varies considerably across the landscape. The two main types of artefact concentrations in this study include surface scatters in the Stud Creek Catchment (or landscape sampling units) and quarries.

Definition of a Quarry

A report undertaken for The Australian Heritage Commission aimed to define a "type profile" for the different types of stone artefact quarries and stone reduction sites in Australia (Hiscock and Mitchell 1990, 1993). Hiscock and Mitchell (1993:19-21) acknowledged that different researchers have different definitions of a quarry. A quarry can be simply a stone source, encompassing any place from where stone was acquired (e.g. Nelson 1987:121; Hiscock 1988:327), has evidence of procurement (e.g. Bird 1985:103; Gould 1977; Witter 1992) or includes both the stone source and its associated artefacts (e.g. Torrence 1986). The variety of definitions led Hiscock and Mitchell (1993:23) to define a quarry as being "the location of an exploited stone source." Sites having a source of lithic raw material and large amounts of production debris, such as cores, flakes and shatter are usually defined as quarries (Barton 1991:148). These definitions have used the site concept to define a specific place or "site" in the landscape. However, what is a quarry and how can it be defined using a non-site distributional approach? For the purpose of this study a quarry is defined as any outcrop with evidence of extraction and core reduction. This definition contrasts with the use of cobbles obtained from the gibber pavement or creek beds which can be defined as a non-quarried stone source. Gould (1977:163) defined a non-quarried stone source as being stone "obtained from the surface of the ground at or near the spot where it was needed for a particular task." The concept of quarry is reevaluated in Chapter 9 to assess whether it is a viable term within a non-site distributional approach.

The lithic resources in the study area fall into two groups: quarried outcrops and the gibber pavement. Quarries tend to be located either along the scarp edge of the Mt Wood Ranges or within the creeks and gullies that dissect the range. Evidence of extraction, material testing, initial core reduction and manufacture can be observed in the vast quantities of flaking debris found on and surrounding the outcrop. In comparison, gibber pavements form an almost continuous surface both on the ranges and the surrounding plains and as isolated patches on the catchment floors. Two quarries, Quarry 27 (Chapter 5 and 6) and Quarry 35 (Chapter 7) and two areas of gibber pavement at Stud 1 and the Upper Stud Creek unit were surveyed (Chapter 8).

Definition of a Landscape Sampling Unit

Three areas within the Stud Creek catchment were intensively surveyed and these form the landscape sampling units. The three units were divided into two topographic zones, the Upper and Lower Stud Creek (Figure 3.5). The Lower Stud Creek units were sub-divided into Stud 1 and Stud 2. Fieldwork at these locations was undertaken in 1996 and 1997 by WNSWAP, under the directorship of Dr Simon Holdaway, while the Upper Stud Creek unit was recorded in 1998 as part of this study.

Artefact concentrations occur in the Stud Creek catchment along the creek margin, where post-European erosion has stripped the overlying soils, lagged the surface and exposed the artefacts (Holdaway *et al.* 2000a, 2004). The Eroding Valley Margin forms an almost continuous strip along the creek, and it is here and on the lagged surfaces that the densest concentrations of artefacts can be found (Holdaway *et al.* 2000a, Figure 3.8). This process of valley floor stripping was observed by Tait (1998), based on LANDSAT satellite imagery, which showed Stud Creek was heavily eroded and choked with post-European alluvium when compared to the nearby Arcoola Creek catchment. Visibility is often constrained by the overlying depositional landform units that have occurred both during pre-European and post-European times. Immediately along the stream margins a post-European alluvium has been redeposited substantially reducing artefact visibility in a ~20 m band on either side of the creek (Figure 3.9). A footslope depositional landsurface and a Tributary Fan depositional landsurface have also hindered artefact visibility. These units were most likely created during the pre-European period and both overlie and incorporate the artefacts within the soil matrix. The gibber pavement is found intermittently throughout the catchment floor and also has a high artefact concentration due to a greater artefact visibility and lack of water movement operating on the surface and removing artefacts. The different landform surfaces were identified during the survey of Stud 1 and used to direct the selection of landscape sampling locations in the following years. Consequently, the survey of Stud 2 focused on the Eroding Valley Margin while the Upper Stud Creek sample primarily focused on this surface and the lagged surface.

3.7 SUMMARY

Despite a difficult environment setting the region is rich in resources. The fundamental problem is the unpredictable and often inadequate rainfall, however, when it does rain a plethora of plants and animals appear making the region highly productive. One of the permanent and predictable resources required by the Late-Holocene occupants of the region is the stone. It is the use of stone sources which is the focus of this study. The sources of stone were divided into two groups, gibber and outcrops, and several raw material type based on the different geological characteristics. The stone properties vary between the sources and influences the patterns of procurement, manufacture, tool use and discard across the landscape. With the removal of large volumes of sediments from the catchment floors vast numbers of artefacts and hearth features have become visible. These areas provide suitable locations to examine the use of the different raw material types and contrast these with the quarry and gibber sources. By understanding the environmental context and the distribution of resources the various factors which influenced the composition of the archaeological record can be identified.

Chapter 4

Field Methodology and Analytical Framework

4.1 INTRODUCTION

This chapter presents the fieldwork methodology and analytical framework used to identify how stone from three different sources was acquired, worked, transported, used and discarded within the study area. To achieve this objective an in-depth study of two quarried outcrops (Quarry 27 and Quarry 35) and three landscape sampling units (Stud 1, Stud 2 and the Upper Stud Creek unit) was undertaken. The chapter is divided into two sections: the fieldwork methodology and the analytical framework. The fieldwork methodology focuses on the objectives, survey method and field recording techniques. The analytical framework describes how the position and characteristics of the smallest unit of analysis, the artefact, is used to build the scales of analysis and ultimately gain an understanding of the organisation of stone technology in the study area.

4.2 FIELDWORK METHODOLOGY

4.2.1 Fieldwork Objectives

The fieldwork objectives were –

- To locate the quarries within the study area by undertaking a field survey.
- To record the geological and archaeological characteristics of each quarry.
- To select the quarries for further analysis.
- To record the geomorphological features found at each quarry and landscape sampling unit.
- To map the location and record the characteristics of individual artefacts found within each quarry selected and each landscape sampling unit by using a non-site distributional approach.

4.2.2 Field Recording Techniques

- **Field Survey for Quarry Location**

The field survey involved a targeted surveying strategy that focused on particular landscape features. Knowledge of the regional geology was vital to developing the survey strategy. Based on the regional geological survey, (Rose *et al.* 1967) and the geological descriptions of the silcrete in the area (Watts 1974, 1978), the search for the silcrete sources focused along the escarpment, and within catchment areas and stream gullies of the Mt Wood Range. In these locations outcrops of silcrete are visible. The survey included the identification of worked outcrops and an assessment of the related geological landforms (Table 4.1; refer Appendix 1 for details). The position of the quarries was overlaid with the position of the landscape sampling units in a GIS to provide an overall view of the available stone resources. Most of the quarries are located in the Twelve Mile Creek Gorge (Figures 3.3, 4.1).

Prior to this study, Witter (1992) undertook two survey transects in the area. In the Mt Wood Gorge transect only one quarry was identified, while two quarries were found along the Thomson's Creek Transect (Figure 3.3). Another large quarry, known to occur in the Mt Wood Range was identified as the most southerly leilera blade quarry (Witter 1992:157; Allen 1996a:371, 1996b:152).

The quarries were described following both geological characteristics and archaeological guidelines to present a general survey report for each source. The geological characteristics are based on Watt's (1974) Ph.D. research on silcretes of the Mt Wood Ranges and Hiscock and Mitchell's (1993) "type profile" of quarry characteristics (refer to Chapter 2). The recording of the quarried outcrops involved the removal of geological samples for petrological analysis, a Global Positioning System (GPS) location, general notes on the setting, recording the variation in the raw material types, the flaking products, its size, the density of the artefacts and a geological description of the outcrop (Table 4.1). Each quarry was given a unique identification number and located on the 1:100 000 Tibooburra orthophotomap (Map 7339, Land Information Centre 1985).

- **Field Survey for the Landscape Sampling Units**

Stud 1 was found as part of a general reconnaissance survey undertaken in 1996. The aim of this survey was to locate and select an area for analysis in June/July of the same year. Stud 1 was chosen as a suitable location due to a high density of artefacts on a large lag exposure providing both an extensive area for surveying and numerous hearths for radiocarbon dating. In June/July of the following year Stud 2 was selected by WNSWAP for further intensive artefact recording. Stud 2 is located on the Eroding Valley Margin, a landform surface which has a high density of artefacts. The survey at Stud 2 aimed to search for artefact variability within a single landscape unit (Holdaway *et al.* 2000a:30). The overall aim of the fieldwork was to determine the types of behavioural information that could be obtained from surface scatters (Holdaway *et al.* 2000a:24). Data from Stud 1 and Stud 2 were used for this study with the permission of WNSWAP.

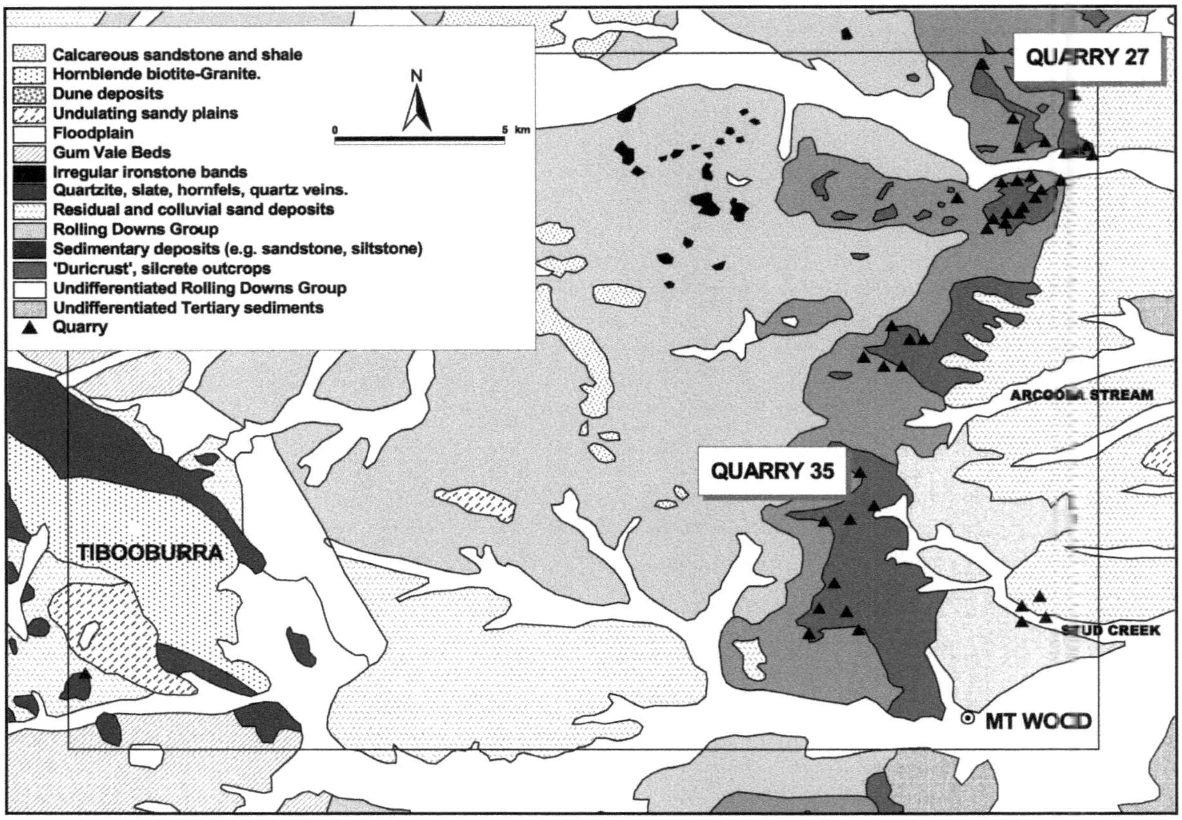

Figure 4.1: The Location of the Quarries in the Study Area

Attribute	*Description*
ID	
Location (GPS coordinates)	
Aids to relocation	Road, creek, geological feature
Raw material	Silcrete, quartzite, quartz, hornfels
Silcrete type	Coarse clast, medium clast, fine clast, amorphous, matrix-dominated, banded, breccia, hornfels
Colour	Light grey, grey, brown ,yellow, pink
Field description and	Massive layer, concretionary conglomerate, columnar
Outcrop form	Pebble, cobble, boulder
Lithology	Sandstone, claystone, siltstone, reworked pebble
Exposure location	Bluff, stream, gully, mesa
Estimated size (m^2)	
Artefact types present	
Artefact density (no. per m^2)	
Topographical features	Ridge, creek, cliff, waterway, gully, gorge
~ size of artefacts	

Table 4.1: Recording Sheet for the Quarry Survey

The Upper Stud Creek Landscape sampling unit, a location where a high concentration of artefacts occurred on a lagged surface, was found during a survey of the area surrounding Quarry 35 in June/July of 1997. I selected the unit for further investigation in 1998. This unit is close to Quarry 35 and was used to further identify the form material removed from a quarried outcrop was discarded and allow comparisons to be made between these artefacts and the reduction of non-quarried stone available in this unit.

4.2.3 Archaeological Surveying at Quarries and Landscape Sampling Units

After selecting the quarries and landscape sampling units for further investigation the surveying of each location was undertaken using a non-site, distributional approach already described in Chapter 2. Below a brief outline of the recording method used at the quarries and landscape sampling units is presented which describes the mapping and recording of individual artefacts found in each location.

To build the survey grid in the Stud Creek Catchment a GPS (Geographic Positioning System) was used by WNSWAP to establish the first datum point. A GPS gives two-dimensional co-ordinates based on the Australian Map Grid and provides an altitude above sea level. Other datums were set-up from the first datum using an electronic total station theodolite (Sokkia SET 5C) and a notebook computer (SDR33 data logger). This method enabled a fixed surveying grid to be created where the individual artefact became the smallest scale of recording with each artefact having a unique ID, associated attributes and co-ordinate data.

A similar surveying strategy was used at both the quarries and the landscape samples, involving three phases of recording. The first phase used arbitrary boundaries to survey an area for artefacts. The area was intensively and systematically surveyed, each artefact larger than 20 mm was marked with a nail and fluorescent flagging. Artefacts smaller than 20 mm were excluded from the survey as they were more subject to post-depositional movement (Holdaway *et al.* 1998, 2000a, 2000b). The second phase recorded the position of each artefact in AMG (Australian Map Grid) coordinates. A team of three people consecutively numbered the artefacts, held the prism target and operated the total station. Each artefact was given a unique identification number provided by the data logger, the number was written on the flagging, and the artefact was then point-provenienced using the total station. Once the position of the artefact was recorded the artefact analysis could begin. With the aid of HP200LX palmtop computers data from each artefact was recorded into *Entrer Trois* (McPherron and Holdaway 1996), a data entry software program loaded with a configuration file written for this specific purpose. This program prompted the attributes to be measured and allowed for the unique identification number to be recorded for each artefact. A similar method was used for the recording of the hearth features in the Stud Creek catchment (Reeves 1997) and other artefact studies within the area (Greenwood 1997; Holdaway *et al.* 1998). This approach is an example of an "on-site" recording scheme that allows each artefact to be mapped, analysed and returned to the original position thereby requiring nothing to be removed from the location.

The data from the data logger and the palmtop computers was downloaded and stored in a relational database (*Microsoft Access*). Using a *Microsoft Access* query, linking the two data files (artefact attributes and AMG coordinates) by their unique identification numbers created text files that incorporated the artefact ID and the artefact's position in space (AMG co-ordinates). This approach enabled point coverages, showing the location of all the artefacts, to be generated in *ArcInfo* and viewed in *ArcView*. Ultimately these coverages were used to search for spatial patterning in the artefact distribution as all attributes of the artefacts could be queried and visually expressed in a GIS.

Artefacts, such as cores and tools, were also photographed either with a Kodac DC 30 digital camera, or a Pentax P30T camera using both colour and black and white film. The core types and tool types were photographed and their unique identification number recorded. General photographs were taken of the survey area, outcrops and geomorphological features.

By using a "non-site" approach a site boundary can not be defined. However, for this study an arbitrary boundary was used to determine the extent of the surveyed area. The boundary has been identified using two criteria: either natural or distributional. A natural boundary was defined through the mapping of geomorphological features. Stream margins, rills, outcrop position and even surface type have all created a decrease in the density of artefacts. A distributional boundary was more difficult to assess. However, by expanding the survey outwards from a dense, central location and measuring the changing density of the artefacts, a distinctive drop in the number of artefacts was identified. The boundary was defined when the artefact density dropped below 1-2 artefacts per metre square. However, further analysis of boundary definitions will be re-examined throughout the analysis.

4.2.4 Geomorphological Surveying at Quarries and Landscape Sampling Units

Geomorphological mapping was undertaken by WNSWAP in order to record the various landform-surfaces and features at Stud 1 and Stud 2 (Holdaway *et al.* 1998, 2000a; Fanning and Holdaway 2000). This approach enabled both a reconstruction of the landscape history, the process of deposition and erosion that has created the present land-surfaces, and provided an understanding of the taphonomic processes, such as the effects of slope and water movement, that influenced artefact visibility and distribution.

Boundaries of landform surfaces, between different geomorphic units, were mapped in three-dimensional coordinates using a theodolite, while the feature type was recorded both in the SDR and a field notebook. The data were downloaded, edited and then stored in *Microsoft Access*. Coverages of the features were then built as polygons in *ArcInfo* and overlaid with the artefact points in *ArcGIS*.

Land-surface features were mapped at two scales: macromorphology and micromorphology (Holdaway *et al.* 2000a:30). The macromorphology (landform elements), based on the position of the landscape surface and the dominant geomorphic feature, and the micromorphology (land surface conditions such as redeposited sands and gravels) were mapped at Stud 1 and Stud 2. Mapping the landscape at these two scales aimed to determine the effects of the land-surface features on the assemblage composition and visibility, and to establish whether post-depositional movement of artefacts was discernible in the assemblages (Fanning and Holdaway in press, a). A similar geomophological mapping regime was undertaken by myself at the Upper Stud Creek landscape sampling unit.

At the quarries features which obviously influenced artefact distribution were mapped by myself at Quarry 27 and by members of WNSWAP at Quarry 35. These features included the outcrop, areas of kangaroo disturbance, landform units, vegetation and topography.

4.3 BUILDING THE ANALYTICAL FRAMEWORK

The analytical framework incorporates aspects of formation processes, technological organisation and non-site distributional archaeology described in Chapter 2. The overall aim of the analysis is to identify how the different stone resources found within the study area were procured, reduced and used as tools.

Four interrelated scales of variation in the archaeological record were defined for the study area (Figure 2.3), each with a series of questions that both direct and focus the analysis (Table 4.2). The first scale of variation is the single artefact (Scale 1), this takes into account aspects of non-site distributional archaeology which aims at moving away from a site based approach to a more comprehensive understanding of the interplay between artefact and the landscape (e.g. Ebert 1992). Information gained from the individual artefacts (Scale 1) was used to characterise each assemblage at the quarries and landscape sampling unit by examining the reduction strategies used in each location (Scale 2). The third scale of variation compares the composition of the two quarry assemblages with the non-quarried stone sources in the three landscape sampling units to define the use of the different types of sources (Scale 3). The last scale builds a reconstruction of the use of the stone resources and determines the role the different stone sources played in the organisation of technology within the study area (Scale 4).

By building the scale of analysis, hierarchically from the smallest unit, the artefact, any spatial variation in the discard patterns of the artefacts, can be found that reflect the different technological strategies employed. Hence, the analysis of individual artefacts represents the lowest scale of inclusiveness and the highest scale of resolution (Steffen *et al.* 1998:135-136). The fieldwork methodology used in this study allows each scale of spatial variation, building from the individual artefact, to be analysed and compared (Figure 2.3). This analytical framework allows for different scales of spatial variation in the distribution of the archaeological record to be compared within an assemblage and between assemblages.

By reducing the archaeological record to the scale of the individual artefact not only can spatial patterns in reduction be found but also the "pull" factors, such as landscape features, that have influenced artefact discard and distribution. By incorporating an inductive use of 'exploratory data analysis', using the artefact as the single unit of analysis, and combining both spatial and contextual data, multiple observations can be used to reinforce patterns and understand the "totality" of the data (Carr 1991:221). As Carr (1991:224) emphasises, an analysis needs to incorporate definitions of measurements and the related inferences, and include the relevant context and attribute data, in order to generate new ideas and hypotheses which can then be tested deductively.

The finest scale of analysis is the recording of the individual artefacts. Each artefact has a unique position in space and a unique set of characteristics. The analytical implications of an artefact's characteristics must firstly be understood before any meaningful interpretation of the reduction strategies employed at the quarries or landscape sampling units can be made.

4.3.1 Scale 1. Definition of Artefact Attributes and their Analytical Application

Classification is the "cornerstone of archaeological methodology" (Odell 1981:321). Classification is a way of organising vast amounts of data into manageable units for comparison within and between assemblages (Andrefsky 1998:59). The artefact definitions and inferences made from the artefacts' attributes are grouped into three main technological categories, cores, flakes and tools. This analysis combines the recording of nominal (morphological characteristics) and ordinal (measurements) values to record the attributes of each artefact. Core and tool typologies were also used to group the assemblages.

Scale 1 – The Individual Artefact

- What are the characteristics of the artefact that reflect core reduction?
- Where was the artefact discarded?
- What type of stone was the artefact made from?

Scale 2 – Characterisation of the Assemblages at the Quarries and Landscape Sampling Units

The Quarry

- What raw material types were available at the quarry?
- How was material extracted from the outcrop?
- What techniques and strategies were used to reduce the cores?
- In what form was material removed from the quarry?
- Does spatial variation in the location of different reduction strategies occur in the quarry assemblage?
- Is there evidence for the selection of higher quality silcrete?

The Landscape Sampling Units

- What raw material sources were available at the landscape sampling units?
- In what form was material introduced into the landscape sampling units from the quarries?
- Were different sources of stone reduced and used in different ways?
- Is there a correlation between the raw material type and the way the cores were reduced?
- Is there a correlation between the raw material type and the type of tools manufactured?

Scale 3 – Comparing the Different Stone Sources

- What similarities and differences occur in the composition of the assemblages at Quarry 27 and Quarry 35?
- Is there a correlation between the landscape features at the quarries and the distribution of the artefacts?
- What factors contributed the formation of the assemblages at the quarries?
- How does the use of the gibber pavement as a stone source compare to the use of the quarried outcrops?

Scale 4 - Establish the Role of Each Stone Source in the Organisation of Lithic Technology within the Study Area

- Compare the strategies used to acquire, reduce and use the material obtained from the gibber pavement, Quarry 27 and Quarry 35.

Table 4.2: Research Questions Related to the Different Scales of Analysis

The characteristics of both the cores and flakes reflect the process of reduction (Sheets 1975:372; Collins 1975:17; Crabtree 1968; Mauldin and Amick 1989; Ingbar *et al.* 1989). The number of attributes each artefact encodes is limitless, hence decisions regarding the selection of particular attributes should aim at answering the research question/s (Rosen 1997:27). The attributes selected for this research provide the fundamental building blocks for the subsequent interpretation of the techniques and strategies used in core reduction.

Recent debate has focused on trying to establish which sets of attributes provide the best indicator of core reduction (Sullivan and Rozen 1985:755; Magne 1989; Mauldin and Amick 1989; Kuhn 1991; Shott 1994; Prentiss and Romanski 1989; Ingbar *et al.* 1989; Larsen 1994). The debate surrounding the most appropriate set of attributes to record prompted Shott (1994:79-81) to propose that weight, cortex, scar count, platform angle, condition (complete flakes or broken flakes) and raw material type should be the minimal attribute set measured as these were more reliable based on experimental studies.

In the following section the influence raw material quality and the size and shape of the available material had on the types of reduction strategies employed is firstly discussed. Secondly, the artefact attributes recorded at the quarries and landscape sampling units are presented and related to the reduction of the cores.

Raw Material Quality

In Chapter 3 the silcrete raw material types were classified as Group I, Group II or Group III. These groups form the largest proportion of the material types found in the research area and form the focus of the quarry study. This strict grouping allowed comparisons to be made between the raw material type and the degree of reduction and the types of tools made from each of the raw material groups.

For blade manufacture and pressure flaking the lithic materials normally favoured are more homogenous and isotropic microcrystalline siliceous lithologies such as silcrete (Cotterell and Kamminga 1987; Domanski *et al.* 1994; Crabtree 1968). Experimental studies of the mechanical properties of these rock types show that they are characterised by low fracture toughness and elastic modulus and generally have a higher compressive and tensile strength (Domanski *et al.* 1994). These characteristics facilitate blade detachment, have greater edge-holding properties and allow retouching and resharpening.

The flaking characteristics of the different types of silcrete are reflected in their mechanical properties, and these in turn are determined by petrological characteristics like mineralogy, grain size and microfabric (Doelman *et al.* 2001). The resistance to fracturing is inversely proportional to the grain size; the larger the grains, the lower the fracture toughness (Farmer 1968:13;

Huang and Wang 1985; Whittaker *et al.* 1992:301-3). For materials with a large variation in grain size, the larger grains exert a controlling influence on fracture toughness (Rice *et al.* 1980, 1981). Witter (1992:34) noted fine retouch on coarse-grained stone is very difficult to achieve, implying that the tool type is dependent on the raw material quality. Fracturing properties, matrix composition and grain size vary between the silcrete groups (Doelman *et al.* 2001). Domanski's analysis of microcrystalline (Group III) and cryptocrystalline (Group II) samples from Tibooburra indicates that the Group III silcretes, with a small grain size, are more consistent in fracturing strength while the Group II silcretes, with a mixed grain size, vary considerably (Doelman *et al.* 2001). This trend implies that Group III silcretes was more predictable during knapping and could be controlled more easily to manufacture tool types that required resharpening or blade manufacture.

The Colour of Silcrete

Silcrete was also classified by its colour, primarily grey, white, yellow and pink, in order to establish the range of colours present at a quarry and to potentially identify the source of the material at the landscape sampling locations. Although heat treatment has been observed in Australia (*cf* Flenniken and White 1983; Hanckel 1985; Akerman 1979) no evidence of this practice was observed in the study area.

The Size and Shape of Cobbles

The cobbles outcropping at the quarries reflect the types of raw material present and the size of the available material. Selection for the reduction of a particular size of cobble and type of raw material may occur as these variables influence the types of reduction techniques used (Dibble and Rolland 1991; Andrefsky 1994b). Consequently, the cobble size, raw material type and amount of cortex present on the surface of the cobbles were recorded at each quarry. The amount of cortex indicates the degree of natural decortification that has occurred as a result of weathering and subsequent spalling.

Core Characteristics that Reflect Reduction Strategies

Core morphology changes through reduction until the core is abandoned. The characteristics of each core reflect the degree of reduction but also the different types of reduction techniques and strategies employed (Table 4.3).

- *Core Type*

Cores are defined either as cobble, uni-platform, bi-platform, bi-directional, multi-platform, prismatic, tranchet, cobble or microblade based on Witter (1992:46-64) and Holdaway *et al.* (1998). These core types were applied to Quarry 27, Quarry 35 and the Upper Stud Creek sampling unit. The only core types recorded at Stud 1 and Stud 2 were uni-platform, bi-platform, microblade and cobble cores. The core type may indicate the flake forms manufactured, the reduction strategy employed, and what was removed from the quarry.

Uni-platform, bi-platform and multi-platform cores refer to the number of platforms and the direction of flake removals. Uni-platform cores have flakes removed from only one platform in the same direction, while bi-platform cores have two platforms and flakes removed from two different directions. A multi-platform core has more than two platforms and with flakes removed from the core in many directions. The number of platforms is often directly proportional to the amount of core reduction, however, uni-platform reduction is also used during blade manufacture. A cobble core refers to a core which has a highly cortical surface and only has one or two scars removed. The removal of only a few flakes from cobble cores and subsequent discard is thought to indicate the testing of the material which was then found unsatisfactory and immediately discarded. This type of core is more likely to occur at a stone source (Root 1992.130; Jones and White 1988:59-61; Binford and O'Connell1984:410).

Prismatic, microblade, bidirectional and tranchet cores represent cores types which are typical of blade producing strategies (Witter 1992:46-53). These core types have parallel flake scars representing the removal of blades. A prismatic core is formed by the removal of flakes from a single platform causing the core to taper into a conical shape. A tranchet core is formed on a large flake where flakes are removed across the ventral surface, from the lateral axis of the flake and is similar to the Redbank A Strategy observed by Hiscock (1993) in the Hunter Valley. A bidirectional core simply refers to blade removals from two opposing platforms. A microblade core is a small blade producing core, usually smaller then 10 cm, with flake scars that extend across the entire face of the core (Holdaway and Stern in press).

- *Cross-section*

Witter (1992:31-32) defines the cross-section of a core as lenticular, square or conical. The type of cross-section often depends on the original shape of the cobble, the reduction strategy employed and the number of flakes removed from the core (Holdaway and Stern in press). To manufacture blades a core can change from having a square, to lenticular and to conical cross section before being discarded (Witter 1992:31-32). If standardised core shapes are found then this could reflect the aim of the production and show whether the core material was maximised by tightly controlling flake removals, such as in blade manufacturing. If flake attributes were noted, such as a bulb of percussion, then the cross-section was simply called flake, indicating that the core was originally a large flake.

Attribute	Value	Definition
Technological type	Complete flake	Has a platform and termination
	Proximal flake	Has a platform and no termination
	Distal flake	A termination but no platform
	Broken flake	No platform or termination
	Complete, broken, distal and proximal tool	As above but with retouch
	Core	Negative scars from the removal of flakes
	Core fragment	A broken core showing evidence of flake removal
	Core tool, core fragment tool	As above but with retouch
	Block	No flake characteristics, cubed shape
	Hammer stone	Cobble shaped with evidence of pecking/ crushing
	Broken hammer stone	Broken hammer cobble, shows pecking/crushing
	Weathered tool	Spooled fragment with retouch
Termination	Feather	Tapering termination
	Abrupt	Non-tapering termination
	Plunge	Curves towards the ventral surface
	Hinge	Curves towards the dorsal surface
	Step	Steps at termination
	Cortical	Termination ends in cortex
Form	Expanding	Proximal end narrower then distal end
	Block	Cubed form
	Blade	Parallel flake scars
	Point	Parallel flakes scars, triangular section and leaf shape
	Intermediate	All other flake forms
Platform type	Unifacial	Struck from a unifacially flaked platform
	Bifacial	Struck from a bifacially flaked platform
	Cortical	Struck from a cortical platform
	Missing	Crushed or missing platform
	Step	Evidence of overhang removal
Tool type	Scraper	Continuous macroscopic scalar retouch
	Backed blade	Geometric microlith
	End-scraper	Scalar retouch in Quadrant 3
	Notch	Retouch forming a cuspate notch
	Denticulate	Retouch form more than one cuspate notch
	Usewear	Continuous edge nibbling
	Adze flake	Tula or burren
	Pirri	Unifacial retouch
	Stepped scraper	Scraper with heavily stepped retouch
	Borer	Backed and tapered to a point
Core type	Uni-platform	Flaking from one direction
	Bi-platform	Flaking from two directions
	Multi-platform	Flaking from more than two directions
	Microblade	Small parallel flakes scars
	Tranchet	Blade flakes removed from a flake axis
	Prismatic	Blade production rotating from a single platform
	Cobble	One or two minor flakes removed, remaining is cortex
	Bidirectional	Flakes removed in opposite directions from two platforms
Flake dimensions	Maximum dimension	Figure 4.4
	Maximum length	
	Maximum width	
	Maximum thickness	
	Platform width	
	Platform thickness	
	Platform angle	
Core cross section	Square	Block form
	Conical	Triangular
	Lenticular	Eye shaped
	Flake	Evidence of a flake attribute
Core dimensions	Maximum length	Figure 4.5
	Maximum width	
	Maximum thickness	
	Length from platform	
	Width at midpoint from length	
	Thickness at midpoint from length	

Table 4.3: Definitions of Artefact Attributes (appended from Holdaway *et al.* 1998)

- *Percentage of Cortex*

The amount of cortex decreases with the amount of core reduction and is often reflective of the distance from the source (Magne 1989:17-18; Hiscock and Mitchell 1993:12-17). At quarries a high percentage of cortex is expected to occur on cores as unprepared cores discarded early in reduction often retain a large amount of cortex on their surfaces (Ahler 1989:90).

- *Terminations on the Flake Scars*

As core reduction continues the likelihood of step or hinge terminations occurring on the flakes being produced increases as the bulk and weight of the core decreases (Hiscock 1988:117). Flakes with step or hinge terminations are considered failures and hinder the removal of further flakes from the core, hence, decisions by the knapper regarding the discard or core rejuvenation must be made (Whittaker 1994:106-109). The number of steps and hinges on the flake scars of each core were counted in order to determine how much reduction the core underwent before it was discarded and why it was discarded.

- *Number of Flake Scars*

The number of flake scars on the surfaces of a core indicates whether the core was heavily reduced, with many flakes removed, or only had few flakes detached before the core was discarded (Arnold 1987b). The number of flakes removed from a core may also hint at a selection for particular raw material types with a greater number of flakes removed from the cores made from higher quality stone. However, similar problems to the number of dorsal scars on flakes can also be attributed to the count of flake scars on cores; often the number of flake scars is proportional to the size of the core (Maudlin and Amick 1989:73).

- *Count of Platforms*

The amount of core reduction can be measured by counting the number of platforms and the presence of cortical platforms on a core. A large number of platforms may imply a greater amount of reduction and core rotation (Flenniken and White 1985:140). In contrast, the presence of a cortical platform on a core shows an early discard of the core. The number of platforms and cortical platforms were counted on the cores at Quarry 27 and Quarry 35.

- *Metrical Attributes*

The size of the cores was measured in two ways, a block measurement (longest length, width and thickness) and a technical orientation measurement. The orientated size measures the flaking axis from the platform to the longest point opposite the platform on the same face; width and thickness are taken at the central point perpendicular to the length (Figure 4.3). The size of the cores reflects the amount of reduction, indicates the size thresholds of reduction strategies and shows the selection of raw material types (Hiscock 1988:23; Kuhn 1991:83).

Measuring the length of the longest flake scar (Figure 4.3) allows comparisons to be made between the flakes removed from the different core types at the quarries and at the landscape sampling units. A size difference will not only indicate the amount of core reduction but also a selection of cores that were suitable for further manufacturing. Core selection may reflect the raw material type as finer-grained material can be reduced further and was preferred for the manufacture of particular tool types (Witter 1992:34; Doelman *et al.* 2001). At Quarry 35 the width of the longest, complete flake scar on the core was also measured to determine the form and size of the flake removed (Figure 4.3).

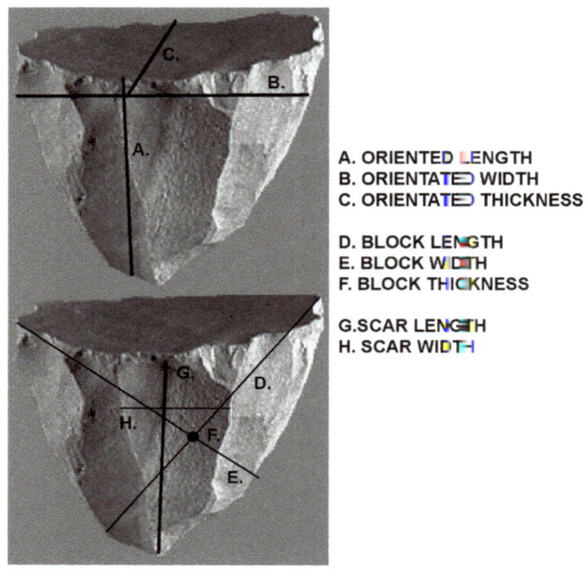

Figure 4.3: The Core Dimensions Measured (adapted from Holdaway and Stern 2004)

The weight of all cores at the quarries was measured. Weight is considered a good indicator of the degree of reduction (Magne and Pokotylo 1981:38).

Flake Characteristics that Reflect Reduction Strategies

- *Technological Type and the Fracture Type*

The technological type firstly refers to whether the artefact was flaking debris or a tool and secondly whether the artefact was complete or broken (Figure 4.4). A tool was defined as having macroscopic edge damage or retouch (Rosen 1991:29). The technological types include; complete flake, platform rejuvenation flake, proximal flake, distal flake, complete split, broken flake or block and their retouched counterparts (Table 4.3).

A complete flake has both a platform and termination while a proximal flake has a platform but no termination

(Figure 4.4). In contrast, a distal flake has a termination but no platform. A platform rejuvenation flake has evidence of an old platform and previous flake removals on the dorsal surface of the flake. The presence of platform rejuvenation flakes in an assemblage reflects a technique used to extend the use-life of a core by removing a spent platform (Holdaway and Stern in press). A block was defined as a cubed fragment with no indication of a conchoidal fracture (Witter 1992:107). A broken flake has a flake appearance but no identifiable proximal or distal end (Holdaway *et al.* 1998). These two types are produced throughout manufacturing but usually occur during initial reduction when a large hammer force has acted on the core (Andrefsky 1998:81-83). Consequently, an area with a high density of blocks and broken flakes may indicate the reduction of a core. Broken flakes can also be formed as a result of treadage in high impact areas (Prentiss and Romanski 1989:97; Hall and Love 1985; Andrefsky 1998:81; Hiscock 1986b) while broken tools may be the result of intensive use or breakage during manufacturing (Shott 1989:17-19).

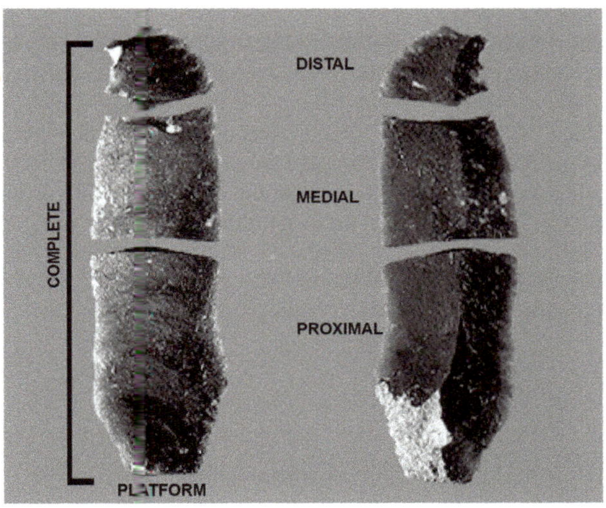

Figure 4.4: The Technological Type and the Position of Breakage(adapted from Holdaway and Stern 2004)

A flake type indicative of the amount of force applied to the core is a complete split. A split occurs when an incorrect or excessive force was used to remove a flake from a core and the impact of the force travels longitudinally through the bulb of percussion splitting the flake in half (Hiscock 1988:365-366). The number of complete splits often decreases as reduction continues, particularly in the manufacture of blades, due to increased levels of flaking control (Clarkson and David 1995:33). Hence, the frequency of complete splits within an assemblage can indicate the amount of core reduction with a higher number of complete splits occurring early in the core reduction process.

- *Flake Form*

The flake form is related to the core face, the force and the angle of strike and the distance from the core edge that the hammerstone impacted on the platform (Pelcin 1997). Expanding, intermediate, block and blade forms were recorded in the assemblages. A flake with a block form is usually produced from a large core with an irregular core face by using a large hammer force hit away from the core edge at a low angle of strike. Flakes with a block form are usually produced early in core reduction when trying to remove cortex from the core surface and reduce the bulk of the core. In contrast, blade and point forms were manufactured after careful core preparation (Witter 1992:63).

A blade is generally twice as long as wide, has parallel lateral margins and parallel dorsal flake scars (Flenniken and White 1985:136). Another flake form identified at Quarry 35 was blades with a point form. A point or converging form was recognised ethnographically as a spear projectile (Thomson 1983). A point has a triangular section and a tapered termination. An expanding flake is usually thin and wide, with a smaller proximal width than the distal end and is produced from a core with a flat, wide face (Holdaway *et al.* 1998). The most common flake form produced during core reduction is the intermediate form. This form is elongated, with a longer length than width, but is relatively thin in comparison to a flake with a block form.

- *Flake Terminations*

The type of termination on a flake can indicate the stage of its removal from the core and the type of force used to detach the flake (Andrefsky 1998:68-87). Hinge and step terminations can reflect the end of a core's use-life, the need to change the manufacturing technique to maximise the material, the skill of the knapper, the presence of flaws or inclusions within the material or the force applied to the core (Hiscock 1988:117; Rosen 1997; Cotterell and Kamminga 1979:105; Whittaker 1994:109; Shelley 1990:191). Abrupt or plunge terminations occur when the knapper tries to extend the use-life of a core by applying a large hammer force to the platform, usually during in the early stages of core reduction (Cotterall and Kamminga 1979:106, 1987:699-701). In a quarry the primary reduction of cores will therefore produce larger numbers of abrupt terminations. Feather terminations are thought to be the desired outcome and occur at all stages of reduction (Cotterall and Kamminga 1987:699; Pelcin 1997:1107).

A platform termination is the result of rotating the core 180° and using a bi-directional flaking technique that results in a flake with a termination that ends on an old platform. This technique is probably undertaken to create a new platform and maximise the use of the material. A cortical termination occurs on a flake that abruptly ends on the cortical exterior of the core and usually occurs early in the core reduction. A cortical termination in conjunction with a cortical platform can be used to calculate the original size of the core. These two termination types were specially defined for this study.

- *Platform Characteristics*

The platform types were classified as missing, step, unifacial, bifacial or cortical based on the research design of Holdaway *et al.* (1998, 2000a). The type of platform can indicate the reduction strategy employed, the stage of reduction and provides information about the face of the core (Crabtree 1972:74; Sheets and Muto 1972:623). A cortical platform combined with a cortical exterior indicates initial reduction, while a cortical platform with no exterior cortex, although still occurring early in the reduction sequence, can be termed secondary flaking (Crabtree 1972:74). A unifacial platform struck from a flat surface can indicate a unidirectional core type and these tend to occur, although not always, early in core reduction (Magne and Pokotylo 1981:36). In comparison, a bifacial platform with two or more previous flakes scars on its surface shows the formation of a secondary flaking platform after core rotation (Andrefsky 1998:94-95; Magne and Pokotylo 1981:36). A flake with a missing or crushed platform implies that the platform collapsed during flake removal. A crushed platform is usually the result of a heavy hammerstone force, a lack of platform preparation, a bipolar reduction technique or an attempt to remove a flake with a small platform area (Bordes and Crabtree 1969:3; Magne 1989:19; Clarkson and David 1995:31; Witter 1992:34). A stepped platform represents overhang removal, a platform preparation technique often used on a core during blade manufacture (Flenniken and White 1985:135; Torrence 1986:195-196).

- *Dorsal Scarring*

The number of dorsal scars and the quadrant direction from where the scar originated were recorded for the complete flakes at Quarry 27 and Quarry 35. The number of dorsal scars is thought to indicate the amount of reduction a core underwent (Johnson 1987b:193; Andrefsky 1998:106; Shott 1994:80). The number of scars increases as the reduction continues in an inverse proportion to the percentage of cortex (Andrefsky 1998:104-106; Bamforth 1992:142). However, the number of flake scars can be influenced by the flake size, the flaking technique used, the raw material type and the type of artefact made (Andrefsky 1998:106; Odell 1989:178; Magne 1989:19; Mauldin and Amick 1989:73).

By using a quadrant system to orientate the flake (Figure 4.5), regularities in reduction can be observed which reflects whether or not core rotation occurred (Baumler 1988:263).

- *The Percentage of Dorsal Cortex*

The amount of cortex present on the dorsal surface of a flake indicates the amount of core reduction (Johnson 1989:127; Andrefsky 1998:101-102; Fullagar and Torrence 1992:36; Hiscock and Mitchell 1993:12-17; Bamforth 1992:142; Shott 1996:10-11). The first flakes removed from a cobble have a completely cortical dorsal surface and a cortical platform (Andrefsky 1998:94). Generally, the presence of any amount of cortex on flakes indicates initial core reduction (Mauldin and Amick 1989:70).
The amount of cortex present on the dorsal surface was divided into four groups (0%, 1-49%, 50-99%, 100%) for the analysis of the artefacts at Quarry 27 and the Lower Stud Creek units. To further refine the amount of cortex present on the dorsal surfaces of the flakes at Quarry 35 and the Upper Stud Creek unit, the percentage of cortex was further sub-divided into 0%, 1-25%, 26-50%, 51-75%, 76-99% and 100% groupings.

Cortex was also divided into two types in the Upper Stud Creek landscape sample. In this locality the type of cortex on the artefacts was used to identify the source of the stone. Weathered cortex, with a white/grey, rough and irregular patina, originates from a quarried outcrop. In contrast, the gibber cobbles have a red/brown, rounded and polished cortex.

- *Metrical Measurements*

Since core reduction is a subtractive process the size of the flaking debris reflects the amount of core reduction; as the core decreases in size so do the flakes removed (Odell 1989; Magne and Pokotylo 1981; Dibble and Pelcin 1995; Andrefsky 1998:96; Bamforth 1992:142). Callipers were used to measure the maximum dimension for all flakes, broken or complete. The axial length, maximum width and thickness at the mid-point, the platform width and thickness were also measured for the complete tools and flakes (Figure 4.5).

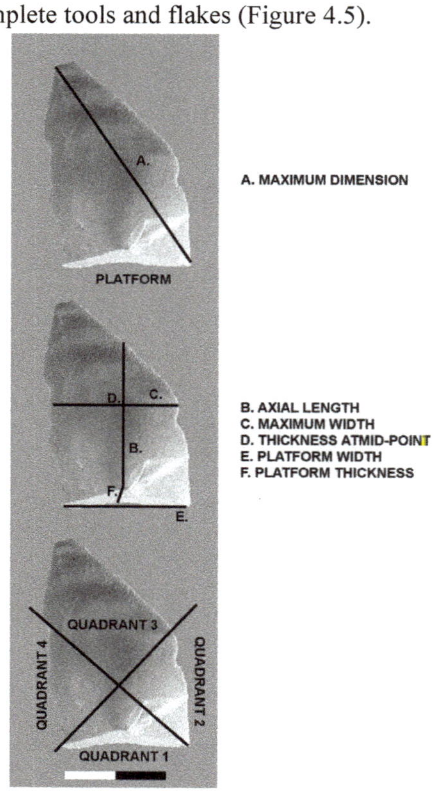

Figure 4.5: The Flake Dimensions Measured and the Location of the Quadrants (adapted from Holdaway and Stern 2004)

Platform size indicates the production strategies employed and the stage of reduction - whether the material was heavily reduced forming flakes with a large platform or finely manufactured with a controlled reduction indicated by thin, small focal platforms and prepared platforms (Magne and Pokotylo 1981; Cleghorn 1986:383; Odell 1989:185; Witter 1992:107; Hiscock 1988:371). Platform size may also reflect the desired flaking outcome and hint at the production of specialised or standardised tools (Andrefsky 1998:88). Small or focalised platforms, from striking the core near the edge with a controlled amount of force, are more common in blade technologies (Hiscock 1988:371; Fullagar and Torrence 1991:133-134). The platform angle also indicates the stage of reduction. As a core is reduced the angle of flake removals becomes more acute (Bamforth 1992:142; Magne and Pokotylo 1981:36; Pelcin 1997). A high platform angle is especially associated with blade manufacture (Witter 1992:48).

Other Technological Types Recorded

Grinding equipment has been grouped into four categories: mill slabs, pestles, mullers and mortars, based on McCarthy (1976). Presumably large grinders are rarely found within the landscape sampling units due to the activities of private collectors in the past. Local landowners, NSW National Parks and Wildlife Service in Tibooburra and the Tibooburra Aboriginal Land Council hold numerous large private collections. Other artefact types rarely found include axes, hammerstones, broken hammer stones, cylcons, ochre and spall tools. Spall tools were made from naturally occurring weathered spalls of silcrete and show retouch along their margins.

Tool Types

The tool types present in the study area are typical of those found during the middle to late Holocene in Australia (Witter 1992:132). These include a variety of flake tools, such as scrapers and notched tools, adze flakes (tulas and burrens), as well as the presence of blade tools, such as unifacial points (pirri) and geometric microliths. Non-retouched flakes with evidence of continuous macroscopic edge damage along a margin were termed usewear (Hiscock 1988:332).

Typological definitions, based on McCarthy (1976), White and O'Connell (1982), Flenniken and White (1985) and Holdaway *et al.* (1998), were used to firstly define the tool type. For the complete tools the number of retouched edges was counted, the quadrant position of the retouched edge recorded (Figure 4.5) and the type of retouch (usewear, notch, scraper, backing) was identified based on Holdaway *et al.* (1998).

The presence of core tools is a much-debated topic in Australia (*cf* Gould 1977; Mulvaney and Joyce 1965; McBryde 1974; Lampert 1981; Hiscock 1988; Witter 1992; Holdaway and Stern in press). For this study a core tool is a core with definite negative scars indicating flake removal plus the presence of retouch along the edges. Often retouch on cores has been confused with overhang removal or step terminations. A core tool was identified as having retouch on the margins of the cores that were not used as a platform.

Artefact Attributes Measured at the Quarries

Artefact recording was based on classification used Holdaway *et al.* (1998) during the TIB13 and the Lower Stud Creek (Stud 1 and 2) field seasons, but expanded to include specific attributes necessary for the interpretation of core reduction strategies at the quarries. The artefact recording system enabled the results from the quarry analysis to be comparable to those from the landscape sampling locations along the upper and lower margins of Stud Creek. The aim of the quarry analysis was to define the process of reduction based on artefact typology, morphological characteristics and size of the artefacts at each quarry location (Table 4.4). The technological analysis should reflect the process of reduction and define the reduction strategies used (Bamforth 1990:75). A fine-grained recording of the core types, flake and core characteristics present at the quarries allows for a detailed analysis of the strategies used to reduce the available stone.

Artefact Attributes Measured at the Landscape Sampling Units

The aim of the analysis of the Lower Stud Creek (Stud 1 and Stud 2) samples was to search for large-scale spatial patterning within each sample (Holdaway *et al.* 2000a). This required the analysis of larger numbers of artefacts over a wide area, consequently fewer attributes were recorded by WNSWAP (Table 4.5).

A more detailed analysis of the Upper Stud Creek landscape sample was undertaken for this study in order to contrast its analysis with the results from Quarry 35. The same core types and termination types were recorded at the Upper Stud Creek landscape sample and Quarry 35. A new cortex type category was included in the analysis to define the type of cortex present on the artefacts (Table 4.5).

4.3.2 Scale 2. The Characterisation of the Quarries and Landscape Sampling Units

The holistic study of different parts of an assemblage, the cores, flaking debris and tools, and their interrelationship with the environmental context forms the basis for the study of technological organisation (Nelson 1991; Larsen 1994:57). The goal of this approach is to ultimately link the characteristics of the artefacts with the problem-solving strategies people undertook regarding how they acquired, reduced, used and discarded material from different stone sources at the quarries and landscape sampling locations. To characterise the assemblages at the quarries and landscape sampling locations two approaches were used: a technological analysis and a spatial analysis.

- *The Technological Analysis*

Core reduction comprises of a series of different techniques and strategies that reflect the overall technology. A technique is defined as a distinctive method or process employed during manufacturing to remove a flake to achieve a particular goal (Flenniken and White 1985:131). Within any cultural system a range of different techniques were known requiring decisions to be made as to which technique should be used when appropriate (Flenniken 1985). A reduction strategy is the overall plan followed in stone reduction (Witter 1995:3) and refers to the general pattern of flake removals from a core (Bamforth 1992:132). A technological strategy, defined in studies of technological organisation and described in Chapter 2, is the sum of different techniques and reduction strategies employed and reflects how people in the past organised their activities by undertaking different problem-solved approaches (Hiscock 1988:22; Binford 1977, 1979; Shott 1986; Torrence 1983; Kelly 1988; Nelson 1991). The techniques and strategies used to reduce the available stone and the amount of core reduction occurring at any particular place influences the composition of the assemblage. Identifying the techniques and reduction strategies used to reduce the stone at the quarries and landscape sampling units enables comparisons to be made between the formation and composition of the different assemblages investigated. To define the 'role' stone from quarried and non-quarried stone sources played in the organisation of lithic technology the stages through which stone passed, described in Chapter 2.2, are compared.

Two main approaches are used to study core reduction and manufacturing techniques. These approaches are broadly divided into the French school (e.g. Perles 1992; Karlin and Julien 1994; Inzan *et al.* 1992; *cf* Jelinek 1991; *cf* Simek 1994) and the development of *chaîne opératoire,* and the American school of reduction sequence studies (e.g. Collins 1975; Bradley 1975).

Reduction sequence studies, although based on earlier observations, are mainly credited to Collins (1975) and Bradley (1975) who aimed at reconstructing a series of discrete stages for core reduction. This approach to lithic reduction makes inferences about the morphological change in the artefact type and form produced from different stages in core reduction (Sullivan and Rozen 1985:773, Magne 1989:89). Reduction stages were generally divided into acquisition, primary processing, tool manufacture and tool use, including rejuvenation and maintenance (Bradley 1975). More recent studies have incorporated metric analysis to "fine tune and gain a more precise reconstruction of stages" (Henry 1989:139). Recently Shott (1994:71) has argued that reduction should be viewed not as a series of discrete stages but as a continuum, including the analysis of both tools and debris. The continuum of core reduction should take into account that humans can respond to a situation in a multitude of different ways and not in a series of stages. However, it is useful when dealing with a large number of artefacts to group the data according to the characteristics which best reflect the stage of reduction.
The use of a *chaîne opératoire* approach to lithic analysis focuses on the process not the product and aims to define the "use history" of an artefact (Kuhn 1995:81). The "use history" is influenced by dynamic processes that involve three main stages through which an artefact passes, from raw material collection, transformation, to the abandonment of tools (Kuhn 1995:15). The goal of a researcher using a *chaîne opératoire* approach is to identify the conceptual framework the prehistoric knapper employed to reduce the available stone and ultimately trace the development of technical skills (Schlanger 1994:148; Karlin and Julien 1994). The approach aims to take into account all aspects of variability including raw material availability, tool use, site function and technological style (Simek 1994:119). A *chaîne opératoire* differs from reduction sequence studies as it includes all aspects of tool use, tool loss and breakage into the reduction scheme to understand the totality of assemblage formation (Simek 1994:119). Whereas researchers undertaking reduction studies have used statistical inference of large numbers of artefacts to define pattern, pattern using a *chaîne opératoire* is usually defined by refitting studies to find not only general stages in lithic technology and the patterning of the individual knapper, but also the cognitive realm of the prehistoric tool makers (Simek 1994:119-120).

The archaeological record is a complex amalgamation of many different and often overlapping knapping episodes occurring at different times and in different places. Consequently, it is frequently impossible to reconstruct the reduction sequence, the series of flake removals from one core, or trace the "flow" (Schiffer 1972) of the material produced from a core throughout the entire cultural system. Another method of assessing the archaeological record is to develop a generalised reduction strategy that describes the typical process of reduction viewed as a common set of knapping behaviours within an assemblage (Hiscock 1988:33-41).

Attributes	***Characteristic***
ID	unique number
Technological Types	complete flake, complete tool, complete split, broken tool, proximal tool, broken flake, proximal flake, core, core fragment, block, hammer, spall tool, coretool, coretool fragment, block, broken hammer
Raw Material	Group I (coarse clast) ,Group II (medium clast, fine clast), Group III (breccia, amorphous, matrix-dominated, banded), quartz, quartzite
Colour	grey, white, yellow, pink, spotty grey, black
Termination type	hinge, feather, step, abrupt, platform, cortical
Cortex	Quarry 27 - 0%, 1-49%, 50-99%, 100%
	Quarry 35 – 0%, 1-25%, 16-49%, 50-75%, 76-99%, 100%
Platform type	unifacial, bifacial, cortical, step, missing
Form	Quarry 27 - expanding, blade, block, intermediate
	Quarry 35 – point, expanding, blade, block, intermediate
Tools	
Tool Type	Adze flake, denticulate, geometric microlith, scraper, end scraper, burin, thumb nail scraper, pirri, notch, backed blade, borer, usewear
Type of retouch	usewear, scraper, step, notch, backing
Number of retouched edges	
Quadrants of retouch	Quadrants 1, 2, 3, 4
Flakes	
Maximum Dimension	All flakes and tools
Length	Complete flakes and complete tools (Figure 4.5)
Width	
Thickness.	
Platform width	
Platform thickness	
Scar number	
Scarring direction	Originating from quadrants 1, 2, 3, 4
Platform angle	Quarry 35
Cores	
Core type	uni-platform, bi-platform, prismatic, multi-platform, microblade, cobble, tranchet, bi-directional
Cross-section	Flake, lenticular, square, conical
Platform count	
Cortical platform	Number
Hinge termination	Number
Step termination	Number
Weight	
Flake scar number	
Length of the longest flake scar	
Width of the longest flake scar (only Quarry 35)	
Block length	Figure 4.3
Block width	
Block thickness	
Technical orientation	Greatest length of the core face from a platform
	Width at a right-angle to length
	Thickness at a right-angle width

Table 4.4: The Artefact Attributes Measured at the Quarries

Attribute	***Characteristic***
ID	unique number
Technological type	complete flake, complete tool, complete split, broken tool, proximal tool, broken flake, proximal flake, core, hammer, spall tool, block, broken hammer, millslab, muller, mortar, pestile, axe, cyclon, ochre
Raw Material	Silcrete;Group I (coarse clast) ,Group II (medium clast, fine clast), Group III(breccia, amorphous, matrix-dominated, banded), sandstone, quartz, hornfels, chalcedony, quartzite, petrified wood, quartz
Colour	Lower Stud Creek - grey, white, yellow, pink, clear
	Upper Stud Creek - black, spotty grey, grey, white, yellow, pink, clear
Termination	Lower Stud Creek - hinge, feather, step, abrupt
	Upper Stud Creek - hinge, feather, step, abrupt, cortical, platform
Cortex type	Upper Stud Creek - gibber, weather, no cortex
Cortex	Lower Stud Creek - 0%, 1-49%, 50-99%, 100%
	Upper Stud Creek - 0%, 1-25%, 26-49%, 50-75%, 76-99%, 100%
Platform	Lower Stud Creek - unifacial, bifacial, cortical, n/a
	Upper Stud Creek - unifacial, bifacial, cortical, step, n/a
Form	Lower Stud Creek - expanding, blade, block, intermediate
	Lower Stud Creek - expanding, blade, block, intermediate, point
Tools	
Tool Type	adze flake, denticulate, backed blade (geometric microlith), scraper, end scraper, burin, thumbnail scraper, pirri, notch, usewear
Number of retouched quads	Numeric
Type of retouch	Usewear, scraper, step, notch or backed
Quadrants of retouch	Quads 1, 2, 3, 4
Flakes	
Maximum dimension	All flakes
Length	Complete flakes and tools (Figure 4.5)
Width	
Thickness	
Platform width	
Platform thickness	
Platform angle	Upper Stud Creek Unit and Stud 2
Cores	
Core type	Lower Stud Creek – uni-platform, bi-platform, microblade, cobble
	Upper Stud Creek - uni-platform, bi-platform, microblade, cobble, tranchet, prismatic, bi-directional, horsehoof, multi-platform
Length of the longest flake scar	
Length	Figure 4.3
Width	
Thickness	

Table 4.5: The Artefact Attributes Measured at the Landscape Samples

For this study core reduction was defined as a continuous process which incorporates all aspects of stone reduction, from initial procurement and manufacture, to tool use and place of discard. In this definition an aspect of *chaîne opératoire,* incorporating the use of tools into the reduction model will help define the general "life history" (Schiffer 1972, 1976) of the different stone sources, rather than individual artefacts, within the cultural system. The broad stages of reduction identified in Figure 2.1 is used as a heuristic device to compare the reduction of cores obtained from Quarry 27, Quarry 35 and the non-quarried stone sources near the landscape sampling units and used to identify the different techniques and reduction strategies employed. Where specific reduction stages can be observed in an assemblage, such as material testing, initial reduction and core preparation or blade manufacture, the characteristics of the artefacts that represent these stages will be discussed. Distinguishing how stone from the different stone sources was extracted, reduced, removed and used can be used to establish their role in the organisation of lithic technology. Contrasting the factors that influence the use of stone sources can aid in gaining an understanding of the underlying technological strategies used by people in the past to successfully live in their environment.

- *The Spatial Analysis*

A GIS provides a powerful tool in the spatial analysis of an assemblage. For this study a GIS analysis was used to establish whether spatial variation occurs in the distribution of artefacts that reflect the different stages of core reduction observed in the quarry assemblages. A GIS was also used to gain an understanding of whether the environmental features present at the quarries influenced the discard of the artefacts.

Without taking into account the spatial variation within an assemblage, and the factors that influence variation, a biased interpretation of the structure of the archaeological record will be presented. Spatial analysis within GIS can be used to place an artefact in 'real' space, using AMG co-ordinates, to compare the artefacts between each other and with the associated landform surface or feature. This method enables measurements of distance from other artefacts or features, thereby allowing a manipulation of any data type in any combination. Within a non-site distributional approach a GIS allows the empirical unit of comparison to be the artefact from which spatial patterning can be investigated. A GIS generates new information about the relationships between elements within the cultural system and is not simply map building but incorporates data manipulation (Savage 1990:22-23). Ultimately a GIS can allow for an examination of cultural processes both synchronically and diachronically within a landscape approach (Savage 1990:29-30).

If the landscape is viewed as having both cultural and environmental variables, then a GIS is an ideal vehicle for determining how these features interrelate across space (*cf* Allen *et al.* 1990; Kvamme 1989). Within GIS overlays of variables can be created and compared to determine spatial patterns. As Zvelebil *et al.* (1992:195) suggest a GIS provides the method for determining spatial patterning by integrating natural and cultural factors, the continuity of human use of space and the importance of 'scale' as it determines both the use of space and the interpretation of its use by archaeologists. GIS allows for space to be explored at many scales, from the single artefact to the wider landscape.

Two techniques within an *ArcInfo/ArcView* GIS, density plots and buffering, were used to search for patterns in the distribution of the artefacts at the quarries . *Arcview* density plots of artefact attributes, such as weight, size and raw material type, were compared in order to find where concentrated areas of different artefact attributes occur within the quarry. A density plot is a visual expression of the distribution of artefact types and their characteristics. It can highlight the spatial differences in reduction by comparing the various core and flake attributes. Plots can be created using *Spatial Analysis*, an extension program of *ArcView*. This program can build a rasta image from vector data, meaning any point map using vector data can then be converted to a density plot giving a rasta image. Specifying the same grid distance and search radius for each plot and then overlying the results within *ArcView* allows the density plots of any artefactual characteristics to be compared. The *Spatial Analysis* program will automatically seek like attributes within the given area and calculate a density. This tool is extremely helpful when the objective is to search for relationships in the distribution of artefact types or attributes.

A buffering technique around selected artefact types was used to support the results of the density plots. This approach was advocated by Johnson's (1984) Local Density Analysis (LDA) and used by Holdaway *et al.* (1998). As an example, particular core types can be buffered using a circular sampling unit of any radius (r) centred on the core. Overlapping buffers show a higher density of the core type and enables all artefacts falling within that radius to be extracted, using the individual ID numbers, and compared statistically between each buffer. By varying the r lengths of the buffer the scale where variation occurs in the quarry assemblages can be identified. Buffering can also be centred on features (landform, outcrop or cobbles) to determine whether particular features influenced artefact distribution. For example, as the distance from the feature increases the characteristics of the artefacts may also vary. *Microsoft Access* queries can be used to examine assemblage variation by comparing the mean size and frequency of the cores and flake with different attributes. The results can be further compared using statistical tests such as chi square tests and Analysis of Variance.

Assemblage comparisons can also be undertaken by measuring the degree of diversity between two assemblages. Diversity measures have a long history in evolutionary studies that focus on adaptive variation and are generally used to characterise the structure of populations (Jones and Leonard 1989:1-2). Archaeology has adopted diversity concepts to measure variation within or between assemblages. *ArcView* density plots may not show artefact diversity or variation but simply present a high artefact density. By measuring artefact diversity, density as an influencing factor in assessing artefact variation was removed. *Diverse*, is a computer program written by Frankel and Holdaway (1993) based on the diversity formulas explored by Bobrowsky and Ball (1989). This test is independent of density in its measurement of an assemblage's degree of richness, evenness and heterogeneity or the "sameness" of the assemblage. As a result, the test can be used to quantify the amount of assemblage variation present within a sampling area and between sampling areas. Measures of diversity are used at Quarry 35 to select a sampling location for the analysis of the flaking debris. The goal of this analysis was to identify an area that has high assemblage diversity and contrast the results of its analysis with two other areas that have low diversity. The degree of diversity was used to assess whether the same reduction strategies or many different reduction strategies were used to create the spatial variation observed in the quarry assemblage. The underlying premise of this argument is that "sameness" of an assemblage would be higher if the artefacts in a particular area were discarded at the same stage of reduction.

The core types and various key core characteristics, such as the percentage of cortex, often indicate the amount of reduction (refer to section 4.2.1). The characteristics of the cores can be shown in space using GIS, which may highlight specific areas where different stages of core reduction took place. The results of the core analysis can then be compared to the analysis of the flaking debris. The quality of the raw material influences both the manufacturing and production of stone tools (Cotterell and Kamminga 1987; Dibble and Rolland 1992). The location of different core types made from different material types may reflect the position of the outcropping material or the selection for a particular raw material type. Buffering the outcrop and cobbles allows comparisons to be made between cores that occur near the outcrop and those that occur at varying distances.

An understanding of the quarry assemblage is fundamental in searching for patterns in distribution of the artefacts that reflect the different stage of core reduction. This methodology incorporates not only clear definitions of a quarry 'boundary' but also the implementation of the appropriate sampling techniques. In this way any spatial or statistical test undertaken can be justified. Furthermore, the use of an individual artefact as the basic unit of analysis, within a non-site distributional approach (e.g. Ebert 1992), allows the scale of analysis to be expanded hierarchically taking into account all the possible variables that influence the spatial distribution of a quarry assemblage. Hence, the artefact represents the highest degree of resolution and smallest scale of comparison. The search for spatial patterns that reflect different stages of reduction can be used to define assemblages at a location, such as at a quarry. Understanding the characteristics of an assemblage at any one location allows comparisons to be made between locations. In this way the quarries can be compared with the sampling locations to define the use of the different raw material sources in the study area.

Relationships between artefacts and features can easily be established using a GIS and can be used to define the entire quarry – the interrelationship of the artefact in space with other artefacts, their characteristics and the associated landscape features. By defining the entire quarry assemblage at Quarry 27 and examining the spatial relationships occurring within the quarry an appropriate sampling methodology was developed for Quarry 35. This approach aimed to establish the quarry 'boundary' and acknowledged the problems of time conflation to assess the quarrying behaviour. A detailed account of the specific field methods used in Quarry 27 and Quarry 35 occurs in the following chapters (Chapters 5 and 7).

- *Creating an Approach to the Analysis of the Quarry Assemblages*

Quarries have been largely ignored by archaeologists for several reasons. Purdy (1984:119) lists these reasons as being no stratigraphic control, the absence of formal artefact types and the problems of dating a quarry assemblage. Often quarries are seen as "a mass of data creating nothing but serious methodological problems" (Ericson 1984:2). However, as Ericson stresses, the quarry is the logical place to begin when studying a stone-tool culture as it helps to define the entire "lithic production system" (Ericson 1984:2). In the mid-1980s a plethora of literature was published on quarrying and mining (Ericson and Purdy 1984; Butler and May 1984; *World Archaeology* overview 1984; Vehik 1985a). These articles presented the leading studies and new directions both in methodology and theory for quarry research at the time. However, little attention was given to the definition of a quarry, the internal spatial structure of quarries or the problems of using different sampling techniques. More recent studies have concentrated on the development of theoretical issues relating to the organisation of stone production (e.g. Montet-white and Holen 1991; Ellis and Lothrop 1989).

A serious oversight in quarry research has been the failure to discuss methodological approaches or take into account the problems of assemblage variation at quarries. Furthermore, there has been a lack of detailed study on the processes and techniques of quarrying (Amaman *et al.* 1992:4). Quarries are notoriously large, with many hundreds or thousands of artefacts, and require appropriate sampling strategies. A variety of methods have been used to record artefact assemblages at quarries. Most quarry studies have undertaken mass collections of artefacts from varying sized sampling units, either systematically or randomly sampled, to record the quarry assemblage (e.g. Singer and Ericson 1977; Singer 1984; Bamforth 1992:134; Byrne 1980; Fullagar and Torrence 1991; Arnold 1987a, 1987b). A number of studies firstly defined areas of high density and either made random collections or recorded artefacts using a variety of methods including grids or transects (e.g. Torrence 1984, 1986; Hartwell and Amick 1993; Wilke and Schroth 1989, Beck and Jones 1994, Jones 1984). Often studies of quarries have either not stated their sampling methodology, making it difficult to assess the quality of the work, been purely descriptive or dealt solely with theoretical issues (e.g. Turner and Bonica 1998; Stevenson *et al.* 1984; McBryde 1973; Clark *et al.* 1978; Reher 1991). Excavations have been used to establish a time depth in the use of quarries and determine whether change through time occurred in the reduction strategies employed (e.g. Elston and Raven 1992; Gramly 1980, 1984; Root 1992; Arnold 1987a, 1987b; Dortch and Gardiner 1976). Refitting studies have also been used as a valuable tool to reconstruct the core reduction strategies (e.g. Jones 1984; Leach 1984; Singer 1984). Few studies have used an "on-site" analysis of the flaking debris found at quarries (e.g. Beck and Jones 1994; Hartwell 1994; Torrence 1984, 1986). At these quarries cores and flakes were generally not collected but recorded at the quarry and returned to their original location.

Among the best methodological approaches to the study of a quarry is Torrence's (1984, 1986) work on the Greek island of Melos. The field methodology was designed to investigate patterns of exchange using data collected from a series of obsidian outcrops and artefact scatters in two locations. The density of the artefacts and the location of the outcrops and their characteristics (e.g. size, shape, condition, and ease of extraction) were recorded. Three sampling methods were used: systematic, "grab" and targeted. After laying a square grid of 25 m^2 across the area of highest artefact density at both quarries, all the artefacts in each quarter m^2 were recorded "on-site". Activity areas, away from the outcrops, were also defined and a number of "grab" samples were taken. A targeted collection of the cores in the survey grid was removed from the quarries for further analysis. Following a similar approach Arnold (1987a) used both a stratified sampling strategy to collect artefacts on different landscape surfaces and excavation to examine craft specialisation at the quarries on Santa Cruz Island, California.

Bamforth's (1992:131) criticisms of previous research on quarries focused on the lack of amalgamation between studies of the distribution of stone and the ways people procured and prepared stone at quarries. By focusing on four quarries in the central Mojave Desert, southeastern California he identified several aspects of raw material variation that influenced human behaviour and needed to be examined in any quarry study. These aspects include the distribution of source areas in space, the accessibility [abundance, density and ease of extraction] of stone at the source and the quality of the stone. During the course of this research these aspects form a guideline placing the two quarries investigated in context. The geological context is an important variable that influences how people in the past used the available stone resources.

The lack of quarry research in Australia prompted Hiscock and Mitchell (1990, 1993) to attempt to define a "type profile" of quarries and reduction sites in Australia. They reviewed past studies, definitions of quarries, theoretical frameworks and methodological approaches to quarries and aimed at classifying quarries according to their characteristics. Hiscock and Mitchell (1993:72-73) found that quarries were largely ignored in Australia because they were thought to have little information potential and were too difficult to analyse. The studies undertaken were criticised as being too descriptive and failing to employ "the full range of methodologies that can be applied to quarries" (Hiscock and Mitchell 1993:73). Hiscock and Mitchell (1993:73-74) also noted a bias in the study of axe quarries, to the detriment of other types of quarries.

The most widely referenced and researched axe quarry is the Mt William Quarry in Victoria. Although considerable work was undertaken on the petrology and distribution of axes from Mt William Quarry little attention was given to the reduction techniques employed, the spatial variation of the discarded artefacts and their characteristics within the quarry (McBryde 1978, 1984a 1984b, 1986; McBryde and Watchman 1976; Binns and McBryde 1972; Watchman and Freedman 1978).

While few studies have focused entirely only on quarries in Australia (e.g. Law 2003; Lamb 1996; Baker 1987; Barker and Schron 1994) data from quarries have formed secondary contextual data for wider distributional studies (e.g. Hiscock 1986b, 1988; Byrne 1980; Bird 1985). As most of the quarry studies in Australia provide only a physical description of the quarry Hiscock and Mitchell (1993:74) concluded that it was impossible to determine the range of variation in the types of quarries, the products manufactured or make inferences about human behaviour.

Quarries have great potential to add to theoretical debates surrounding the organisation of technology. As Purdy (1984:119) notes, "quarries are not subject to routine inquiry." Furthermore, many of the problems acknowledged in surface archaeology are often ignored in quarry studies. Few studies have dealt with the problems of sampling, or overprinting, admitted to time conflation and lack of a strict chronology, defined boundaries or assessed the density distributions and the spatial variation of artefacts. The lack of quantifiable research focusing on quarries in Australia has created a gap in our understanding of an important part of the archaeological record.

A number of valuable ethnographic accounts of quarrying behaviour in Australia that provide useful information on the techniques used to reduce stone have been undertaken (Jones and White 1988; Binford and O'Connell 1984;

Gould 1977, 1980; Horne and Aiston 1924; Aiston 1928; Thomson 1983). Simple ethnographic analogies are often used to interpret quarrying behaviour without any attempt at explaining the actual complexity of the formation, distribution and composition of a quarry assemblage. Quarry methodologies need to incorporate an understanding of the entire quarry to establish where and how spatial variation occurs and to justify the sampling method used. Making these steps enables a more accurate interpretation of quarrying behaviour.

According to Hiscock and Mitchell (1990:30) to characterise a quarry assemblage three main criteria of assessment should be followed; defining the raw material procurement strategy, the nature of the reduction, and the stages of reduction present within the quarry. A detailed appraisal of procurement incorporates an examination of the raw material type and the method of extraction. The nature of reduction includes an analysis of the reduction techniques and strategies employed at the quarry. The stages of reduction can be seen in the changes cores underwent at the quarry and the assessment of the product or products made and removed from the quarry.

For this study a detailed examination of the geological characteristics of the quarry (outcrop type and size, original cobble or cobble size and raw material types available) and the features that contributed to artefact displacement was undertaken. Analysis of the artefacts from the quarries was undertaken in two stages; a technological analysis of the cores, flakes and tools found in the quarry and a spatial analysis of the distribution of the artefacts that reflect the reduction strategies employed (described below). The analysis aims to answer the questions posed in Table 4.2.

A general description of the entire quarry assemblage focuses on the frequency of the different technological types and types of raw material available. The quarry assemblage is then broken down into the three main components: cores, flaking debris and tools. The characteristics of each group was then described in terms of the frequency of the different attributes, using tests of association, such as a chi-square test, and tests of population differences on ordinal values, such as t-tests and analysis of variance (*cf* Fletcher and Lock 1994). The results of the analysis were used to determine the reduction techniques and strategies employed and to identify the form of the material that was removed from the quarry.

- *Approach to the Analysis of the Landscape Sampling Units*

The aim of analysing the landscape sampling units was to establish how the different raw material sources, quarried outcrops and gibber cobbles were acquired, reduced and discarded within each of the units. The questions related to this scale of analysis are in Table 4.2. Each unit was described in terms of the location, the available raw material sources and the post-depositional factors which influenced the spatial integrity of the assemblage.

To assess how similar the assemblages in the three units were, a comparison of the frequency of different technological types was undertaken. At the Upper Stud Creek Unit the type of cortex, representing the source of the stone, was identified. The characteristics of the cores from the Upper Stud Creek Unit were used to contrast the cores made from quarried outcrops and gibber cobbles using various statistical tests to compare their frequency and size. A series of predictions were made about the characteristics of the cores that reflect the source. As the type of cortex was not recorded at the Lower Stud Creek units these predictions were used to describe the cores and extrapolate the results to the stone sources used at Stud 1 and Stud 2. The characteristics of the flaking debris were then described and the frequency of the different attributes compared. The analysis of the tools aimed to determine whether different tools were made from different sources of stone and different qualities of silcrete. Many researchers in Australia have observed differences in the types of tools made from different raw materials (e.g. Gould 1977). The results are used to establish the reduction strategies employed at the landscape sampling units for the stone obtained from quarried outcrops and non-quarried local sources.

4.3.3 Scale 3 - Comparing the Different Stone Sources

Each individual quarry may vary according to any number of characteristics, such as the raw material type, the amount of "on-site" reduction, the product removed or the location, consequently there is no such thing as a 'typical quarry'. However, common features, such as the type of outcrop, are present at each quarry location and collectively these may influence the spatial distribution of the artefacts. In order to define these features and their influence, and to quantify how *different* two quarries might be, a statistical comparison of the metrical and morphological characteristics of the cores and flakes from Quarry 27 and Quarry 35 was undertaken. In turn, this comparison helped to establish how best to sample a quarry and to redefine a quarry. These results could then be compared with the acquisition, reduction and use of the non-quarried stone resources to further understand the role of that the available stone resources had in the organisation of lithic technology.

The technological analysis of the artefacts at the quarries allowed comparisons to be made between the assemblages. The ways material was extracted, reduced and removed from each of the quarries was compared using a variety of methods. The amount or "intensity" of core reduction can be reflected in the size of the artefacts and the frequency of different technological types and their characteristics (Kuhn 1991:82). The characteristics of the artefacts also reflect the ways cores were reduced at each quarry and in what form material was removed from the quarry. Statistical tests of association and tests of difference were used to compare the nominal and ordinal data from the flakes and cores in the assemblages. Reduction charts, developed by Witter (1992:101-107), were used to establish a profile of the size and form of the complete flakes in the quarry assemblages. Ratios of elongation of the complete flakes in the two quarries were also used to define the form of the flakes removed from the cores (Clarkson and David 1995:31). The results of the analysis were used to compare the techniques and strategies of core reduction used at both of the quarries.

Assemblage Comparisons

The frequency of different artefact types can reflect the amount of core reduction, the intensity of use and the type of use, such as a quarry or occupation location (Torrence 1986:181-189; Hiscock and Mitchell 1993:28-32; Witter 1992:132; Hiscock 1988:81-84; Chatters 1989:359-36). By describing the frequency of different artefact types, such as cores, tools and grinding equipment, within an assemblage an idea can be gained about the different types of activities that were undertaken at a particular location (Binford 1978a, 1978b). However, assemblage variability in an organised system can also be used to identify the technological strategies employed by people to deal with their associated "venture and contingency risks", already described in Chapter 2 (Elston 1990:161). It is useful to compare the ratios of various technological types within an assemblage as this method focuses on the composition of the assemblage and not the number of different technological types. Although artefacts were transported away from a stone source, discarded throughout the landscape, used many times and sometimes recycled, it is the discard behaviour that is important – where and why artefacts were discarded in particular locations. It is this reason why assemblage comparisons are meaningful, reflecting the ways people organised their technology.

- *Complete Flake to Core Ratio*

The ratio of cores to flakes can be used to measure the amount of core reduction occurring at any location. If there is a low ratio of flakes to cores then this may indicate that the cores were not heavily reduced and were discarded early in reduction (Kuhn 1991:82). However, if a high number of flakes were present in relation to the number of cores then more flakes were removed from each core.

- *Broken Flake to Complete Flake Ratio*

The number of broken flakes reflects the amount of breakages during core reduction and the affects of trampling on an assemblage (Gifford-Gonzalez *et al.* 1985; Prentiss and Romanski 1990:97; Hall and Love 1985; Hiscock 1986b). A high number of broken flakes to complete flakes suggest that considerable manufacturing or trampling had occurred in that location.

Complete Flake to Block Ratio

Blocks usually occur early in core reduction when a large hammer force has acted on the core (Torrence 1986:181-189; Andrefsky 1998:81-83). Consequently, a high ratio of blocks to complete flakes not only indicates core reduction but early core reduction.

- *Complete Flake to Complete Tool Ratio*

A high ratio of complete tools to complete flakes may indicate a later stage of reduction and determine the functional characteristics of the sampling location (Binford 1978a, 1978b). A high diversity of tool types may imply a more permanent, long-term occupation or frequent, episodic occupation, particularly if the tools are highly curated (Shott 1997). High numbers of tools made from local stone may also suggest a decrease in mobility and an increasing reliance on stone that was immediately available (Elston 1990:162).

- *Complete Flake to Complete Split Ratio*

The frequency of complete splits in an assemblage can be used as a marker for the stage of core reduction (Clarkson and David 1995:33). A high number of complete splits to complete flakes suggest that more early-stage core reduction had occurred in the assemblage.

The Influence of Spatial Variation and Time on the Artefacts within the Quarry Assemblages

A number of common features were present in both quarries. These features include the type of outcrop, the raw material types available, the topographic, landform surfaces and geomorphic location. To investigate whether these features influenced the composition and distribution of the different reduction strategies employed the results of the spatial analysis from each quarry assemblage (Scale 2) were compared to establish the similarities and differences between the two quarries. Each feature and the artefacts associated with it were compared from each quarry to examine whether patterns emerged in the distribution of the assemblage.

The degree of spatial differentiation of core reduction activities has been equated with the amount of specialist production undertaken at a quarry (Arnold 1987a, 1987b; Torrence 1984, 1986; Fullagar and Torrence 1992:135). If a high degree of segregation of different reduction activities occurred at a quarry then this was due to specialist knappers who were highly organised and controlled the working and distribution of stone away from the quarry. The degree of spatial variation in the characteristics of the artefacts that reflect reduction is reflected in the 'strength' of the patterns. If distinct areas exist where different reduction activities were concentrated then this may indicate a more structured use of the quarry. However, these patterns may also be a repeated response to the distribution of the landform features and/or the time invested in the reduction of stone at a quarry. The many interacting factors, such as time processes, reduction strategies and environmental context, which contribute the formation and distribution of quarry assemblages need to be teased apart. In Chapter 9 these factors are explored and used to further define the role of the stone sources in the organisation of lithic technology (Scale 4).

What is a Quarry and how does it Compare to the Reduction of the Non-Quarried Gibber Cobbles?

By comparing the two quarries, and gaining some understanding of the variability in the distribution, formation and composition of a quarry assemblage, it was possible to contrast the strategies used to reduce the quarried stone with the reduction of the cobbles obtained from a non-quarried stone source. The results of the analysis of the landscape sampling units was used to compare the ways cores obtained from the different stone sources were reduced, manufactured into tools and what types of tools were made from each of the sources. The definition of a quarry, identified by the organisational implications behind the procurement and reduction of this type of stone source, could then be re-examined by comparing the results of quarry analysis with the use of stone obtained from the gibber pavement.

The use of the different qualities of silcrete was compared between the landscape sampling units and the quarries in order to assess whether the type of silcrete (defined in Chapter 3) influenced the tool form and use. By comparing the tool types and the raw material throughout the sampling locations patterns may emerge that can be related to the use of the different stone sources.

4.3.4 Scale 4 - Establishing the 'Role' of Each Stone Source in the Organisation of Lithic Technology within the Study Area

Studies using an organisation of technology approach have often analysed the distribution of artefacts over wide areas – the landscape (e.g. Kelly 1988; Ebert 1992; Nelson 1991; Ricklis and Cox 1993). Kelly (1994:133) argued that to find organisational patterns in the distribution of artefacts a wide region must be investigated to define the relationships between "activities, raw material, tool manufacture, use and discard that occur in different places". This study attempts to define the 'role' of the different stone resources in the organisation of technology within an area rich in lithic resources: the Mt Wood Ranges. The scale of the study is not large, the study area does not cover the entire distribution of stone obtained from the ranges nor does it encompass the entire pattern of settlement-subsistence thought to exist for prehistoric hunter/gatherers (*cf* Binford 1980). However, it does provide valuable insights into the procurement, manufacture and use of the different stone resources that can be used to make wider inferences about the ways people organised themselves in the landscape.

To fully investigate strategies of stone reduction the key is to understand the stone sources (Ericson 1984). This understanding is the first step and provides valuable information about the ways stone was procured and reduced. A detailed analysis of the available stone sources needs to be undertaken or only wide, often simplistic generalisations can be made which limit the information potential of the archaeological record. In the study area an abundance of high quality material is available at quarried outcrops while the gibber pavement provides an almost continuous source of material. These two types of sources represent the range of stone resources available in the region. By undertaking a detailed analysis of these two types of stone sources the results allow inferences to be made about their 'role' in the organisation of lithic technology.

The investigation of the available stone resources is the "baseline" from which technological patterns in stone use can be examined (Bamforth 1992:150). After teasing apart the multiple factors which interact and influence the formation, composition and distribution of artefacts in the study area a better understanding of the types of technological strategies employed can be gained.
Petrological examination was used to source stone in Australia (Binns and McBryde 1972; McBryde 1979; McBryde and Harrison 1981). This method is not possible within the study area as outcrops of similar silcrete types can be found in very different locations. However, if stone reduction is dependent on type of source and raw material type, and this is reflected in the end-product discarded at the landscape sampling units, it does not matter which outcrop the stone came from as it is the patterns in the discard of the raw material sources that are important. These patterns reflect whether different procurement and manufacturing strategies occurred for the different types of stone sources.

The data obtained from Quarry 35 and Quarry 27 provides an ideal opportunity to investigate the reduction strategies used at the quarries and to compare these results with data collected from the Upper and Lower Stud Creek samples. This comparison serves two purposes; to characterise the reduction of non-quarried stone and to identify how material obtained from a quarried stone source was discarded. At this scale, factors such as the distance from the quarry (although small) and raw material type available, can be compared between the Upper and Lower Stud Creek sampling units in order to define which factors have influenced the composition of these assemblages. In this way the reduction processes can be established for each type of stone source.

4.4 SUMMARY

This chapter presented the fieldwork methodology and analytical framework used to interpret patterns of variation in the distribution of artefacts within the study area. The field methodology provides the structure of the data collection on which the analysis was based. The combination of the artefact's characteristics and spatial location allows the archaeological record to be reduced to the finest scale of analysis - the individual artefact. From an individual artefact, with a single point in space, data can be examined at any scale to search for pattern and to ultimately answer the research question; what 'role' did the different stone sources play in the organisation of lithic technology.

The analytical approach firstly focuses on the artefact and the influence its characteristics have on the interpretation of the reduction process. The approach combines a technological analysis with the use of a GIS as an analytical tool to search for patterns of raw material use. Four scales of interpretation were examined ranging from the artefact, to the role of the raw material sources in the organisation of lithic technology.

In the following five chapters the analytical framework provided in this chapter is used to organise and analyse the data. By examining the characteristics and position of each artefact within the quarries and landscape sampling units the assemblages can be characterised and compared to search for patterns in the use and distribution of the available raw material sources.

Chapter 5

Artefact Analysis of Quarry 27

5.1 INTRODUCTION

At Quarry 27 cobbles of high quality Group III silcrete were used as a source of stone. These cobbles were found only in a small, localised area within a larger outcrop of Group II silcrete. The characteristics and position of all the cores, flaking debris and tools greater than 20 mm in maximum diameter and surrounding the cobbles of Group III silcrete were recorded. The aim of recording of an entire quarry assemblage was to identify how stone from a high quality raw material source was extracted, reduced and the form in which material was removed from the quarry. By recording all the artefacts at Quarry 27 any sampling bias in the results and interpretation were avoided, allowing the characterisation of the entire quarry assemblage.

Quarry 27 was selected for a detailed analysis for several reasons: firstly, the outcrop of Group II silcrete was easily visible within the landscape as a single exposure surrounded by gibber pavement (Figure 5.1). The cobbles of Group III silcrete within this outcrop were the focus of quarrying activity and these are concentrated in only one area of the larger Group II outcrop. Most of the flaking debris are not widely dispersed but were discarded around the Group III cobbles. Secondly, the relatively small size, and medium density of the quarry, made it feasible to record within a single field season. Lastly, mapping the areas of disturbance in the survey grid could assess the spatial integrity of an artefact's position.

5.2 DESCRIPTION OF THE QUARRY

Quarry 27 lies on the eastern slope of the Mt Wood Range approximately 120 m ASL here the range gradually descends towards the Bulloo Embayment (Figure 4.1). The outcrop is a single, isolated exposure located on a small tributary of the Twelve Mile Creek, approximately two kilometres north of the Twelve Mile Creek Gorge. During the Quaternary water erosion formed a small gully and subsequent erosion has exposed the outcrop on the eastern face of this gully (Figure 5.2). Surrounding the gully are extensive gibber plains.

The assemblage recorded at Quarry 27 includes 3383 artefacts and 166 Group III silcrete cobbles. Most of the artefacts fall within the main survey grid (n=3179), which centred on the quarried outcrop, and only 204 artefacts were found in the survey transects (Figure 5.1). The aim of surveying these transects was to define the quarry boundaries. The survey transects were established to determine whether any artefact concentrations occurred beyond the survey grid and to further define the boundary of the quarry. The density of artefacts in the survey grid is 7 per m^2 while a density of only 0.1 per m^2 artefacts was found in the transects.

5.2.1 Geological Characteristics

Most of the silcrete available at Quarry 27 was derived from Tertiary sandstones that silicified to form Group II silcrete. The silicification of finer material, such as a siltstone or claystone, resulted in a small pocket of Group III silcrete being formed within the massive layer of Group II (medium clast) silcrete. The total outcrop, mainly composed of Group II silcrete, stretches 104 m from north to south and 60 m from east to west. Only in a small area of the outcrop, 21 m from north to south and 17 m from east to west, were cobbles of Group III silcrete found (Figure 5.1). The nearest outcrops of Group III silcrete occurs downstream within the Twelve Mile Creek Gorge (Figure 4.1). No other outcrops of Group III cobbles can be found along this tributary of Twelve Mile Creek.

The outcrop has weathered to the surface as large, densely packed cobbles. The mean cobble length is 270 mm (n=166, Standard Deviation, SD=116 mm). Where the cobbles have been subject to exposure, and subsequent weathering, natural spalling of the stone has occurred, decreasing the amount of cortex present on their surfaces. This process substantially reduces the amount of cortex not only on the cobbles but also on the cores and flaking debris.

At Quarry 27 the lithology of the outcrop ranges from a low quality, medium clast silcrete (Group II) to a high-quality, white porcellanite (Group III) silicified from a fine-grained homogenous claystone (refer to Chapter 3 for descriptions of the raw material types). Most of the outcrop is composed of medium clast silcrete but only limited amounts of this material type are artefacts (n= 176, 5%). Most of the artefacts recorded at the quarry (Table 5.1) were made from the high quality Group III silcrete (n=3200, 95%).

There is little variation in the colour of the cobbles. Most of the artefacts at Quarry 27 made from the Group III silcrete are grey (n=2779, 82%) and are derived from the adjacent outcrop. Only two cobbles are white porcellenite or pink, Group III silcrete (Table 5.1).

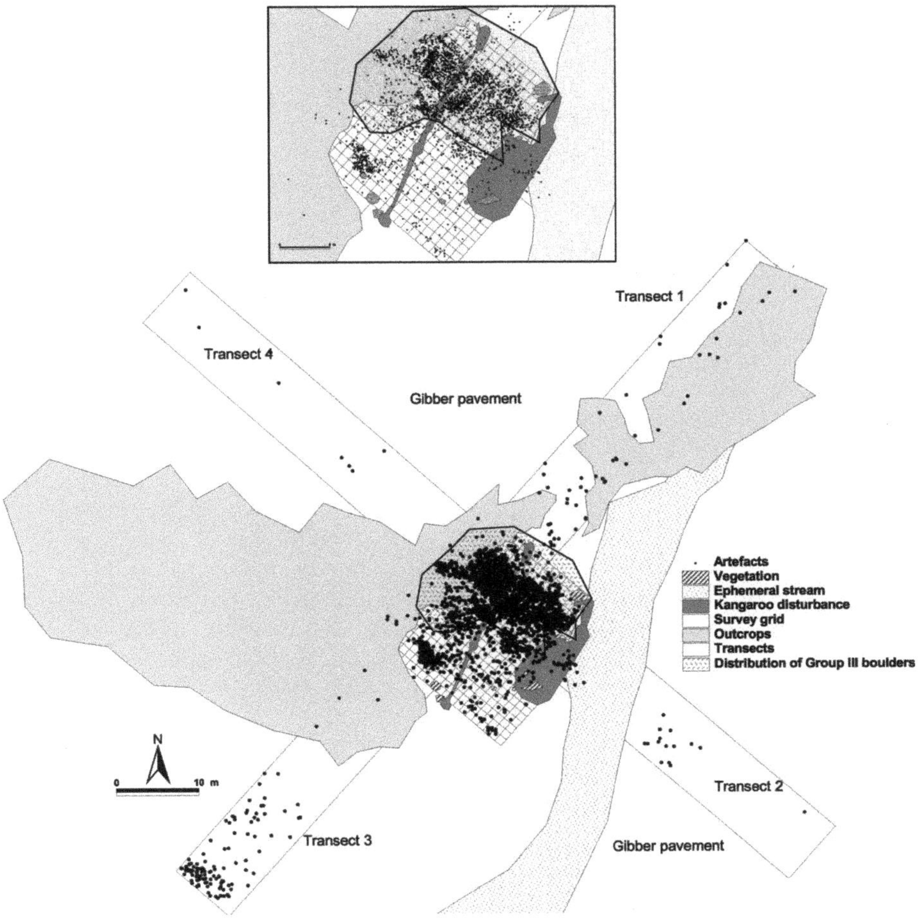

Figure 5.1: Quarry 27

Material Group	Material Types	Count	%	Raw Material Colour			
				Grey	Yellow	Pink	White
Group III silcrete	Amorphous	361	11	357	3	1	
	Banded	148	4	148			
	Breccia	39	1	38		1	
	Matrix Dominated	2637	78	2624		13	
	Porcellanite	15	0.4				15
	Total	3200	95	2779	3	15	15
Group II silcrete	Fine Clast	4	0.1	3			1
	Medium Clast	176	5	141	15	20	1
	Total	180	5	144	15	20	
Other	Quartz	1	0.03				1
	Quartzite	2	0.06			2	
	Total	3	0.09			2	1

Table 5.1: Raw Material Types found at Quarry 27 and the Survey Transects

Some of the artefacts in the quarry assemblage were made from other stone non-local sources; these include quartzite, milky quartz and either a fine or medium clast Group II silcrete (Table 5.1). The yellow silcrete (n=18, 1%) was also obtained from another source and discarded at Quarry 27. Most (n=15) of the yellow artefacts are hammerstones made from gibber cobbles. Two quartzite hammerstones and several pink or yellow, medium clast hammerstones made from gibber cobbles were also found in the quarry assemblage.

5.2.2 Identifying Artefact Displacement in the Quarry

The geomorphological processes that created the modern landscape of the study area were discussed in Chapter 3, and a detailed description of the impact of the various geomorphological processes on the spatial integrity of the quarry assemblage was presented. During the fieldwork, features that might have influenced the movement of artefacts were mapped to provide a basis for assessing the extent of artefact disturbance. These features include places of dense vegetation growth, kangaroo tracks and sleeping areas, and low-lying areas subject to water wash. In the small, ephemeral stream (refer to Chapter 3.4.4) sporadic water flow has dispersed the artefacts located in the northeast corner of the surveyed area (Figure 5.1). The burying of artefacts by colluvium, and the downslope movement of artefacts, has also influenced the spatial integrity of the Quarry 27 assemblage.

Substantial artefact movement and burying is unlikely to have occurred except in areas where dense vegetation occurs (43 m^2), kangaroo tracks or sleeping areas (68 m^2), and wash along the stream margins (26 m^2), covering a total area of 137 m^2. Artefacts could have been removed from these areas by water movement and/or animal disturbance. Often these areas overlap, indicating that several different processes might have affected the locations the artefacts. Where the northeastern corner of the survey grid and the stream intersect only a few artefacts (n=10) are present in the survey grid (Figure 5.1 and 5.5). In this low-lying area it is likely that water movement has size-sorted artefacts so that only large artefacts remain. Large artefacts, with a maximum dimension of 55 mm (SD=28 mm), cluster at the stream margin and indicate that water flow has removed the smaller artefacts and redeposited the larger flakes. Previous experimental studies, and three-dimensional computer simulations of artefact movement and size-sorting in the study area, supports the idea that sheet-wash has contributed to the movement of artefacts with a maximum dimension of less than 30 mm (Webb 1996; Schick 1987a, 1987b; Pigdon 1996; Holdaway *et al.* 1998; Fanning and Holdaway in press, a).

Most of the artefacts in the survey grid are part of a surface scatter. However, some artefacts may have been buried by colluvium since fine silts have been deposited down the slope. The redeposition of colluvium resulted from post-European erosion of the landscape (Holdaway *et al.* 1998). Smaller artefacts may have been incorporated into the soil matrix leaving the larger artefacts on the surface. Without excavating the colluvium in the survey grid the degree of artefact burial is unknown.

From the outcrop to the stream there is a vertical drop of approximately three metres resulting in a gradient of nine degrees (Figure 5.3). The slope raises the possibility that gravity has caused down-slope movement of the larger artefacts (Rick 1976). The maximum dimension of the complete flakes is large (57 mm) but their size and flat shape may have hindered substantial down-slope movement of the artefact assemblage (Pigdon 1996; Greenwood 1997; Fanning and Holdaway in press, a; Reid and Frostick 1985:147-148). Gravity is more likely to have influenced the location of the cores since their larger size and spherical shape makes them more susceptible to downslope movement (Reid and Frostick 1985:147-148). The cores were discarded in the area where the Group III cobbles outcrop (Figure 5.1). If considerable downslope movement of the cores can occurred they would not be concentrated on the outcrop but would have been dispersed along the slope or clustered at the lowest point of the slope.

The lack of any substantial disturbance of the surface assemblage is supported by the presence of artefact conjoins in the Stud Creek Catchment (Greenwood 1997). Two discrete areas of conjoins could be observed within the survey grid at Quarry 27 (Figure 5.4), recognised due to the different colour and material type of the flaking debris in each case. Two clusters of pink (n=12 conjoins) and white (n=11 conjoins) silcrete artifacts were refitted; these comprise of cores and flakes. The bulk of the flaking debris in each area of conjoining artefacts clusters within a 2.5 m diameter (Figure 5.4). With this in mind, it is possible to suggest that only a limited degree of artefact movement has affected the spatial patterning of debris discarded at the quarry.

Figure 5.2: View of Quarry 27 (looking west)

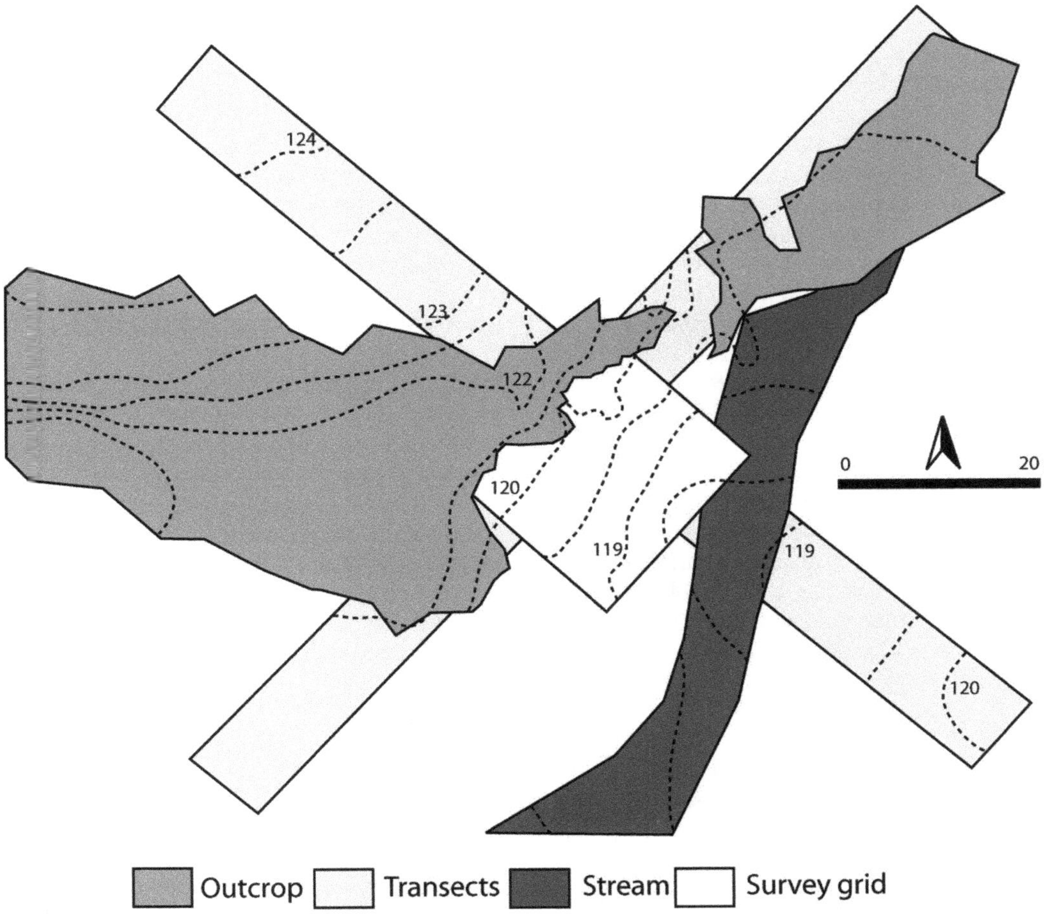

Figure 5.2: Topographic Map of Quarry 27

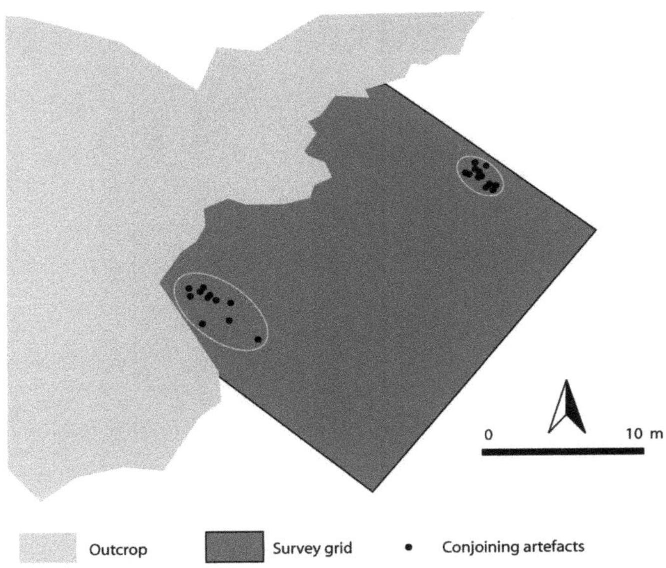

Figure 5.3: The Location of Two Areas of Conjoins at Quarry 27

5.3 SPECIFIC FIELDWORK METHOD

At Quarry 27 a central position for a surveying datum was selected and a series of readings using a Geographic Positioning System (GPS) provided its three-dimensional coordinates. The average reading was used to determine the datum position in Australian Grid Map coordinates (AGM) and also to orientate the electronic theodolite. A marker was randomly thrown onto the outcrop and its position used as the central reference point to begin surveying. Around this point, unit Area 1 (A1) was strung as a 1 m square. The eight squares surrounding A1 were then gridded to produce a 3 by 3 m square labelled as unit A2. This method was used to grid the next 16 units, each extending the grid one metre, and labelled consecutively (Figure 5.5). Surveying continued until there were less than two artefacts per metre square. The number of artefacts was recorded for each 1 m square in order to assess the density and distribution of the quarry assemblage. The expansion of the survey grid followed the squares with the greatest artefact concentrations. The survey grid (Figure 5.5) covered an area of 22 by 22 metres (484 m^2). The mapping and artefact recording methods described in Chapter 4 were applied to entire quarry.

After the artefact analysis was completed, the perimeter of the survey grid was mapped and transects were established. Four 50 m transects were surveyed from unit A3 (Figure 5.1, 5.5). The artefacts within these transects were also pegged and recorded. Geomorphological (Figure 5.1) and topographic maps were completed using an electronic theodolite (Figure 5.3). The mapping recorded the position of the outcrop, the location of the Group III cobbles within the outcrop, the location of the survey grid and transects, and the features which might have disturbed the artefact locations (Figure 5.1).

5.4 ANALYSIS OF THE ARTEFACTS

The quarry assemblage is the result of many overlapping knapping events. Without data drawn from replication experiments or conjoining analysis individual reduction sequences cannot be reconstructed. Nevertheless, the use of different knapping strategies influences the composition of an assemblage (Collins 1975:17). By studying the technological characteristics of the artefacts the core reduction strategies can be identified. The core and flake attributes that provide such technological information are described in Chapter 4.

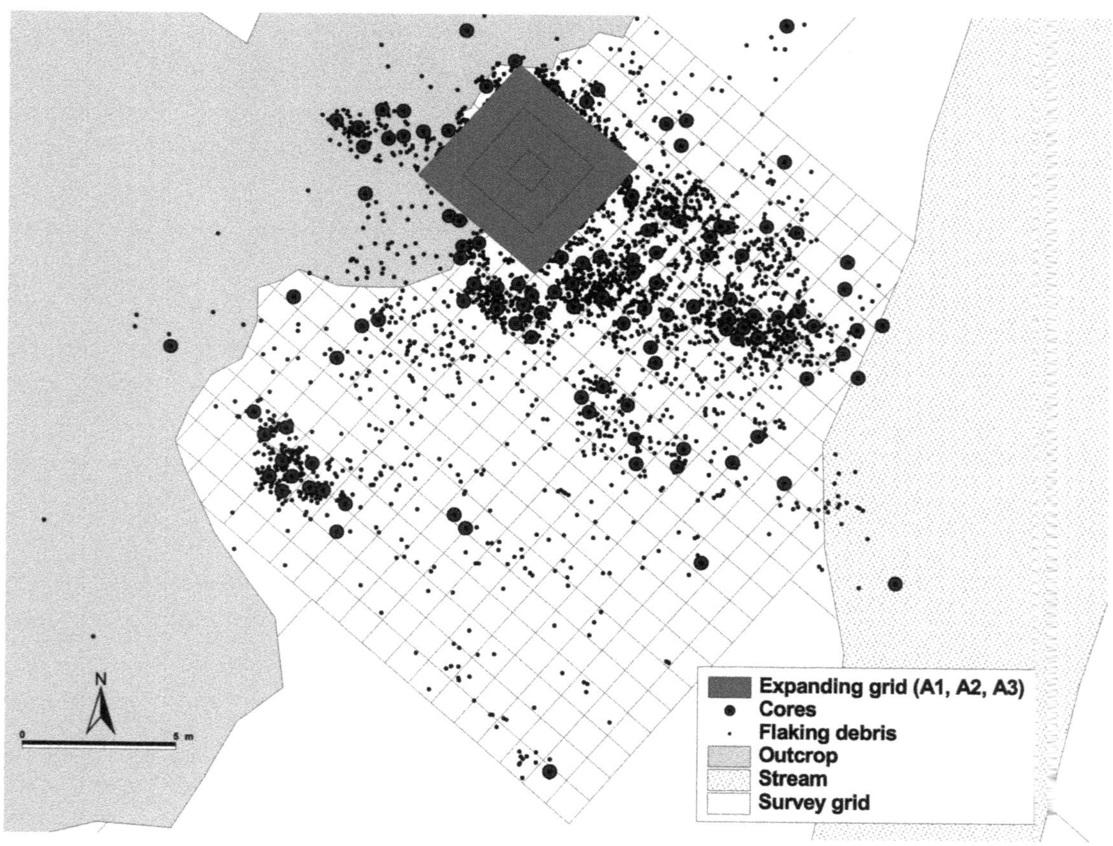

Figure 5.5: The Expanding Survey Grid and the Extent of the Survey Grid at Quarry 27

The first stage of the analysis describes the characteristics of the cores and flaking debris found within the quarry assemblage. This data was used to identify the reduction strategies that contributed to the formation of the Quarry 27 assemblage. The aim of this analyse was to answer four specific questions already presented in Table 4.2:

- How was material obtained from the outcrop?
- How was the stone worked?
- In what form was material removed from the quarry?
- What tools were manufactured and/or discarded at the quarry?

5.4.1 General Assemblage Characteristics

A significant proportion of the quarry assemblage comprises of block fragments, hammerstones, cores, core fragments, broken flakes, complete flakes, complete splits and platform rejuvenation flakes (Table 5.2), artefact types that are generally regarded as characteristic of quarry assemblages (Hiscock and Mitchell 1993:12-17; Hiscock 1988:107).

Only 116 tools (4%) were recorded in the quarry assemblage. Most of the tools were made from complete flakes, but some were also made from naturally occurring silcrete spalls (Table 5.2).

	Artefact Type	Count	%
Cores	Core	208	6
	Core Fragment	116	3
Non-retouched flakes	Complete flake	1358	40
	Platform Rejuvenation Flake	28	1
	Complete Split	331	10
	Distal Flake	249	7
	Proximal Flake	64	2
	Broken Flake	607	18
Retouched flakes	Complete Tool	79	2
	Distal Tool	14	0.4
	Proximal Tool	2	0.1
	Broken Tool	16	1
Retouched Spall	Spall Tool	5	0.1
Other	Hammer Stone	50	1
Debris	Block	256	8
	Total	**3383**	

Table 5.2 The Technological Types Present at Quarry 27

5.4.2 Available Raw Material Types

Cobbles of Group III silcrete (n=166) were found only in one discrete location within the larger outcrop of Group II, medium clast silcrete (Figure 5.1). The weathering and spalling of the cobbles helps to identify the types of silcrete present. Most of the Group III cobbles in the outcrop are still wedged within the substrate and overlain by colluvium (Table 5.3). Of the 166 Group III cobbles recorded in the survey area 70% (n=117) were partially visible on the surface. Consequently, the extraction of flakeable material at Quarry 27 simply involved the removal of the cobble from the surrounding colluvium. Buried cobbles were significantly larger than those cobbles already exposed at the surface (Table 5.3; t=3.56, df=109, p<0.0001). These exposed cobbles are more similar in length to the cobble cores at 157 mm in maximum length (SD=77 mm). However, a significant size difference occurs between the exposed cobbles and the cobble cores indicating that smaller exposed cobbles were selected as cores (t=3.9, df=73, p<0.0001). Only 44 cobbles of Group III silcrete were fully exposed, ranging from 135-495 mm in length, and only 16 of these were less then 200 mm in maximum diameter.

	Count	Mean Length (mm)	SD (mm)
Buried cobbles	117	291	112
Exposed cobbles	49	210	103
Total	166	267	116

Table 5.3: The Size of the Exposed and Buried Cobbles in the Survey Grid

A large number of gibber hammerstones (n=48) and two quartzite hammerstones were discarded in the survey grid. The gibber cobbles were probably obtained from the gibber pavement surrounding the quarry. Quartzite was not available in the immediate vicinity of the quarry but can be found as cobbles in a few streams located in the Mt Wood Ranges.

5.4.3 Analysis of the Cores

Of the 208 cores recorded in the survey grid and transects only six different core types were identified (Table 5.4). Core types, and the attributes recorded for them, were described in Chapter 4 and their relationship to the process of core reduction was discussed. Looking only at the survey grid, 189 (97%) cores were made from Group III silcrete and only three cores were made from medium clast (n=2) or fine clast (n=1) Group II silcrete (Table 5.5). It is the Group III cores found in the survey grid that are the focus of the core analysis (Figure 5.1). The different types of cores, their specific characteristics, and their relationship to the reduction strategies employed are discussed in the following section. The multi-platform, bi-platform and uni-platform cores have similar characteristics and are discussed together.

Core Type	Total Count of all Cores Recorded		Group III Silcrete Cores		Transects	
	Count	%	Count	%	Count	%
Cobble	37	18	31	17		
Uni-platform	83	40	76	41	2	29
Bi-platform	61	29	53	28	4	57
Multi-platform	25	12	24	13		
Microblade	2	1	1	1	1	14
Prismatic	1	1	1	1		
Total	208		189		7	

Table 5.4: The Different Core Types in the Survey Grid and Transects

	Material Type	Survey Grid	Total Count in Survey Grid and Transects
Group II	Fine Clast	1	1
	Medium Clast	2	8
	Total	3	9
Group III	Banded	21	22
	Amorphous	7	8
	Matrix-Dominated	159	167
	Porcellanite	2	2
	Total	192	208

Table 5.5: The Raw Material Types of the Cores in the Survey Grid and Transects

- **The Cobble Cores**

The characteristics of the cobble cores, such as the size, weight and amount of surface cortex, are indicative of a limited amount of core reduction. The 31 cobble cores found in the survey grid have large amounts of cortex on their exterior surfaces (Table 5.6). In comparison to the other core types the cobble cores are large and heavy, and had a few, relatively small flakes removed from them. The small flake scars, in comparison to the orientated platform length of the cobble cores, indicates that the entire face was not used to remove flakes (Table 5.7). That only one or two small flakes were removed from these cores suggests that flaking was undertaken simply to remove the cortex and view the quality of the underlying material.

The 'testing' of the cobbles to identify the material type was necessary as cobbles of Group II and Group III silcrete were present in the outcrop. Since most of the artefacts at Quarry 27 were made from Group III silcrete a selection for the reduction of cobbles made from Group III silcrete is evident (Table 5.5). The cobble cores that were not selected and reduced further may have flaws or fractures in the rock, the size and shape of the cobble was not suitable or they could not be extracted from the surrounding substrate (Bamforth 1992:132)

The cobble cores have few platforms and a large number of these are cortical compared to the other core types (Table 5.7). The presence of cortical platforms and the lack of core rotation suggest a limited amount of reduction. Cobble cores that do not have any cortical platforms (46%, n=16) indicate that flakes were removed from an already exposed platform surface created either by weathering or bifacial flake removals (Witter 1992:54).

The number of flake scars with step or hinge terminations on the face of a core is considered indicative of the amount of core reduction (Hiscock 1988:117). As a core undergoes further reduction, and its weight decreases, the number of flakes produced with these termination types increases. The cobble cores in this assemblage have few flake scars with step or hinge terminations (Table 5.8).

Core Type	Percentage of cortex	Total		Scar Length (mm)		Weight (g)		Length (mm)	
		Count	%	Mean	SD	Mean	SD	Mean	SD
Bi-Platform	0%	1	2	34	6	100	57	70	16
	1-49%	30	57	51	13	344	273	93	24
	50-99%	22	42	57	17	477	324	112	36
	Sub-total	53	28	53	15	388	299	100	31
Multi-Platform	0%	2	8	63	18	210	191	72	29
	1-49%	18	75	56	28	555	350	106	28
	50-99%	4	17	61	14	611	273	115	21
	Sub-total	24	12	57	25	536	335	105	28
Cobble	0%								
	1-49%	2	7	24	6	854	204	188	101
	50-99%	29	93	45	26	707	336	148	73
	Sub-total	31	17	43	25	781	372	157	77
Uni-Platform	0%	5	6	33	8	126	70	86	21
	1-49%	42	58	44	18	288	234	97	25
	50-99%	29	36	63	30	582	338	120	41
	Sub-total	76	41	53	35	396	317	109	43
Prismatic	0%								
	1-49%	1	100	46		350		104	
	50-99%								
	Sub-total	1	1	46		350		104	
Microblade	0%	1	100	44		50		45	
	1-49%								
	50-99%								
	Sub-total	1		44		50		45	
	Total	189							

Table 5.6: The Characteristics of the Different Types of Cores made from Group III Silcrete in Relationship to the Amount of Cortex on their Surfaces

		Core Type					
		Bi-Platform	Multi-Platform	Cobble	Uni-Platform	Microblade	Prismatic
Length (mm)	Mean	100	105	157	109	45	104
	SD	31	28	77	43		
Width (mm)	Mean	81	89	133	95	45	104
	SD	30	24	81	45		
Thickness (mm)	Mean	59	67	95	63	37	74
	SD	22	25	55	31		
Orientated Platform Length (mm)	Mean	63	72	81	60	45	56
	SD	26	35	59	34		
Length of Flake Scars (mm)	Mean	52	57	43	53	44	46
	SD	15	25	25	35		
Flake Number	Mean	4	6	2	2	2	4
	SD	2	2	2	1		
Number of Cortical Platforms	Mean	0.3	0.2	0.9	0.1		
	SD	0.6	0.5	2	0.3		
Number of Platforms	Mean	2	3	1	1	2	1
	SD		0.7	0.4			

Table 5.7: The Characteristics of the Group III Cores in the Survey Grid

Number	Microblade		Uni-Platform		Cobble		Prismatic		Multi-Platform		Bi-Platform		Total	
	Hinge	Step	Hinge	Step	Hinge	Step	Hinge	Step	Hinge	Step	Hinge	Step	Hinge	Step
0	1	1	48	50	26	21			7	9	27	27	109	108
1			24	16	6	6		1	11	4	17	11	58	38
2			4	2	1	4	1		4	4	15	9	14	19
3			1	2		2			2	3	4	3	8	10
4				1		1				1				3
5										1				1
6										1				1
7														
8										1				1

Table 5.8: The Number of Group III Cores with Step or Hinge Terminations in the Survey Grid

- **Uni-platform, Bi-platform and Multi-platform Cores**

The most common core types in the quarry are uni-platform and bi-platform cores (Table 5.4). If cores maintain a large size and weight when discarded then this usually indicates that they were not reduced to any considerable degree (Magne and Pokotylo 1981:38). The similar sizes and weights of the multi-platform, uni-platform and bi-platform cores suggest that these core types were not substantially reduced but were discarded at a similar stage of reduction (Table 5.6).

The size and weight of the multi-platform, uni-platform and bi-platform cores indicates that they are considerably smaller than the cobble cores and have undergone further reduction (Table 5.6, 5.7). However, unlike the uni-platform cores more flakes were removed from the bi-platform and multi-platform cores than the cobble cores (Table 5.7). An average of two flakes were removed from each uni-platform and cobble core, while an average of four and six flakes respectively were removed from the bi-platform and multi-platform cores (Table 5.7). The number of cortical platforms on the multi-platform, uni-platform and bi-platform cores is less than on the cobble cores and hints at the removal of cortex to form a new platform (Table 5.7).

There is a significant difference in the length (Figure 4.3) of the cobble cores compared to the multi-platform, uni-platform and bi-platform cores (f=11.289, df=3, p<0.0001) but no significant difference in length between the multi-platform, bi-platform and uni-platform cores (f=1.07, df=2, p=0.3). The sizes of these later core types are similar even though the bi-platform and multi-platform cores have undergone rotation to establish new platforms and have had more flakes removed (Table 5.7).

The uni-platform and bi-platform cores generally have a high percentage of cortex (50-99%) on the core surface (Table 5.6). Most of the bi-platform cores (42%, n=22) have more than 50% of their surfaces covered in cortex. A similar pattern emerges for the uni-platform cores (Table 5.6). It is clear that there is a relationship between the percentage of cortex on the core surface, the mean scar length, the weight and the block length for these core types (Table 5.6). Smaller, lighter uni-platform and bi-platform cores have less cortex on their surfaces and shorter mean flake scar lengths (Table 5.6). The cores with no cortex underwent more reduction before they were discarded (Clarkson and David 1995:31). Only nine cores in the survey grid were completely decortified indicating that the core surface was prepared only enough to allow the removal of a few non-cortical flakes before the core was discarded (Table 5.6).

The multi-platform cores have more platforms, a greater number of flakes removed and less cortex on their surfaces than the other core types (Tables 5.5, 5.7). Of the 25 multi-platform cores in the survey grid 75% (n=18) have between 1-49% of their surfaces covered in cortex, while only 17% (n=4) have more than 50% of their surfaces covered in cortex (Table 5.6). It seems evident that the size of the original cobble used to manufacture the multi-platform cores was large allowing further platform rotations and a greater number of flake removals. Regardless, the multi-platform cores were discarded when they were the same size as the uni-platform and bi-platform cores, although more flakes were removed from them.

By comparing the lengths of the flake scars to the orientated platform length of the multi-platform, uni-platform and bi-platform cores it is evident that the entire core face was used to produce the maximum flake size possible (Table 5.7). In contrast, the flake scars on the cobble cores are considerably smaller than the orientated platform length (Table 5.7). Not only were the flake scars large but no significant difference in scar length (f=0.955, df=3, p=0.415) was found between the bi-platform, uni-platform and multi-platform cores. It seems likely that these cores were discarded when they could no longer produce the desired flake size. Based on the length of the flake scars large flakes, approximately 50 mm in length, were produced before the core was discarded (Table 5.7).

The bi-platform and multi-platform cores have a larger number of flake scars with a step or hinge termination on the core surface than do the uni-platform or cobble cores (Table 5.8). The presence of steps or hinges on a core makes the further removal of flakes more difficult and often leads to the discard of the core (Whittaker 1994:106-108). Since more flakes were removed from the bi-platform and multi-platform cores the likelihood of errors occurring during manufacture is greater for these core types. However, the overall low numbers of hinge and step terminations on the cores in the survey grid indicates that most of the cores, independent of type, were discarded early in reduction (Table 5.8). Of the 189 Group III cores in the survey grid most (n=108, 58%) have flake scars that end with feather terminations and only 35 (18%) cores had two or more flake scars with step terminations (Table 5.8). The flake scars with hinge terminations show a similar pattern, with only 23 (12%) cores having two or more hinge terminations present on flake scars.

Most of the cores in the survey grid (73%, n=135) have retained the square cross section of the original cobble (Table 5.9). The uni-platform cores that underwent greater reduction have a conical cross section (Witter 1992:47-48). If flakes are continually removed from all the margins of a single platform the core begins to taper and forms a conical cross section. Cores with a conical cross section are smaller, have small flake scars and have less cortex on their surfaces than the other core types (Table 5.9). The ventral surfaces of large flakes (n=20) were also used to remove small flakes (Table 5.9). The size of the flake scars made on these cores are significantly smaller than the cores with a square or lenticular cross section (f=2.6, df=3, p=0.04) but not significantly smaller than the flake scars on the cores with a conical section (t=1.328, df=33, p=0.1932). The length of the uni-platform cores with a conical or flake cross section are also not significantly different (t=6.831, df=33, p=0.4993). These results indicate that the uni-platform cores with a conical or flake section underwent not only the same amount of reduction but the greatest amount of reduction.

From the above comparisons it is apparent that there are few differences between the multi-platform, uni-platform and bi-platform cores with respect to most of the core attributes. The most notable similarities between these three core types are the lengths of flake scars, the orientated platform length and the size of the cores. It is these three features which strongly suggest that the aim of core reduction at Quarry 27 was to manufacture large flakes. This necessitated the testing and selection of suitable cobble cores and the removal of cortex to prepare the core face before large, decortified flakes could be manufactured.

Section	Square		Conical		Lenticular		Flake	
Core Type	**Count**	**%**	**Count**	**%**	**Count**	**%**	**Count**	**%**
Bi-platform	37	70	6	11	7	13	3	6
Multi-platform	16	67	2	8	5	21	1	4
Cobble	31	100						
Uni-platform	51	67	5	7	4	5	16	21
Prismatic			1	100				
Microblade			1	100				
Total	135 (73%)		15 (8%)		16 (9%)		20 (11%)	
			Cortex					
0%	2	1	3	20			4	20
1-49%	61	45	11	73	10	63	13	65
50-99%	72	53	1	7	6	38	3	15
	Mean	**SD**	**Mean**	**SD**	**Mean**	**SD**	**Mean**	**SD**
Flake Scar Length mm)	52	26	46	13	60	27	39	17

Table 5.9: The Characteristics of the Cores made from Group III Silcrete with Different Cross Section Types in the Survey Grid

- **Blade-Producing Cores**

There are only three blade-producing cores in the survey grid and the transect assemblages (Table 5.4). Two blade cores were discarded in the survey grid and one prismatic core was discarded in Transect 3 (Figure 5.1). Clearly, blade production was not the main working strategy employed at the quarry. The blade cores were substantially smaller and lighter than the other core types (Table 5.6, 5.7) but still had a considerable uselife shown by the lack of flakes scars with hinge or step terminations (Table 5.8), the block size and the length of the flake scars (Whittaker 1994:109).

Cores with a conical cross section indicate greater core reduction and the use of a specific reduction strategy to remove flakes usually associated with blade production (Witter 1992: 47-48). Some cores in the survey grid have a conical section but only two, the prismatic and microblade cores, were used to manufacture blades (Table 5.9).

5.4.4 Analysis of the Flaking Debris

- **Description of the Flake Assemblage in the Survey Grid**

The non-retouched flakes of Group III silcrete found in the survey grid are the focus of the flake analysis. The flake assemblage from this grid contains 1221 complete flakes made from Group III silcrete and only 50 made from Group II silcrete (Table 5.10). A count of the platforms from the platform rejuvenation flakes, complete flakes and tools, proximal flakes and tools and half the number of the complete splits indicates that at least 1533 flakes were discarded in the survey grid.

A large number of complete splits, blocks and broken flakes are present in this assemblage (Table 5.10). These three types of artefacts usually occur during initial core reduction and indicate that a considerable force was used to remove flakes (Andrefsky 1998:81-83). The blocks are large and have large amounts of cortex on their surfaces, another feature consistent with early core reduction. In comparison to the blocks, the broken flakes are generally small and have little cortex on the dorsal surfaces (Table 5.10). Flakes can be broken during manufacture or the result of trampling (Prentiss and Romanski 1990:97; Hall and Love 1985; Andrefsky 1998:81). Based on the small amount of cortex on the broken flakes it is evident that they were manufactured later in reduction.

- **The Description of the Complete Flakes in the Survey Grid Assemblage**

The first flakes struck from a cobble generally have a cortical platform and a dorsal surface completely covered in cortex (Andrefsky 1998:101-102). In the survey grid only 27 (2%) of the complete flakes have these characteristics. The length of the complete flakes with an entirely cortical dorsal surface is significantly smaller (t=2.49, df=120, p<0.0001) than the complete flakes with 50-99% cortex on the dorsal surface, and these are the same size as complete flakes with 1-49% cortex on their dorsal surfaces (Table 5.11). The fact that only a few, small flakes with cortical dorsal surfaces and cortical platforms were identified in the survey grid assemblage indicates that these may have resulted from the 'testing' of the cobbles.

The mean length of the complete flakes with both a cortical termination and a cortical platform provides an indication of the orientated length of the cobble as these flakes traverse the entire length of the core face. The lengths of the flakes (n=8) range from 19 to 91 mm but average at 56 mm (SD=19 mm), a similar length to the flake scars on the uni-platform, bi-platform and multi-platform cores (Table 5.7).

The largest flakes in the survey grid generally have a block or intermediate form, a large, cortical platform and a large maximum width and thickness at the mid-point (Tables 5.11, 5.12). Large flakes have comparatively greater numbers of abrupt, plunge, cortical or platform terminations and a higher proportion of cortex on the dorsal surface than the smaller flakes (Table 5.12). Figure 5.6 illustrates the relationship between the termination type and the length of complete flakes. The largest flakes have plunge, abrupt or cortical terminations indicating that these flakes were the first to be removed from the core while the smallest flakes have a step, hinge or feather termination and were removed from the core last. The large, complete flakes with plunge, abrupt or cortical terminations traveled the entire length of the core face and were probably used to decortify the surface of the core.

	Artefact Type	Total		Maximum Dimension (mm)		Percentage of Cortex							
						0%		1-49%		50-99%		100%	
		Count	%	Mean	SD	Count	%	Count	%	Count	%	Count	%
Unretouched Flakes	Complete Flake	1221	44	57	25	347	28	560	46	256	21	58	5
	Platform Rejuvenation	25	1	76	23	8	32	10	40	7	28		
	Complete split	315	11	51	19	105	33	143	45	57	18	10	3
	Distal flake	223	8	45	19	87	39	91	41	31	14	14	6
	Proximal flake	60	2	40	15	26	43	27	45	7	12		
	Broken flake	566	20	38	14	258	46	206	36	74	13	28	5
	Sub-total	2410	86			831		1037		432		110	
Retouched Flakes	Complete tool	68	2	65	24	19	28	43	63	6	9		
	Distal tool	10		67	21	3	30	4	40	3	30		
	Proximal tool	2		64	47	1	50	1	50				
	Broken tool	13		60	28	3	23	10	77				
Spall	Spall tool	3		111	23			1	33	2	67		
	Sub-total	96	2			26		59		11			
Other	Block	249	9	61	24	26	10	107	43	107	43	9	4
	Total	2755		51	23	857		1144		539		119	

Table 5.10: The Different Artefact Types made from Group III Silcrete in the Survey Grid

Attribute	Total		Percentage of Cortex							
			0%		1-49%		50-99%		100%	
Termination	Count	%	Count	%	Count	%	Count	%	Count	%
Abrupt	178	15	48	14	92	17	35	14	2	3
Cortical	225	19			116	21	90	35	19	33
Feather	580	48	209	61	245	44	99	39	28	48
Hinge	11[illegible]	9	48	14	46	8	16	6	2	3
Platform	10[illegible]	1	4	1	6	1		0		
Plunge	38[illegible]	3	8	2	21	4	8	3	1	2
Step	66[illegible]	5	27	8	27	5	6	2	6	10
Total	120[illegible]		344		553		254		58	
Blade	13[illegible]	1	11	3	1		70	27		
Block	19[illegible]	16	22	6	93	17	28	11	9	16
Expand	14[illegible]	12	42	12	68	12			9	16
Intermediate	86[illegible]	71	272	78	398	71	158	62	40	69
Total	122[illegible]		347		560		256		58	
Bifacial	73	6	19	5	35	6	17	7	2	3
Cortical	14[illegible]	11			78	14	43	17	19	33
Missing	22[illegible]	19	66	19	99	18	52	20	11	19
Step	25	2	16	5	7	1	1		1	2
Unifacial	75[illegible]	62	246	71	341	61	143	56	25	43
Total (%)	122[illegible]		347		560		256		58	
Length (mm)	Mean	SD	Mean	SD	Mean	SD	Mean	SD	Mean	SD
	45	23	34	16	47	22	55	28	47	22

Table 5.11: The Characteristics of the Complete Flakes made from Group III Silcrete in Relationship to the Amount of Cortex on their Dorsal Surfaces

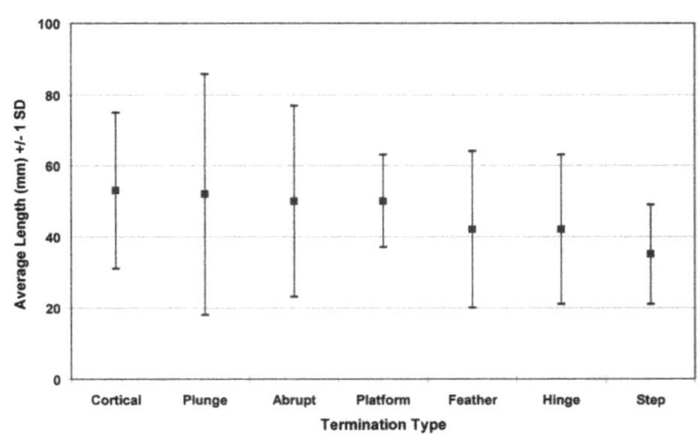

Figure 5.6: The Relationship Between the Termination Types and Length of the Complete Flakes made from Group III Silcrete in the Survey Grid

Attribute	Platform Width (mm)		Platform Thickness (mm)		Length (mm)		Width (mm)		Thickness (mm)	
	Mean	SD	Mean	SD	Mean	SD	Mean	SD	Mean	SD
Form										
Blade	15	8	5	3	47	12	23	19	7	4
Block	30	20	13	10	63	27	54	21	26	10
Expand	17	14	6	5	39	20	46	20	10	5
Intermediate	22	15	8	6	41	19	35	17	10	5
Platform										
Bifacial	36	19	13	7	54	24	44	22	14	11
Cortical	29	15	11	8	48	25	43	19	14	9
Missing					46	25	39	22	12	8
Step	15	9	4	4	40	18	32	14	10	6
Unifacial	27	13	10	6	43	22	39	19	12	8
Termination										
Abrupt	25	16	9	7	50	27	44	20	16	11
Cortical	25	19	10	8	53	22	47	22	16	10
Feather	22	16	8	7	42	22	37	18	10	7
Hinge	21	16	8	7	42	21	39	18	11	7
Platform	20	8	8	7	50	13	51	11	22	10
Plunge	16	18	6	6	52	34	39	19	14	12
Step	18	18	6	7	35	14	32	16	9	6
Flake Scars										
0	24	17	9	8	47	25	45	19	14	8
1	23	17	9	7	45	24	40	20	12	9
2	22	17	9	7	43	21	38	19	12	8
3	20	17	7	6	43	24	37	19	12	9
4	23	16	8	7	49	24	41	20	13	8
5	23	17	7	6	56	29	46	26	16	11
6	27	17	10	7	50	18	46	25	17	10
Total	22	17	12	9	45	23	39	20	12	9

Table 5.12: The Characteristics of the Complete Flakes made from Group III Silcrete in the Survey Grid

Most of the complete flakes (45%, n=571) have dorsal scars originating from Quadrant 1 indicating that the cores were often not rotated (Table 5.13; Figure 4.4). These flakes have a mean length of 47 mm (SD=25 mm) and are considerably smaller than flakes with four or more scars on the dorsal surface (Table 5.12). Few complete flakes (11%) have more than four scars on the dorsal surface (Table 5.12). The rotation of a core to create a second platform is reflected on the dorsal surface of the complete flakes. These flakes have scars on their dorsal surfaces originating from either Quadrants 1 and 2 or Quadrants 1 and 4. Although most of the cores in the survey grid assemblage were only reduced from one platform, 16% of the complete flakes were produced from bi-platform cores (Table 5.13). That few cores are rotated 180 degrees is indicated by the small number of complete flakes with scars originating from Quadrants 1 and 3 (Table 5.13). Cores were generally not rotated more than twice as only 11% of the complete flakes had scars originating from three or more quadrants (Table 5.13).

These results are consistent with the numbers of platforms on the cores (Table 5.7).

Quadrants with Flake Scars	Number of Complete Flakes		Length (mm)	
	%	Count	Mean	SD
No Scars	9	114	47	25
1	45	545	44	23
1, 2	8	98	38	20
1, 3	5	67	49	21
1, 4	8	97	48	24
1, 2, 3	2	22	57	30
1, 2, 3, 4	2	21	47	23
1, 2, 4	4	53	45	28
1, 3, 4	2	25	41	22
2	3	40	43	20
2, 3	1	9	36	17
2, 4	3	42	41	22
2, 3, 4	1	10	43	18
3	1	17	55	29
3, 4	1	8	44	23
4	5	53	46	23

Table 5.13: The Complete Flakes made from Group III Silcrete with Flake Scars Originating from Different Quadrants in the Survey Grid

Only a small number of platform rejuvenation flakes (n=25) were found in the survey grid indicating that platform rejuvenation was not undertaken very often. The characteristics of the cores indicate that they were discarded early in reduction, hence there was no need to extend the life of a core by undertaking platform rejuvenation. Compared to the complete flakes the platform rejuvenation flakes are large but have a relatively low percentage of cortex on their dorsal surface (Table 5.10). The absence of cortex on the platform rejuvenation flakes suggests that this technique was used after the core was decortified.

Most of the complete flakes (45%, n=560) have between 1-49% of their dorsal surfaces covered in cortex (Table 5.11). These flakes are significantly smaller (t=20.31, df=430, p<0.0001) than the complete flakes with between 50-99% of their dorsal surfaces covered in cortex (Table 5.11). The complete flakes with less cortex on their dorsal surface have proportionately more step, hinge, plunge or abrupt terminations, fewer cortical terminations and either an intermediate or a block form (Table 5.11). A high number of the split flakes also have between 1-49% cortex on their dorsal surfaces (Table 5.10) and are the same size (45 mm, SD=18) as the complete flakes (t=0.718, df=701, p=0.479) with comparable amounts of cortex on their dorsal surfaces (Table 5.11). It seems likely that the splits were produced at the same stage of core reduction as the complete flakes with 1-49% cortex on their dorsal surfaces. If considerable force was used on the core, with the aim of removing a large flake, a misdirected strike of the hammerstone on the core platform could result in a split flake (Hiscock 1988:365-366). Consequently, the presence of split flakes with cortex on their dorsal surface may reflect the need to remove large flakes to decortify the cores.

Platform termination was a category identified during the analysis and used to distinguish when a core was rotated 180° to create a new platform opposite the original platform. Flakes with platform terminations are accidentally produced when trying to manufacture large flakes and the force used to remove the flake is too large. These flakes remove part of an old platform at the base of the core. Hence, the length of these flakes represents the size of the orientated platform length of the core (refer Figure 4.4). Only 10 complete flakes in the survey grid have a platform termination. The mean length of these flakes is 50 mm (Table 5.12; Figure 5.6) and relatively little cortex is found on their dorsal surfaces (Table 5.11). The presence of platform terminations in the assemblage indicates that at least some of the cores in the quarry were reduced using a bi-directional technique to extend the life of a core.

Few complete flakes with bifacial platforms or cortical platforms are present in the Quarry 27 assemblage (Table 5.11). The type of the platform a flake has is usually a good indicator of the amount of core reduction (Andrefsky 1998:92-96). Most of the flakes have a unifacial platform indicating that little core rotation or reduction occurred at Quarry 27 (Table 5.11). The complete flakes with a cortical dorsal surface either have a unifacial or cortical platform (Table 5.11) indicating that they were removed early in core reduction before the core was decortified or rotated (Johnson 1989; Andrefsky 1998:93-94,101-102). In contrast, the complete flakes with no dorsal cortex have proportionately more step platforms, indicating overhang removal, and bifacial platforms (Table 5.11). These platform characteristics are associated with later core reduction when cores have been rotated and their platforms needed to be maintained (Andrefsky 1998:95; Whittaker 1994:101-104). However, complete flakes with a bifacial platform are not significantly small, (t=1.3, df=223, p=0.[illegible]9) than the flakes with a unifacial platform (Table 5.12). Consequently, the complete flakes with a unifacial or bifacial platform were probably produced at a similar stage of core reduction. As expected, the complete flakes with a cortical platform are significantly larger than the flakes with unifacial platforms (t=2.5[illegible]4, df=933, p=0.00118) yet surprisingly there is no difference found in the lengths of the flakes with a cortical or bifacial platforms (t=1.736, df=229, p=0.0839). These results indicate that flakes with bifacial or cortical platforms were manufactured early in core reduction. Hence, if core rotation occurred, it took place early in core reduction.

Of the complete flakes present in the survey grid, 28% (n=347) have no cortex on their dorsal surfaces (Table 5.11). Most of these flakes have feather terminations, intermediate forms, unifacial platforms and are significantly smaller (t=9.59, df=905, p<0.0001) in length than flakes with 1-49% cortex on their dorsal surfaces (Table 5.11). Based on the low number of flakes with cortex on their dorsal surfaces it seems likely that most of

these flakes discarded in the survey grid were produced after initial core reduction and platform preparation.
As a core is reduced the amount of cortex on the dorsal surface decreases and number of flakes with step and hinge terminations generally increases (Hiscock 1988:117). There are only a small number of complete flakes with step (n=66, 5%) or hinge (n=112, 9%) terminations in the survey grid (Table 5.12). However, more step and hinge terminations are found on the complete flakes with no cortex (Table 5.11). Complete flakes with hinge or step terminations are significantly smaller (f=11.68, df=6, p<0.0001) in mean length than complete flakes with plunge, cortical, abrupt or platform terminations (Figure 5.6, Table 5.12). Since the flakes with feather terminations are also small and similar in size to those with hinge or step terminations, this may indicate that the larger flakes with feather terminations were removed from the quarry. A histogram showing the lengths of complete flakes with a feather termination type indicate that most of the flakes are less than 60 mm in length (Figure 5.7). The small number of flakes above 60 mm may indicate that large flakes, presumably with no cortex, were selected and subsequently removed from the quarry.

The number of complete and broken flakes with a blade form is very small (n=19), suggesting that blades were not the primary product manufactured at the quarry. The flakes with a blade form were more likely to have no cortex on the dorsal surface but were relatively large in size compared to most of the other flake forms (Tables 5.11, 5.12). Blade manufacture may represent a single visit to Quarry 27 when a different tool type was required.

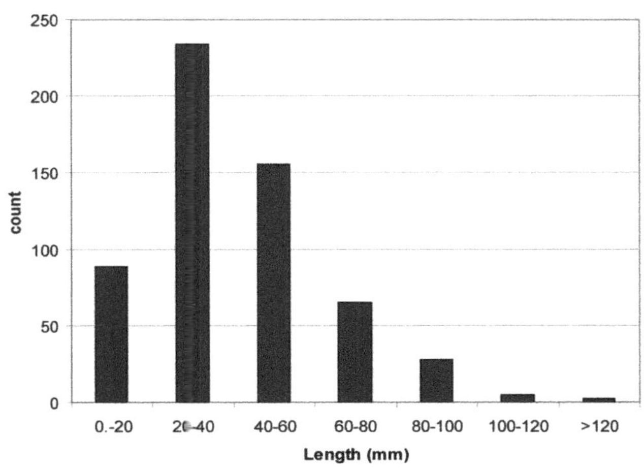

Figure 5.7: The Length of the Complete Flakes with Feather Terminations made from Group III Silcrete

5.5 TOOL TYPES IN THE SURVEY GRID

Only 116 complete or broken tools were present in the survey grid. This number represents only 3% of the total assemblage. Most of the tools have usewear damage, denticulate, notched or scraper retouch (Table 5.14). The tools with denticulate, notched or scraper retouch and the two saws are large in comparison to those with usewear damage (Table 5.10). The endscraper and adze flake are also small (Table 5.14).

Most of the tools were made from Group III silcrete (95%) and only six (5%) are made from Group II, medium clast silcrete (Table 5.14). The proportion of tool types made from Group II and Group III mirrors the proportions of the silcrete types found in the total assemblage. No significant difference (χ^2= 0.08, df=1, p=0.929) was found between the type of tool (with five or more of each type) and the type of material (Table 5.1).

Most of the tools were made on complete flakes (Table 5.10) and these are significantly larger than the non-retouched complete flakes (t=2.3, df=76, p<0.0001). The broken tools are also larger than the equivalent non-retouched components (Table 5.10). These results indicate a selection for a large tool size, even if the flake was already broken.

The presence of tools at the quarry may be a result of two different discard patterns: the immediate and expedient use of tools made from material found at the outcrop and the abandonment of curated tool types of non-local, imported material (Andrefsky 1994b:379; Shott 1989). The use of 'expedient' or 'situational' (Binford 1973, 1979) tools is indicated by the presence of the tools made on naturally occurring spalls of weathered silcrete that have been retouched as either notches (n=2) or denticulates (n=3). There are only five of these tools in the assemblage but they do indicate a selection for a convenient size and concave shape (Figure 5.8). In contrast, a small number of discarded tools from exotic, non-local stone were also found, namely a tula and two scrapers (Figure 5.9, 5.10). These curated tool types are considerably smaller than those made from the local material and may indicate retooling (Gramly 1980; Torrence 1989b; Binford 1979).

Tool Type	Total		Material Type		Maximum Dimension (mm)	
	Count	%	Group II	Group III	Mean	SD
Denticulate	23	20	2	21	82	27
Notch	23	20	1	22	73	26
Point	1	1		1	78	
Scraper	40	34	2	38	66	22
Stepped scraper	2	2		2	60	13
Endscraper	1	1		1	37	
Borer	2	2		2	51	6
Saw	2	2		2	83	
Adze flake (tula)	1	1		1	40	
Usewear	21	18	1	20	44	16
Total	116		6	110		

Table 5.14: The Characteristics of the Different Tool Types

Figure 5.8: Artefacts found at Quarry 27. A. Spall Tool with Denticulate Retouch; B. Medium Clast Blade in Transect 3; C. Tula made from Group III Silcrete; D. Medium Clast Scraper

5.6 THE SURVEY TRANSECTS

Only 202 artefacts were found in the survey transects. Most of these artefacts were made from material obtained from the outcrop and only two refitting artefacts were made from a gibber cobble. The source of the Group III artefacts in the survey transects was probably Quarry 27 (Figure 5.1). Most of the artefacts were found in the southern (Transect 3, n=127) and the northern transects (Transect 1, n=47) which ran parallel to the creek. Only 28 artefacts were found in the two transects, Transect 2 and Transect 4, that crossed the gibber pavement (Transect 2, n=8, Transect 4, n=20). In Transect 1, Transect 2 and Transect 4 the "drop-off" rate reduces within the first 20 m and no increase in density occurs as the distance from the survey grid increases (Figure 5.1).

The high number of discarded artefacts in the transects along the creek (Transect 1 and Transect 3) suggests not only a concentration of activities along the creek but also the lack of activities occurring on the gibber pavement. It is interesting to note that the density of artefacts in Transect 3 sharply decreases away from the main quarry area until at a distance of approximately 24 metres from the survey boundary the density again increases (Figure 5.1). It is probable that the flaking debris found at the end of Transect 3 represents non-quarrying activities and may even reflect a discrete knapping event. To test this assumption the artefacts found in this transect were compared to the artefacts found in the main quarry area.

The results of a comparison between the artefacts found in Transect 3 and the quarry show a number of differences indicating that the Transect 3 assemblage is less indicative of quarrying behaviour. These differences include the relative proportions of artefact types, the amount of cortex on the complete flakes, the types of raw materials represented, the form of the flakes and platform characteristics (Tables 5.1, 5.2, 5.6, 5.12, 5.15-19). A substantially lower ratio of cores to complete flakes was found in Transect 3 (1:14) when compared to the survey grid (1:6). The complete flakes in Transect 3 generally have little cortex on their dorsal surfaces, fewer cortical terminations and more feather terminations (Table 5.12, 5.17). The lengths of the complete flakes are also significantly greater (t=9.79, df=3, p<0.0001) in the transect than within the quarry with a mean length of 55 mm (SD=19 mm) (Tables 5.12, 5.17). A greater range in form and material type was also found in T3 assemblage with more blades and more artefacts made from Group II, medium clast silcrete compared to the survey grid (Table 5.17, Figure 5.11). These factors, in conjunction with the large flake size and the low number of step (n=0) and hinge (n=4) terminations on the complete flakes, indicate that the cores in T3 (Table 5.17) were not substantially reduced before they were discarded. The analysis of the artefacts in T3 suggests that most of the flaking debris may be the product of a core preparation phase before blade removal.

Several differences in the flake characteristics and their material type were observed between the T3 assemblage and the survey grid. The complete flakes made from Group II silcrete are small and have little cortex on their dorsal surfaces (Table 5.17). Flakes with these characteristics indicate that the cores were underwent considerable reduction before they were discarded (Andrefsky 1998:96-102). However, the cores of Group II silcrete are large in comparison to the cores of Group III silcrete indicating that the Group II cores were not substantially reduced and abandoned early in reduction (Table 5.18). Blades were made from both Group III and Group II silcrete in T3, however, only one blade core of Group III silcrete was found. It is likely that at least one blade core made from Group II silcrete was used to produce the majority of the Group II flakes in Transect 3. This core was subsequently removed from the survey transect and discarded elsewhere in the landscape.

Technological Type		Total		Maximum Dimension (mm)	
		Count	%	Mean	SD
Non-retouched flakes	Complete Flake	59	46	63	21
	Platform Rejuvenation	1	1	60	
	Complete Split	9	7	59	18
	Distal Flake	14	11	50	19
	Broken Flake	17	13	46	13
	Total	100	78		
Retouched flakes	Complete Tool	3	2	67	6
	Distal Tool	2	2	81	15
	Broken Tool	3	2	95	32
	Total	8	6		
Other	Block	3	2	73	38
				Length (mm)	
	Core	8	6	115	71
	Core Fragment	8	6	75	26

Table 5.15: The Technological Types found in Transect 3

	Material	Count	%
Group III	Amorphous	2	2
	Banded	18	14
	Breccia	2	2
	Matrix-Dominated	76	60
	Sub-total	98	78
Group II	Medium clast	28	22
Other	Quartz	1	1
	Total	127	101

Table 5.16: The Raw Material Types found in Transect 3

	Total		Group III				Group II			
					Length (mm)				Length (mm)	
	Count	%	Count	%	Mean	SD	Count	%	Mean	SD
Percentage of Cortex										
0%	26	44	16	39	54	17	10	56	51	14
1-49%	29	49	21	51	58	25	8	44	59	11
50-99%	4	7	4	10	43	23				
Form										
Blade	6	10	4	10	60	9	2	11	48	18
Block	14	24	12	29	69	25	2	11	66	2
Expand	1	2	1	2	44					
Intermediate	38	64	24	59	47	18	14	78	54	13
Platform Type										
Cortical	6	10	5	12	55	17	1	6	59	
Missing	10	17	8	20	60	26	2	11	69	4
Unifacial	43	73	28	68	53	22	15	83	52	13
Termination										
Abrupt	17	29	12	20	67	26	5	28	58	9
Cortical	5	8	5	8	49	20				
Feather	32	54	20	34	49	19	12	67	52	15
Hinge	4	7	3	5	49	10	1	6	66	
Plunge	1	2	1	2	68					
	59		41		55	19	18		54	13

Table 5.17: The Characteristics of the Complete Flakes in Transect 3

Material Type	Core Type	Count	Length (mm)		Cortex		
			Mean	SD	0%	1-49%	50-99%
Banded	Bi-platform	1	58		1		
Matrix-Dominated	Bi-platform	1	67	13		1	
	Microblade	1	98			1	
Group III	**Total**	3	77	20	1	2	
Group II	Bi-platform	2	103	9		1	1
	Cobble	1	145				1
	Uni-platform	2	131	1		1	1
	Total	5	123	19		2	3

Table 5.18: The Material Type and Length of the Cores in Transect 3

Few cores were found in Transect 3 (n=8). Half (n=4) of the cores are bi-platform and considerably smaller than those found in the quarry area (Table 5.10, 5.19). In contrast, the scar length on the uni-platform cores is similar in size in both areas (Table 5.10, 5.19). Only one core has a cortical platform indicating that initial reduction and decortification were already undertaken on most of the cores. Since no outcrop is present within the southern region of the transect it is probable that cores were removed from the quarry and further worked in this location (Figure 5.1). It is interesting to note that five of the eight cores present in this area are made from Group II, medium clast silcrete, a material that generally was not reduced within the survey grid. It is possible that these cores were from another quarry source, transported to T3 and further reduced, however, the large size of these cores and high percentages of cortex may indicate otherwise (Table 5.18). In contrast to the Group II cores, the Group III cores are smaller and have less cortex (Table 5.18). While only one microblade core was found in the transect it was made from the same Group III, matrix-dominated silcrete found at the quarry. The core was not heavily reduced with a length of 98 mm and still has cortex (1-49%) on its surface.

Core Type	Count	Block Length (mm)		Scar Length (mm)		Flake Number (mm)	
		Mean	SD	Mean	SD	Mean	SD
Bi-Platform	4	85	23	52	12	3	1
Microblade	1	98		81		4	
Cobble	1	145		41		1	
Uni-Platform	2	131	1	56	16	4	3

Table 5.19: The Count and Dimensions of the Different Core Types in Transect 3

Several different manufacturing strategies seem to have occurred within Transect 3. The flaking debris represents a small working area where blades were produced, and then removed, combined with the initial reduction of medium clast cores. Consequently, the primary goal was to produce a small number of blades within this area, as opposed to the quarry. At least one medium clast core was used to manufacture blades and one blade core made from Group III silcrete was found in Transect 3. As no blade core made from Group II silcrete was found in this transect it is probable that the core was removed.

5.7 REDUCTION STRATEGIES INFERRED FROM THE QUARRY 27 ASSEMBLAGE

The quarry assemblage is the result of a series of overlapping events and activities. Individual knapping events were identified by the presence of conjoins and the limited production of blades found in two discrete areas. However, the bulk of the assemblage represents the outcome of many knapping episodes. Nevertheless, the characteristics of the artefacts in the quarry assemblage reflect a core reduction process that indicates its role in the organisation of lithic technology.

The aim of this section is to draw together information obtained from the analysis of the cores and flaking debris in order to make inferences about the reduction strategies implemented at Quarry 27. Based on the characteristics of the Group III cores and flakes in the survey grid three major reduction stages were identified at Quarry 27:

material selection and testing, initial core reduction and preparation, and the manufacture of large flakes. Figure 5.9 provides a schematic diagram of the general core reduction process and the formation of the quarry assemblage at Quarry 27.

The large number of uni-platform and bi-platform cores, and their large size, indicates that little reduction or core rotation occurred in the quarry (Andrefsky 1998:101). That cores were not reduced to any great degree is reflected in the small number of flakes scars with hinge or step terminations, the limited presence of cores with a conical cross sections and the substantial amount of cortex remaining on the cores (Hiscock 1988:117; Witter 1992:31-32; Mauldin and Amick 1989:70). The characteristics of the flaking debris also indicate that little core reduction took place at the quarry. The flakes have several characteristics indicative of a limited amount of core reduction including a large size, the presence of cortex on their dorsal surfaces and generally do not have hinge or step terminations (Andrefsky 1998:85-104).

5.7.1 Material Selection and Testing

A single outcrop of silcrete can grade from high to poor quality depending on the parent lithology of the material. At Quarry 27 a small, discrete 'patch' of extremely homogeneous, high quality Group III silcrete was found within a massive exposure of Group II silcrete. The cobbles made from Group III were almost exclusively the focus of quarrying activity. Consequently most of the artefacts are made from Group III silcrete and were discarded near these cobbles.

Small cobbles of Group III silcrete, already exposed on the surface, seem to have been mainly tested and selected for further reduction. There was no need to extract these cobbles from the substrate or to split open the larger cobbles. The cobble cores and the exposed cobbles were more similar in size than the cobbles that remain buried. However, a few large cobbles were split on anvils as several of these were observed at the quarry.

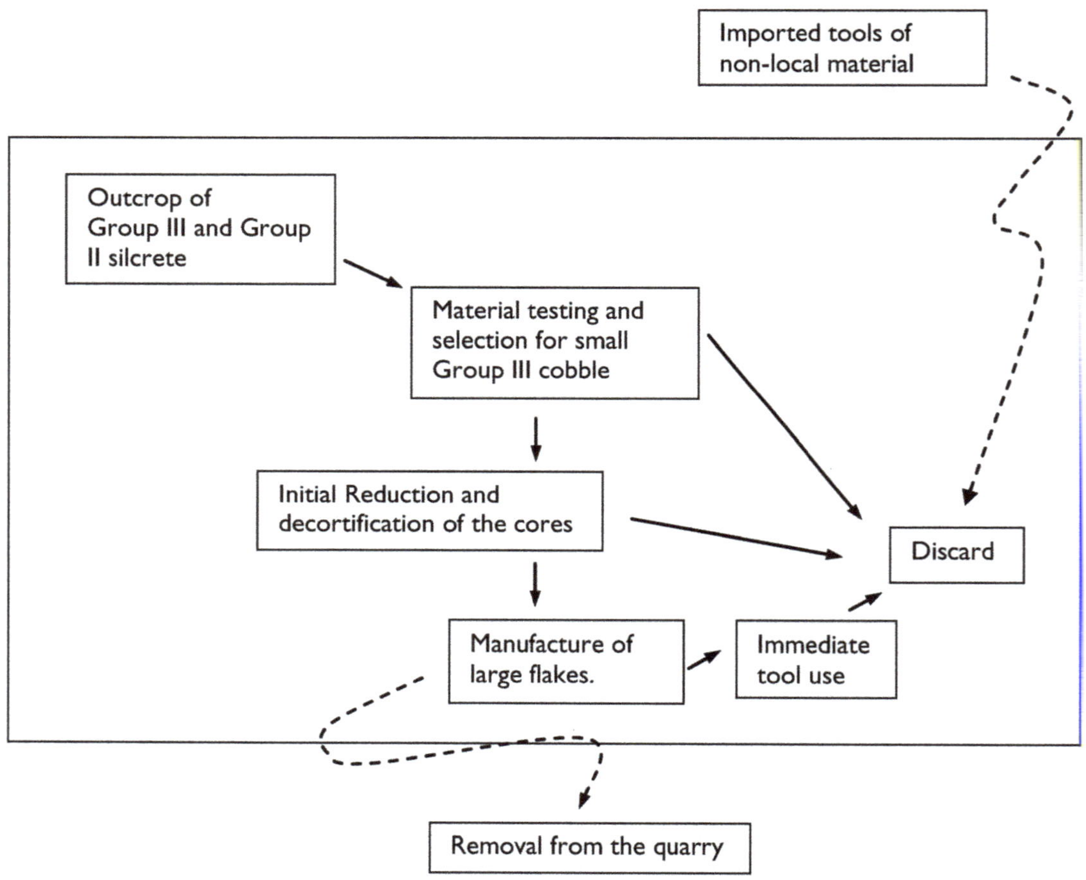

Figure 5.9: Schematic Diagram Showing the General Process of Core Reduction Observed at Quarry 27

The characteristics of the cobble cores are indicative of material testing. The aim of testing was to assess the material type and not to remove large quantities of cortex, produce large numbers of flakes or reduce the overall size of these cores. Consequently, only a few, small flakes were removed from the cobble cores and they remain large and almost completely covered in cortex. At Quarry 27 it was only necessary to 'test' the cobbles and assess the quality of the material if they were completely covered in cortex. It is likely that most of the cobbles tested were selected and reduced as only 31 cobble cores were discarded in survey grid. The cobble cores that were rejected may have had fractures in the rock, not been a suitable size or were buried in the colluvium.

Small flakes with cortical platforms and dorsal surfaces indicate that these flakes were produced during material testing. The low number of flakes with these characteristics suggests that material testing was a strategy not often required at the quarry. The limited amount of material testing may be due to the natural weathering of the cobbles which removed the cortex and exposed the underlying material. The weathering and spalling of the cobbles is advantageous at it provides natural platforms for the removal of flakes, limiting the amount of platform preparation required and providing the acute angle needed to remove flakes (Flenniken and White 1985:134). The use of naturally occurring platforms on the cobbles is one reason why few cortical platforms were seen on the cores and complete flakes.

5.7.2 Initial Reduction and Core Preparation

Once an appropriate core was selected for reduction, initial working and decortification of the core occurred. This strategy is reflected in the bi-platform, multi platform and uni-platform cores with more than 50% cortex of their surfaces. During core preparation cortex is often removed using a free-hand percussion method (Flenniken and White 1985:134). The aim was to decortify the face of a core resulting in more flakes being removed from each core. Consequently, the flake scars are also larger and traverse the entire face of the core. Accordingly, the largest flakes in the Quarry 27 assemblage have plunge, abrupt or cortical terminations indicating that they travelled across the face of the core and removed the base of the core. These flakes also tend to have a block form and large amounts of cortex on their dorsal surfaces, characteristics indicative of early core reduction (Andrefsky 1998:85-104).

5.7.3 The Manufacture of Large Flakes

The manufacture of large flakes was the last act of reduction before the core was discarded. Large flakes with no cortex were removed from the multi-platform, uni-platform and bi-platform cores. These cores are smaller, have little cortex or no cortex remaining on the surface and sometimes have a conical cross section.
It seems likely that large flakes with feather terminations were removed from the quarry. This inference is supported by a drop in numbers of complete flakes greater than 50 mm in length and a similar scar length on the cores. The size of the flake scars on all the cores, bar the cobble cores, is the same across all the core types and the length of these flake scars extends across the entire length of the core face. The similar size of the flake scars on the cores also indicates that large, standardised flakes were manufactured and removed from the quarry. Hence, the cores were discarded when a large flake could no longer be produced. No significant size difference occurs between core types, based on the block measurements or the orientated platform measurements on the uni-platform, bi-platform and multi platform cores. The similar lengths of the orientated platform measurements on the uni-platform, multi-platform and bi-platform cores suggest that the length of the core face was important to produce the maximum flake length possible. Platform rejuvenation sometimes occurred to extend the life of a core, but this technique was not used often.

The material at Quarry 27 would have been highly suitable for use as a tulas, endscrapers or any other tool types requiring fine retouch (Flenniken and White 1985:132-133). It is probable that large flakes of this material were used in such a fashion.

5.7.4 Blade Manufacture

Few blade cores or flakes with a blade form occur in the survey grid indicating that blade manufacture was not the focus of core reduction at the quarry. However, the presence of knapping debris characteristic of blade production suggests that the quarry may have been visited more than once with different reduction strategies undertaken during each visit.

5.7.5 Tool Use

The presence of tools, made from the material found at Quarry 27, indicates that some tool manufacture and use occurred at the quarry. The tools are significantly larger than the non-retouched counterparts indicating a selection for size rather than whether the flake was broken or complete. Due to the size of the broken flakes it seems unlikely that these tools were broken during use but were simply selected for their convenient size and form. A selection for high quality Group III silcrete for tool use can also be observed in the survey grid as few tools were made from Group II silcrete. The tools made from Group II silcrete were probably manufactured from stone obtained from another source as Group II silcrete was generally not worked at Quarry 27.

Non-local tools, such as the tula, were discarded at Quarry 27. Tulas are considered that most curated tool type of the Australian Small Tool Tradition (Hiscock and Veth 1991). Curated tools that were transported, resharpened and replaced at Quarry 27 may provide an example of retooling (*cf* Gramly 1980) or may have resulted from accidental loss (*cf* Shott 1989:17-18).

Expedient or situational tools indicate the manufacture of a tool for an immediate use usually prompted by an unplanned need (Binford 1979). Based on the tool types little effort was invested in the manufacture most of the tools found in the survey grid. The expedient use of tools is highlighted by the use of the naturally occurring weathered spalls. These tools, which were mainly notched, denticulate and scraper types suggests that they were made and used, and discarded immediately (Andrefsky 1994b).

It is doubtful that the tools identified in the assemblage were the primary manufacturing goal at Quarry 27, but rather a side-product made in response to an immediate demand. Tools are often made at quarries but removed and used elsewhere in the landscape (Torrence 1986; Root 1992; Arnold 1987a, 1987b). The use of tools at a quarry seems to be atypical of quarrying behaviour and does not reflect the bulk of the manufacturing debris or the product removed.

5.8 SUMMARY

Only a limited amount of core reduction occurred at the quarry, implying both a minimal labour input and a limited amount of time was invested in the reduction of cores at Quarry 27. Although the quarry assemblage represents a number of overlapping knapping events only a simple set of reduction strategies was employed to reduce the cores and this can be interpreted as small-scale, large flake production (Figure 5.9). While Quarry 27 is a high quality source, the limited amount of reduction implies that procurement was unlikely to be a result of a direct access procurement strategy but embedded in the acquisition of another resource. The isolated location of the quarry and the small size of the outcrop, small size and density of the quarry and little time or effort was invested in the reduction of cores indicates that stone procurement was opportunistic. However, the similar size of the flakes removed from the multi-platform, uni-platform and bi-platform cores indicates that a certain degree of planning influenced the manufacture of large flakes to a particular size. These flakes were removed from the quarry for later use.

Chapter 6

Spatial Analysis of Quarry 27

6.1 INTRODUCTION

The sheer numbers of artefacts at a quarry makes the recording of a quarry assemblage extremely difficult to undertake without creating a sampling bias in the results (Ericson 1984). Quarries are repeatedly visited and sometimes used for a long duration leading to a spatially immense, dense volume of discarded cores and flaking debris. During one visit to a quarry a knapper can produce vast quantities of waste material using many different techniques and reduction strategies. Researchers undertaking quarry studies have often failed to understand the formation and distribution of a quarry assemblage. Furthermore, inappropriate sampling strategies distort the results and create a somewhat limited view of the complexity of a quarry assemblage (e.g. Lamb 1996).

It has been widely acknowledged that the characteristics of the discarded cores and flakes reflect the different reduction strategies employed and the stage of core reduction (e.g. Collins 1975). Yet, whether spatial variation in the placement of different strategies or stages occurs in a quarry assemblage usually remains unexplored. In addition, the underlying factors that influenced the distribution of artefacts in a quarry assemblage are rarely investigated. This

Chapter explores the distribution of the Quarry 27 assemblage to establish whether the artefacts resulting from different stages of core reduction occur in spatially discrete locations at the quarry. Spatial variation in the assemblage is observed through the presence/absence and relative abundance of the core and flake characteristics that reflect the stages of core reduction identified in Chapter 5

In Chapter 4 a brief overview of previous approaches to the analysis of a quarry assemblage was presented. Quarries are often sampled using random sampling squares (e.g. Bamforth 1992:134; Byrne 1980). Problems arise with the use of random sampling squares as the entire distribution of the assemblage, and how much spatial variation occurs within that assemblage, is not taken into account during the analysis. Furthermore, depending on the size of sampling unit it may encompass one knapping event or several knapping events that reflect different strategies or stages of core reduction. Not knowing where spatial variation occurs, or even if it occurs, in the distribution of different stages of reduction is a fundamental problem in quarry analysis. A more useful approach is to firstly defining the boundary of the quarry, based on the density of artefacts, and use transects or a grid system to systematically survey the entire area (e.g. Torrence 1986). By using this method there is a better chance of discovering whether or not spatial variation in the assemblage occurs. However, the researcher is still faced with the problem of what sized sampling unit and how many sampling units should be used to gain the best results depending on the research questions asked. By using the artefact as the unit of analysis these problems are removed from the analysis.

In the following chapter the characteristics of the cores and flakes that reflect their place in the reduction process are located in space and used to search for spatial variation in the quarry assemblage. In addition, not only can an artefact's characteristics and position be compared in a GIS analysis but also its relationship with the various landscape features, such as the location of the outcrop, found at the quarry. Consequently, the constraining or "pull" factors that have influenced the discard of artefacts are examined in this chapter by comparing the landscape features and the artefact distribution. By defining factors that influenced the distribution of the artefacts within Quarry 27 a better understanding of the formation and composition of a quarry assemblage can be gained. The results of this chapter will be used to justify the sampling method applied to Quarry 35 (Chapter 7).

6.2 SPATIAL VARIATION IN THE DISTRIBUTION OF ARTEFACTS

The results of Chapter 5 are used to establish which key attributes, based on the characteristics of the artefacts, highlight the location of the different stages of core reduction identified in the quarry assemblage. These attributes were used to direct the search for spatial variation in the distribution of artefacts at Quarry 27 (Table 6.1). By viewing the location of these key attributes within *ArcView,* a GIS program, it is possible to establish whether different reduction activities occurred in different locations within the survey grid and the transects of Quarry 27.

Several different methods, already described in Chapter 4, were used to search for spatial variation in the distribution of the artefacts. In the following analysis, *ArcView* Density Plots and Buffers were used to search for patterns in the distribution of artefacts.

Extraction, Material Selection and Testing	
Characteristics of the Cores	*Characteristics of the Complete Flakes*
• Cobble cores • Large Size • Heavy • 1 or 2 flakes removed • Small flake scar • Cortical Platforms • 1 or 2 Platforms	• 100% cortex • Cortical platform • Small length
Initial Reduction and Core Preparation	
• Bifacial or uni-platform cores • Large flake scars • 50-99% cortex • Large size (length) • Heavy • > 2 flakes removed	• 50-99% cortex • Cortical termination • Block form • Large size (length) • Uni-platform Platform
Secondary Reduction and Manufacturing the Product	
• Multi platform cores • Blade cores • 1-49% or 0% cortex • Step and hinge terminations • Conical section • Small size (length) • Light weight	• Blade form • No cortex • Small size (length) • Increase in step and hinge terminations

Table 6.1: The Artefact Characteristics Indicative of the Different Stages of Core Reduction Observed in the Quarry Assemblage

6.2.1 Results of the Spatial Analysis

Density plots were created for the flake and core characteristics that best represent the process of core reduction (Table 6.1). Each density plot represented either the nominal values (the different attributes recorded) or size groupings of the ordinal measurements. All the density plots had a grid size of 50 cm^2. Hence, the results in Figures 6.1-6.4 represent the densest area of each attribute per 50 cm^2. The size groupings of the ordinal measurements, such as the length of a complete flake, were established by taking the range of values, from the minimum to the maximum number, and dividing the range into equal size categories.

By overlaying the density plots showing the different characteristics of the artefacts at Quarry 27 that reflect the stage of reduction, and comparing the areas of greatest density, the location of the three stages of core reduction were identified (Table 6.1, Figures 6.1-6.4). Although little blade manufacturing occurred at the quarry the placement of the cores and flakes that reflect this strategy was also identified.

- **The Location of Material Testing and Selection**

The characteristics of the cores discarded near the outcrop in the survey grid reflect an area where material testing took place. The cobble cores in this location are large and heavy, generally have only one cortical, platform and one flake removed before they were discarded (Table 5.6). The density plots of these characteristics tend to cluster in a single, discrete location near the outcrop (Figure 6.1).

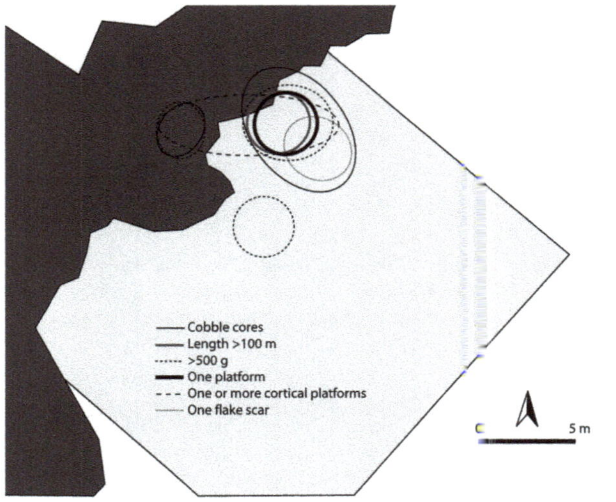

Figure 6.1: Results of the Density Plots Showing the Characteristics of the Cores Indicating Material Testing and Selection at Quarry 27

The complete flakes with characteristics indicative of early reduction have the same distribution as the cobble cores at Quarry 27. The density plots of the complete flakes with cortex covering their entire dorsal surfaces, no dorsal scars, cortical platforms and cortical terminations concentrate in the same location as the cobble cores (Figures 6.1, 6.2). These characteristics typically occur on flakes produced early core reduction (e.g. Andrefsky 1998:88-107) and suggest that material testing probably occurred in this location.

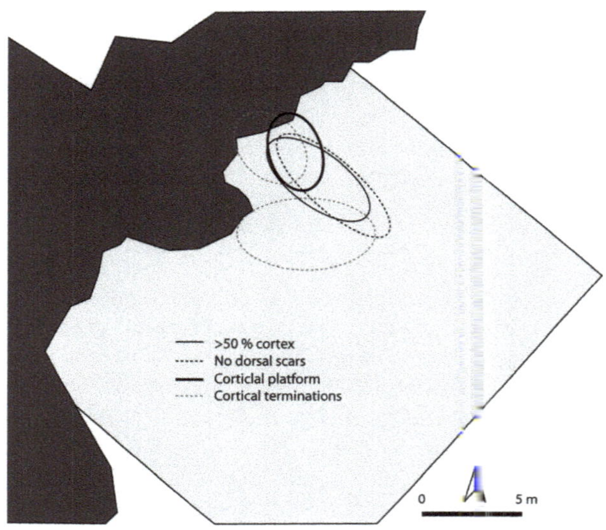

Figure 6.2: Results of the Density Plots Showing the Characteristics of the Complete Flakes Indicating Material Testing and Selection at Quarry 27

- **The Location of Initial Reduction and Core Preparation**

Based on the results of Chapter 5, initial core reduction and preparation can be observed through the discard of uni-platform and bi-platform cores with large amounts of cortex still remaining on their surfaces. The primary goal of initial reduction was to remove large flakes and decortify the core.

Two areas, near the outcrop, have the densest concentration of uni-platform cores still retaining between 50-99% cortex on their surfaces (Figure 6.3). The northern area also has the cores with the largest flake scars (Figure 6.4). This pattern is consistent with the results of Chapter 5 which shows that uni-platform cores generally have large flake scars (Table 5.10). In contrast to the uni-platform cores, the bi-platform cores with 50-99% cortex concentrate in two main areas; one small area near the outcrop, overlapping with a cluster of uni-platform cores with the same amount of cortex, and another nearer to the creek (Figure 6.3).

Density plots show that the complete flakes near the uni-platform cores in the northern cluster have large amounts of cortex on their dorsal surfaces, few dorsal scars, a block form and are large (Figures 6.2, 6.4).

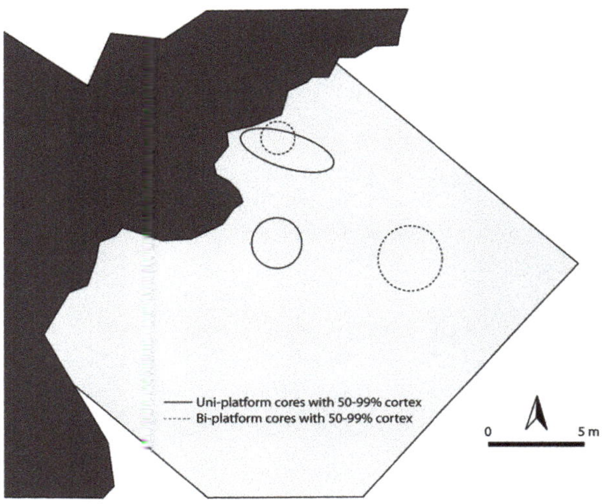

Figure 6.3: Results of the Density Plots Showing the Densest Location of the Uni-platform and Bi-platform Cores with 50-99% Cortex at Quarry 27

By creating a 0.5 m buffer around each of the uni-platform and bi-platform cores four concentrations were identified (Figure 6.5). The artefacts found in the buffers were compared to assess whether different stages of core reduction occur in each location. These comparisons show only a few differences were found between the buffers that may reflect the amount of core reduction. It seems likely that the flakes produced from the reduction of the uni-platform and bi-platform cores were made during the same stage of reduction, further supporting the results of Chapter 5.

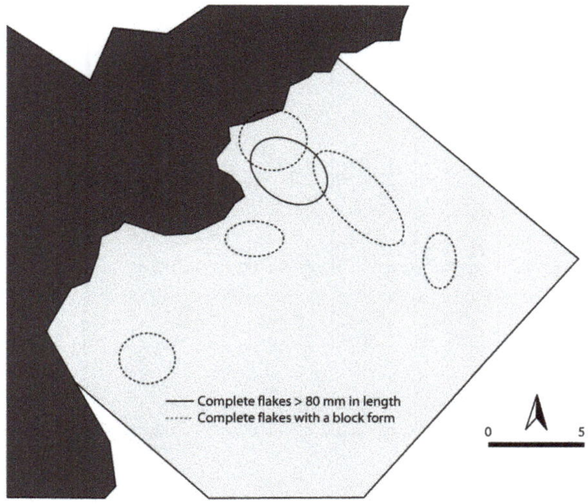

Figure 6.4: Results of the Density Plots Showing the Densest Location of the Complete Flakes with a length >80 mm and a block Form at Quarry 27

While proportionately more complete splits were found within the bi-platform core buffer and more blocks and broken flakes in the uni-platform buffer no significant difference was found in their frequencies (Tables 6.2, 6.3). Likewise, chi square tests of the frequency of the different characteristics of the complete flakes between the two buffers show only one significant difference (Table 6.2). The form of the complete flakes in each buffer differed significantly with more flakes having expanding forms in the bi-platform buffer.

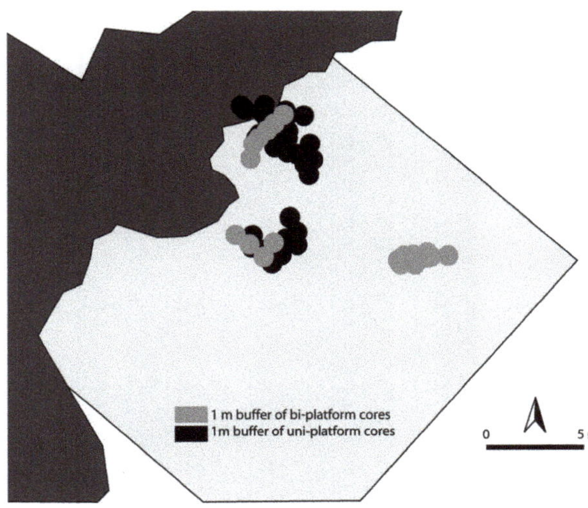

Figure 6.5: 1m Buffers Surrounding the Bi-platform and Uni-platform Cores at Quarry 27

The size of the flakes removed from a core usually reflects how much reduction a core underwent before being discarded (e.g. Andrefsky 1998:96). A significant result was found by comparing the size of the complete flakes in the buffers (Table 6.4; t=2.482, df=115, p=0.0145). The smaller size of the complete flakes in the uni-platform buffer suggests a further reduction of the cores in this buffer compared to those in the bi-platform buffer. It was expected that the flakes in the bi-platform buffer would be smaller but this was not the case. Furthermore, the average number of flakes removed from each core also indicates a further reduction of the cores in the uni-platform buffer (Table 6.5). However, the size of the uni-platform (t=0.4444, df=115, p=0.6631) and bi-platform cores (t=0.3084, df=10, p=0.7641) in each buffer is not significantly different indicating that they were reduced to the same size before being discarded (Table 6.3).

Test	X^2	df	p	Result
Technological type (block, complete flake, complete split, broken flake, core, distal flake)	3.0	4	0.555	Not significant
Complete Flakes				
Cortex (0%, 1-49%, 50-99%)	1.6	2	0.45	Not significant
Termination (step, hinge, feather, cortical, abrupt)	4.2	4	0.379	Not significant
Scar Number (n=0, n=3)	1.4	1	0.713	Not significant
Platform (n/a, unifacial)	3.6	1	0.03	Significant *phi*=0.1775
Form (block, expanding, indeterminate)	6.8	2	0.058	Not significant

Table 6.2: Chi Square Tests of Artefact Characteristics in the Uni-platform and Bi-platform Core Buffers

	Technological Type	Uni-platform Core Buffer		Bi-platform Core Buffer	
		Count	**%**	**Count**	**%**
Core	Core	17	9	14	13
	Core fragment	4	2	3	3
	Sub-total	21	11	17	16
Non-retouched flakes	Complete flake	77	41	40	38
	Platform Rejuvenation	0	0	1	1
	Distal flake	15	8	6	6
	Proximal flake	4	2	2	2
	Broken flake	41	22	16	15
	Complete split	7	4	7	7
	Sub-total	144	77	72	69
Retouched flakes	Broken tool	1	1	1	1
	Complete tool	2	1	2	2
	Distal tool	0	0	1	1
	Proximal tool	0	0	0	0
	Weather tool	0	0	1	1
	Sub-total	3	2	5	5
Other	Hammer	4	2	4	4
	Block	14	7	6	6
	Total	186		104	

Table 6.3: The Technological Types in the Uni-platform and Bi-platform Cores Buffers

	Uni-Platform Buffer		Bi-platform Buffer	
	Count	**%**	**Count**	**%**
Form				
Block	14	20	1[illegible]	8
Expanding	5	7	1[illegible]	11
Intermediate	50	72	11[illegible]	81
Platform Type				
Bifacial	1	2	3	3
Cortical		0	9	10
Missing	13	30	1[illegible]	13
Step	2	5	8	9
Unifacial	28	64	5[illegible]	65
Scar Number				
0	6	8	3	8
1	15	19	1[illegible]	30
2	18	23	8	20
3	25	32	1[illegible]	30
=>4	13	17	5	13
Termination				
Abrupt	17	14	10	18
Cortical	14	12	8	14
Feather	62	51	20	35
Hinge	18	15	6	11
Plunge	1	1	2	4
Step	7	6	6	11
Percentage of Cortex				
0%	30	39	[illegible]2	30
1-49%	31	40	21	53
50-99%	12	16	7	18
100%	4	5		
Length (mm)				
Count	77		40	
Mean	38		50	
SD	23		28	

Table 6.4: The Characteristics of Complete Flakes in the Uni-platform and Bi-platform Cores Buffer

		Core Type	Bi-Platform	Cobble	Uni-Platform
Uni-platform buffer	Length (mm)	Count	3	2	12
		Mean	98	116	110
		SD	29	69	33
Bi-platform buffer		Count	9		5
		Mean	103		103
		SD	23		17
Uni-platform buffer	Scar number	Mean	5	1	3
		SD	2		2
Bi-platform buffer		Mean	3		1
		SD	1		1

Table 6.5: A Comparison of the Size and Scar Number of the Different Core Types in the Uni-platform and Bi-Platform Core Buffers

- **The Location of Flake Manufacture**

Based on the results of the density plots the flaking debris in only one area highlight where the greatest amount of core reduction occurred. This area is located a short distance, approximately 4 m south, from the area where most of the material testing and initial core reduction took place. The cores in this location are small, have a high number of hinge terminations on their flake scars, a small flake scars and a large number of flakes removed (6-7 flakes; Figure 6.6). The presence of these characteristics on a core indicates further reduction (Hiscock 1988:117; Kuhn 1991:83). Furthermore, the characteristics of the complete flakes in this area are also indicative of greater amount of core reduction (*cf* Andrefsky 1998:88-107). These flakes have little cortex on their dorsal surfaces, large numbers of step and feather terminations, small lengths and large numbers of dorsal scars (Figure 6.7).

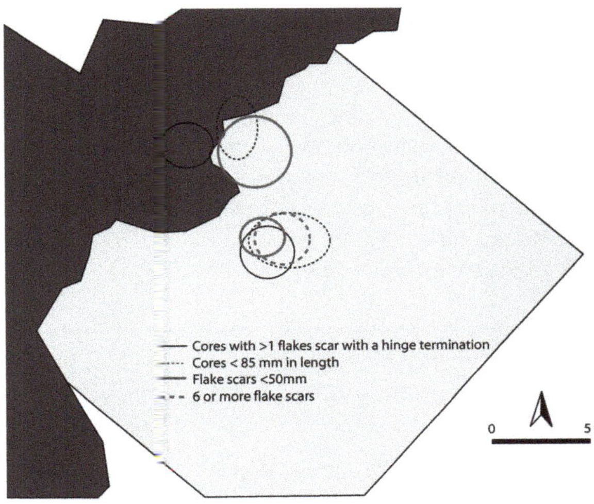

Figure 6.6: Results of the Density Plots Showing the Cores with Characteristics indicating later stages of Reduction at Quarry 27

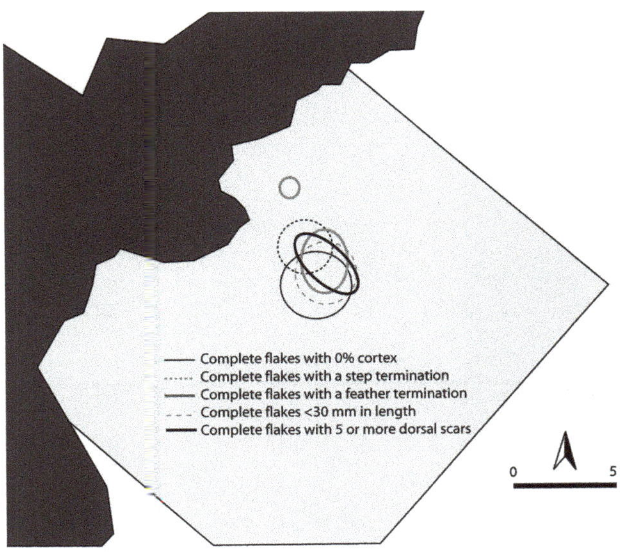

Figure 6.7: Results of the Density Plot Showing the Complete Flakes with Characteristics indicating later stages of Reduction at Quarry 27

- **Blade Manufacture**

As already noted in Chapter 5 little blade manufacture occurred at Quarry 27. The manufacture of blades occurred in two localised areas and could be the result of a discrete overprinting event. A small number of blades were found in the survey grid and along the southern boundary of Transect 3 (Figure 6.8). In the survey grid two blade producing cores are closely associated with the highest density of flakes with a blade form (Figure 6.8). In contrast, within Transect 3 no blade cores were found near the flakes with a blade form. The only blade core in Transect 3 was found 15 m from these flakes indicating that the core was worked, removed and subsequently discarded away from the main area of blade manufacture.

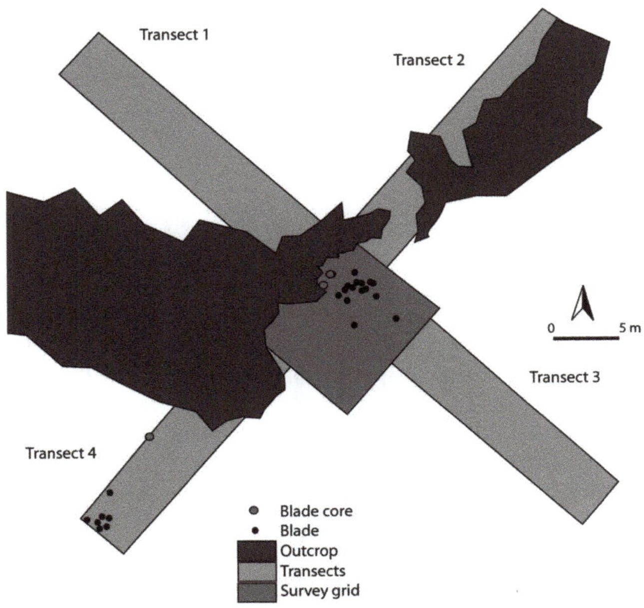

Figure 6.8: Location of Flakes with a Blade Form and Blade Cores at Quarry 27

6.2.2 Buffering Tests of the Cobbles and Cores at Quarry 27

Artefacts or features can be buffered to any diameter in a GIS. The characteristics of the artefacts found in these buffers can be compared to assess whether any spatial variation in the amount of core reduction occurred between the buffers.

- **Assemblage Proportion to Buffer Size**

The density plots show a strong spatial relationship between the location of the discarded artefacts and the degree of core reduction. An analysis designed to clarify the spatial relationship between the cores and the surrounding artefacts compares the proportion of artefacts in three buffers ranging from 0.25 m to a 1 m radius around each core (Figure 6.9). This analysis defines which technological type has the strongest or weakest spatial relationship to the cores. A similar analysis was undertaken for the Group III cobbles and compared to the results of the core buffers to determine whether the position of the cobbles influenced the distribution of the artefacts.

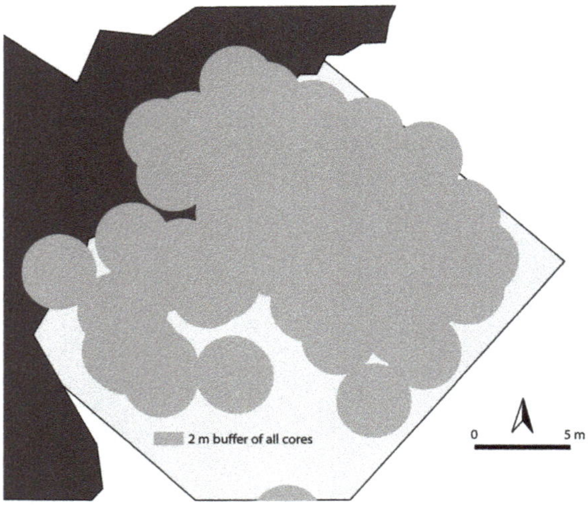

Figure 6.9: Buffers of Core Location at Quarry 27

A strong spatial relationship occurs between the cores and particular kinds of flaking debris (Table 6.6). A strong association was found between the cores, and the blocks, broken flakes, complete flakes and platform rejuvenation flakes with the 0.25 m buffer (Table 6.6). In contrast, the weakest spatial relationship is between cores and cobbles (Table 6.6). The limited number of cobbles near to the cores may hint at the transportation of cores away from the outcrop. By comparing the number of artefacts within the core buffers the results show that 26% of the total assemblage was found within the first 0.25 m of the cores. This result implies that the distribution of the flaking debris was directly associated with the distribution of the cores (Table 6.7).

	Technological Type	0.25 m		0.5 m		1 m		Total
		Count	%	Count	%	Count	%	
	Cobble	6	4	8	5	11	7	166
Core	Core fragment	22	19	46	40	76	66	116
Non-retouched flakes	Complete flake	322	24	756	56	1074	79	1358
	Complete split	70	21	195	59	264	80	331
	Distal flake	53	21	127	51	197	79	249
	Proximal flake	14	22	43	67	58	91	64
	Broken flake	142	23	342	56	501	83	607
	Platform rejuvenation	9	32	16	57	22	79	28
Retouched flakes	Complete tool	14	18	43	54	61	77	79
	Distal tool	1	7	6	43	6	43	14
	Proximal tool	1	50	1	50	2	100	2
	Broken tool	6	38	8	50	9	56	16
Retouched spall		1	20	4	80	4	80	5
Other	Hammer	9	18	26	52	40	80	50
	Block	61	24	142	55	222	87	256

Table 6.6: The Technological Types in the Buffers around the Cores

Buffer (m)	Cores		Cobbles	
	Count	% of total assemblage	Count	% of total assemblage
0.25	919	26	283	8
0.5	1949	55	904	25
1	2700	76	2057	58
1.5	3041	86	2617	74
2	3144	89	2816	79

Table 6.7: The Proportions of the Quarry Assemblage in the Core and Cobble Buffers

In comparison to the core buffers there is little spatial relationship between unworked cobbles and the discarded artefacts. Only when the cobbles are buffered to 1.5 m do the assemblage proportions of most technological types reach 75% (Table 6.7). The most striking difference between the core and the cobble buffers was seen within first 0.25 m (Table 6.7). Only 8% of the total assemblage in this buffer size surrounded the unworked, Group III cobbles, whereas 26% of the total assemblage was found in the 0.25 m core buffer.

While the position of the cobbles does not directly influence the composition of the surrounding assemblage it has certainly influenced the position of the quarrying activity. Very few artefacts occur beyond a 2 m buffer of the cobbles indicating that core reduction was focused on the area where the cobbles outcrop (Table 6.3).

Technological Type	0.25 m		0.5 m		1 m		1.5 m		2 m	
	Count	%	Count	%	Count	%	Count	%	Count	%
Core	22	11	55	26	119	57	156	75	164	78
Core Fragment	9	8	26	22	56	48	77	66	80	69
Complete flake	108	8	346	25	824	61	1046	77	1131	83
Complete split	33	10	84	25	194	59	236	71	259	78
Distal flake	17	7	64	26	143	57	108	76	196	79
Proximal flake	8	13	21	33	42	66	56	38	58	91
Broken flake	54	9	188	31	399	66	499	32	538	89
Platform rejuvenation	3	11	9	32	18	64	20	71	23	82
Complete tool	4	5	15	19	38	48	60	76	62	78
Distal tool					5	36	8	57	9	64
Broken tool	2	13	7	44	10	63	11	59	12	75
Proximal tool			21		72	100	72	100	72	100
			2	40	2	40	4	30	4	80
Block		9		30		70		33		91
Hammer		2		18		52		30		88

Table 6.8: The Technological Types in the Cobble Buffers

- **Buffering Cores to 0.5m**

Since the location and characteristics of the cores was thought to reflect the location of different core reduction processes, another buffering test was used to search for spatial patterns in the distribution of artefacts. All the cores, independent of type, were buffered to 0.5 m using *ArcView*. Where the core buffers overlap a single polygon was formed (Figure 6.10). The largest distributions of cores were found by selecting the polygons in *ArcView* with a perimeter greater than 17 m. Artefacts within each of these polygons were then selected, exported as ID numbers and their characteristics compared. The P value represents the perimeter of the buffered area with P49 having a perimeter of 49 m. By comparing the flakes and cores from each of the core buffers it was hoped that different stages of core reduction could be identified in different parts of the quarry.

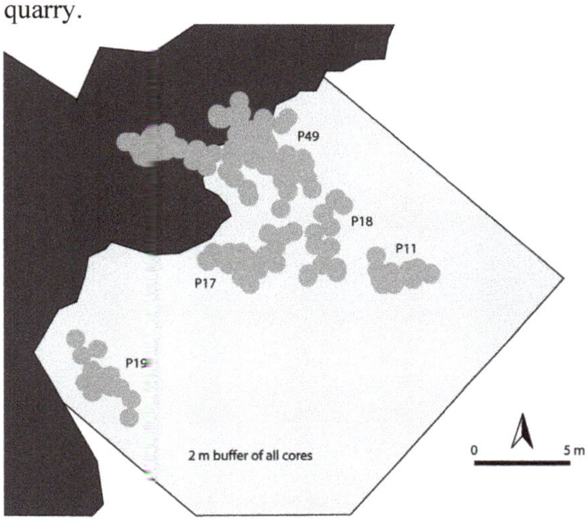

Figure 6.10: The Location of Buffers of 0.5 m around the Cores in the Survey Grid at Quarry 27

Typically, the size of a core and the length of its flake scars indicate how much reduction the core underwent before being discarded. These values are shown in Table 6.9 and were used to determine whether different amounts of core reduction occurred in the four buffers. However, no significant difference in the length of the cores was found between the buffers (Table 6.0; *f*=1.113, *df*=3, *p*=0.3471). This result suggests that overall little difference in the amount of core reduction occurred at Quarry 27; the cores underwent the same amount of reduction before being discarded. However, other core and flake characteristics indicate that the cores in P17 and P19 were reduced further than those in P19 and P49.

The high number of cobble cores in P49 (Table 6.9) shows that material is being tested at this location (Figure 6.19). Furthermore, most of the cores in P49 are uni-platform indicating that little core rotation occurred in this buffer (Table 6.9). Overall fewer flakes were removed from the cobble and uni-platform cores in this buffer compared to those in P17, P18 or P19, again indicating a less core reduction (Table 6.9).

Core Type		Bi-platform	Multi-platform	Cobble	Uni-platform	Prismatic	Mean mm
Length (mm)							
P17	Mean	98	89	116	107		104
	SD	29	35	69	33		33
	Count	3	3	2	13		21
P18	Mean	72	92	136	97		88
	SD	12	13		20		23
	Count	5	2	1	4		12
P19	Mean	93	120	118	99		104
	SD	35	42		16		30
	Count	4	3	1	3		11
P49	Mean	103	107	110	100	104	104
	SD	23	35	25	32		28
	Count	17	8	14	32		71
Scar Length (mm)							
P17	Mean	51	56	31	42		44
	SD	17	40	8	13		18
P18	Mean	33	44	63	39		39
	SD	10	13	9			12
P19	Mean	62	88	60	49		66
	SD	12	45		9		28
P49	Mean	55	58	36	47	46	48
	SD	13	19	15	20		18
Number of Scars							
P17	Mean	5	6	1	3		3
	SD	2	2		1		2
P18	Mean	4	6	2	3		4
	SD	2	4		2		2
P19	Mean	3	5	1	3		3
	SD	2	2		2		2
P49	Mean	6	5	2	2		3
	SD	9	3	1	1		5
Number of Cores							**Total**
P17	Count	3	3	2	13		21
	%	14	14	10	62		
P18	Count	5	2	1	4		12
	%	42	17	8	33		
P19	Count	4	3	1	3		11
	%	36	27	9	27		
P49	Count	17	8	14	32	1	72
	%	24	11	19	44	1	

Table 6.9: The Characteristics of the Cores in the Core Buffers

Analysis of the flaking debris within the P49 buffer indicates that most of the artefacts resulted from material testing and initial core reduction. This result is shown by the high number of cores, complete splits, broken flakes and blocks found in the buffer (Table 6.10). A significant difference was found in the number of these artefact types in the four buffers (χ^2=28.934, *df*=9, *p*=0.0007, *phi*=0.221). High frequencies of these artefact types are generally associated with an early stage of core reduction observed at quarries (Hiscock and Mitchell 1993:20-21).

	Technological Type	P19		P49		P17		P18	
		Count	%	Count	%	Count	%	Count	%
Core	Core	11	7	72	10	21	8	12	6
	Core fragment	6	4	22	3	5	2	4	2
	Sub-total	17	11	94	13	26	10	16	8
Non-Retouched flakes	Complete flake	65	42	281	40	100	36	74	38
	Complete split	26	17	63	9	26	9	27	14
	Platform rejuvenation	3	2	8	1			1	1
	Distal flake	15	10	37	5	19	7	11	6
	Proximal flake	2	1	18	3	6	2	3	2
	Broken flake	15	10	120	17	60	22	41	21
	Sub-total	126	82	527	75	211	76	157	82
Retouched flakes	Complete tool	3	2	10	1	9	3	6	3
	Distal tool	1	1						
	Proximal tool			1					
	Weather tool							1	1
	Broken tool			1		1		3	2
	Sub-total	4	3	12	1	10	3	10	6
Other	Block	5	3	60	8	23	8	11	6
	Hammer	3	2	8	1	4	1		
	Total	155		701		274		194	

Table 6.10: The Technological Types in the Core Buffers

Test	*Chi square*	*df*	*p*	Result
Termination (abrupt, cortical, feather, hinge)	21.260	9	0.0115	Significant *phi*²=0.099
Percentage of Cortex	11.988	12	0.4467	Not significant
Number of dorsal scars on complete flakes (0-3)	9.869	9	0.3612	Not significant
Platform Type (cortical, unifacial)	6.664	9	0.6720	Not significant
Flake form (expanding, intermediate, block)	11.296	6	0.796	Not significant

Table 6.11: The Results of the Chi Square Tests Comparing the Characteristics of the Complete Flakes in the Core Buffers

The characteristics of the complete flakes are very similar in the four buffers showing few significant differences (Table 6.11). However, a significant difference was found in the frequency of the different termination types (6.11). In P49 abrupt terminations are more common than in the other buffers associated with a lower frequency of feather terminations (Table 6.12). High numbers of abrupt terminations indicate an early stage in core reduction (e.g. Whittaker 1994:106-109).

Complete flakes with a cortical platform and more than 50% of their dorsal surfaces covered in cortex indicate that they were manufactured early in core reduction (e.g. Andrefsky 1998:93:101-102). A high frequency of complete flakes with these characteristics occurs in P49 but these were not significantly different from the other buffers (Tables 6.11, 6.12).

	P17		P18		P19		P49	
Termination	**Count**	**%**	**Count**	**%**	**Count**	**%**	**Count**	**%**
Abrupt	17	11	14	12	9	8	72	18
Cortical	23	15	18	15	27	24	61	15
Feather	75	49	66	55	63	56	177	44
Hinge	21	14	11	9	6	5	45	11
Missing	3	2	2	2			7	2
Platform	1	1			1	1	6	2
Plunge	2	1	5	4	3	3	15	4
Step	12	8	3	3	4	4	16	4
Percentage of Cortex								
0%	35	35	28	38	18	28	66	23
1-49%	42	42	30	41	30	47	135	48
50-99%	17	17	12	16	11	7	65	23
100%	6	6	4	5	5	8	15	5
Platform Type								
Bifacial	6	5	4	5	2	3	23	7
Cortical	11	10	8	10	12	17	36	12
Missing	19	17	9	11	11	16	53	17
Step	9	8	5	6	1	1	3	1
Unifacial	70	61	57	69	44	63	195	63
Form								
Blade	1		2	1			5	1
Block	19	9	12	7	22	17	66	12
Expanding	20	9	10	6	6	5	39	7
Intermediate	181	82	143	86	102	78	429	80

Table 6.12: The Characteristics of the Complete Flakes in the Different Core Buffers

There is a significant difference (*f*=12.40, *df*=3, *p*<0.0001) in the lengths of complete flakes in each of the core buffers. P49 is close to the outcrop and has the largest complete flakes (Table 6.13). A similar pattern emerges in the size distribution for most of the other technological types, with larger artefacts again found in P49 (Table 6.14). These results suggest that more initial core reduction, producing large flakes, occurred in the P49 buffer.

Groups	Count	Mean Length (mm)	SD (mm)
P49	281	51	24
P19	65	44	2[illegible]
P18	74	37	15
P17	100	39	2[illegible]

Table 6.13: The Mean Length of Complete Flakes in the Core Buffers

	Technological Type	P49		P19		P18		P17	
		Mean mm	SD mm	Mean mm	SD mm	Mean mm	SD mm	Mean mm	SD mm
Non-retouched flakes	Complete split	55	21	50	17	47	22	44	14
	Platform rejuvenation	79	23	90	37	74			
	Distal flake	50	23	41	13	33	5	34	9
	Proximal flake	43	16	60	25	45	15	40	15
	Broken flake	39	14	41	16	37	11	36	16
Retouched flakes	Complete tool	69	22	67	36	62	29	48	14
	Broken tool	68				47	9	69	
	Proximal tool	31							
	Distal tool			54					
	Weather tool					110			
Other	Block	66	24	41	14	61	18	70	29

Table 6.14: The Mean Maximum Dimension of the Technological Types in the Core Buffers

Further rotation of the cores in P19 is shown by an increase in the number of bi-platform and multi-platform cores in this buffer compared to P49 (Table 6.9). However, the bi-platform cores in P19 are not significantly smaller than those in P49 (t=0.7119, df=19, p=0.4852). Furthermore, the large size of the multi-platform and uni-platform cores, the low number of flakes removed from the cores and their size indicates that these cores were discarded early in reduction (Table 6.9). Indeed, most of the cores in this buffer have between 50-99% cortex on the surface, yet another indication of an early core discard (Table 6.15).

Percentage of Cortex	P17		P18		P19		P49	
	Count	%	Count	%	Count	%	Count	%
0%	2	10					2	3
1-49%	11	52	10	83	4	36	39	54
50-99%	8	38	2	17	7	64	31	43
Total	21		12		11		72	21

Table 6.15: The Amount of Cortex on the Cores in the Core Buffers

In P19 a similar stage of core reduction to P49 was identified in the assemblages even though only 155 artefacts were found in the buffered area. Indeed, greater numbers of the complete flakes within P19 have cortical platforms, cortical terminations or completely cortical dorsal surfaces occur than the other three buffers (Table 6.12). Furthermore, the large size of the complete flakes indicates only a limited reduction of the cores occurred in P19 (Table 6.12). The high number of complete splits and flakes with a block form (Table 6.10) in this buffer indicates that a large force, indicating initial reduction, was used on the cores (e.g. Hiscock 1988:365-366).

Some of the characteristics of the cores discarded in P17 and P18 indicate that they underwent further reduction than the cores in P19 or P49. The smallest multi platform cores (n=2) occur in P17, surprisingly combined with the largest uni-platform cores (Table 6.9). The multi-platform cores in P17 and P18 have a large number of small flake scars indicating that the cores in these buffers underwent considerable reduction (Table 6.9). The bi-platform cores in P18 are considerably smaller than in the other three buffers and these are significantly smaller than the bi-platform cores in P49 (t=2.866, df=20, p=0.0095). A significant difference in the size of the flake scars was also found between the buffers (f=4.672, df=3, p=0.0041) with the smallest flakes scars occurring on the cores in P17 and P18.

Further reduction of the cores in P17 was evident by the increase in the number of step and hinge terminations on the complete flakes (Table 6.12) and their small size (Table 6.13). These characteristics in combination with the high number of complete flakes with no cortex (Table 6.12) may indicate the manufacture of large flakes for removal (described in Chapter 5). An increase in the number of flakes with an expanding form is also evident and suggests a more controlled removal of flakes to prepare the core face before large flakes could be produced. Interestingly, a higher number of tools were found in this buffer when compared to those found in P19 or P49.

6.3 ARTEFACT DISTRIBUTION BY LANDFORM FEATURE

The goal of this section was to explore whether any of the various landscape features associated with Quarry 27 have influenced the spatial distribution of artefacts. By comparing the number and density of artefacts on the different landform units it is possible to examine whether the associated landform feature influenced where the artefacts were discarded (Table 6.16).

The lowest density of artefacts is found on the gibber pavement in Transect 2 and Transect 4 (Figure 5.1). The gibber forms a continuous pavement away from the stream gully where Quarry 27 is located. The low number of artefacts on the gibber pavement may be due to the lack of activities or movement of people across the gibber pavement. Most of the artefacts were found on the sandy depositional landform surface within the stream gully. Even away from the quarry, within Transect 1 and Transect 3, the artefact count is considerably higher than in the two transects which cross the gibber pavement. Not only do Transect 1 and Transect 3 follow the sandy depositional landform unit, near the floor of the gully, but these transects also run parallel to the stream. Hence, based on the increased discard of artefacts along the stream it seems that the stream gully was the main area of concentrated activity away from the outcrop.

Landform Unit	Area Surveyed (m²)	Count	Artefact Density (per m²)
Outcrop 1	278	195	0.7
Outcrop 2	185	19	0.1
Sandy Depositional (Transect 1)	214	32	0.2
Sandy Depositional (Transect 3)	227	123	0.5
Sandy Depositional (Quarry)	389	2895	7.5
Gibber (Transect 4)	398	8	0.02
Gibber (Transect 2)	374	15	0.04
Stream bed	87	19	0.2

Table 6.16: The Mean Density of Artefacts in the Landform Units

The reduction of the high quality Group III cobbles found at Quarry 27 represents the draw card, the main incentive that initiated the procurement of material and the subsequent concentrated discard of artefacts in this location. Consequently, most of the artefacts (72%, n=2420) were found within the area of outcropping Group III cobbles (Figure 5.1). Outcrop 2 has only Group II, medium clast silcrete present resulting in a low density of artefacts while Outcrop 1 overlies the area of Group III cobbles and subsequently has a higher density of artefacts (Table 6.16).

6.3.1 Buffering the Outcrop

To determine whether the outcrop influenced the spatial distribution of the artefacts in Quarry 27 a series of buffers were created using *ArcView* and the characteristics of the cores and flaking debris within each buffer and on the outcrop were compared. The buffers are 1 m apart and extend for 6 m away from the outcrop (Figure 6.11).

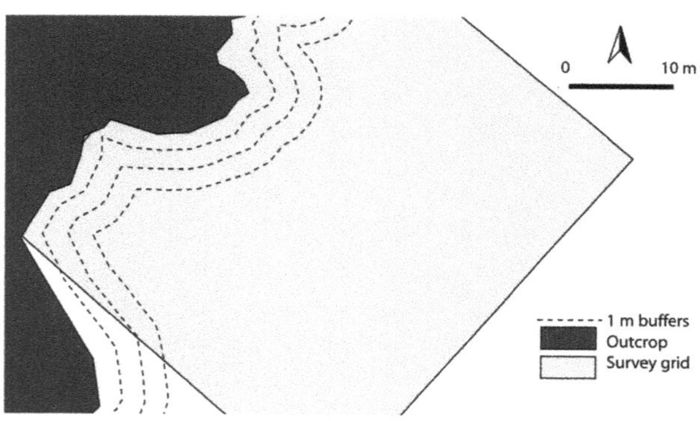

Figure 6.11: Examples of Buffers around the Outcrop at Quarry 27

- **Flaking Debris**

A comparison of the flaking debris between each of the outcrop buffers indicates that initial core reduction occurred near the outcrop. This pattern was shown in the technological types, flake form and the size of the complete flakes (Table 6.17).

A simple comparison of the proportions of different technological types found in each buffer indicates that the distance from the outcrop influences only a few types of artefacts. The number of blocks, platform rejuvenation flakes and cores are proportionately higher nearer to the outcrop, while the frequency of complete tools tends to increase in the buffers away from the outcrop (Table 6.17). High numbers of blocks, platform rejuvenation flakes and cores are commonly found in quarry assemblages (Hiscock and Mitchell 1993:21-22). Other artefact types, such as complete flakes and complete splits, remain proportionately the same across all the buffers away from the outcrop (Table 6.17) A chi square comparing the frequency of cores, complete splits, broken flakes, complete tools and blocks on the outcrop and in Buffer 6 gives a significant result (χ^2=23.219, *df*=4, *p*=0.00001, *phi*=0.327).

The form of the complete flakes undergoes some change as the distance increases from the outcrop. The most notable difference in the flake form is the high proportion of block flakes on the outcrop, as opposed to the 6 m buffer, again indicative of initial core reduction (Table 6.18). A significant difference was in the frequency of the complete flakes with expanding, block and intermediate forms in Buffer 6 and on the outcrop (χ^2=10.498, *df*=2, *p*=0.0053, *phi*=0.176). Furthermore, the decrease in the size of the complete flakes indicates that a greater core reduction occurred away from the outcrop (Table 6.19). The complete flakes in Buffer 6 are significantly smaller than those on the outcrop (*t*=3.278, *df*=174, *p*=0.0013).

- **Cores**

While there are more cores near the outcrop (Table 6.17) no change in the type of cores occurs between the buffers (Table 6.20). Bi-platform, uni-platform and multi platform cores tend to be consistent in frequency across the quarry and may reflect the limited amount of core reduction occurring at the quarry (Table 6.20). The greatest number of cores occurs within the 2 m buffer and not on the outcrop. This may suggest that once the cobbles were extracted from the outcrop and tested, the selected cores were generally reduced elsewhere in the survey grid.

It was expected that cores with more cortex would occur on or near the outcrop, however, most of the cores discarded in this location have between 1-49% of their surfaces covered in cortex (Table 6.20). Only in the 4 m buffer do most of the cores have more than 50% of their surfaces covered in cortex. These results may indicate that once a core was selected it was usually always reduced to manufacture the large flakes required.

		0 m		1 m		2 m		3 m		4 m		5 m		6 m	
		Count	%	Count	%	Count	%	Count	%	Count	%	Count	%	Count	%
Core	Core	23	10	18	6	36	8	24	6	23	6	10	3	8	3
	Core fragment	11	5	7	2	15	3	13	3	17	4	7	2	6	3
	Sub-total	34	15	25	8	51	12	37	9	40	11	17	6	14	6
Non-retouched flakes	Complete flake	95	42	103	35	172	39	152	37	173	46	129	44	82	34
	Complete split	17	8	33	11	45	10	43	10	30	8	31	11	32	13
	Distal flake	15	7	19	6	31	7	35	9	28	7	18	6	11	5
	Proximal flake	4	2	3	1	10	2	10	2	7	2	5	2	6	3
	Broken flake	22	10	74	25	70	16	81	20	57	15	56	19	59	25
	Platform rejuvenation	5	2	2	1	4	1	6	1	1		5	2		
	Sub-total	158	71	234	79	332	76	327	80	296	78	244	84	190	79
Retouched flakes	Complete tool	6	3	6	2	6	1	7	2	16	4	8	3	11	5
	Distal tool	2	1			1		1		1				1	
	Proximal tool			1		1									
	Broken tool	1		1		3	1	1						4	2
	Weather tool							2						1	
	Sub-total	9	4	8	3	11	3	11	3	17	4	8	3	17	7
Other	Block	18	8	29	10	37	8	29	7	23	6	21	7	17	7
	Hammerstone	5	2	1		6	1	7	2	4	1			2	1
	Total	224		297		437		411		380		290		240	

Table 6.17: The Technological Types in the Outcrop Buffers

	0 m		1 m		2 m		3 m		4 m		5 m		6 m	
Form	Count	%	Count	%	Count	%	Count	%	Count	%	Count	%	Count	%
Blade					2	1	2	1	3	1	2	1	5	3
Block	31	20	24	11	40	12	44	14	43	15	21	9	15	8
Expand	11	7	14	6	23	7	35	11	22	8	17	7	10	5
Intermedi te	114	73	186	83	258	80	236	74	220	76	187	82	157	84

Table 6.18: The Characteristics of the Complete Flakes in the Outcrop Buffers

Outcrop Buffer	Count	Length	
		Mean (mm)	SD
0 m	94	53	28
1 m	103	48	22
2 m	172	48	25
3 m	152	45	21
4 m	173	42	22
5 m	129	40	22
6 m	82	41	19

Table 6.19 The Mean Length of the Complete Flakes in the Outcrop Buffers

Core Type	Buffers (m)						
	0	1	2	3	4	5	6
Bi-Platform	6	4	9	8	5	1	2
Microblade		1					1
Multi	5	2	3	3	5	2	1
Prismatic			1				
Cobble	4	2	7	2	1	3	1
Uni-Platform	8	9	16	11	12	4	3
Percentage of Cortex							
0%	1	3			3		
1-49%	14	8	18	15	9	5	7
50-99%	8	7	17	9	11	5	1
Total	23	18	36	24	23	10	8

Table 6.20: The Types of Cores and Amount of Cortex on the Cores in the Outcrop Buffers

Generally, the size of the bi-platform, uni-platform cores has decreased by the 6 m buffer (Figure 6.12). However, the size of the cobble and multi-platform cores tend to be more variable in the six buffers and on the outcrop. An overall decrease in the size of the flake scars on these cores is consistent with the results of the core size; as the distance from the outcrop increases the mean size of the flake scars also decreases (Figure 6.13).

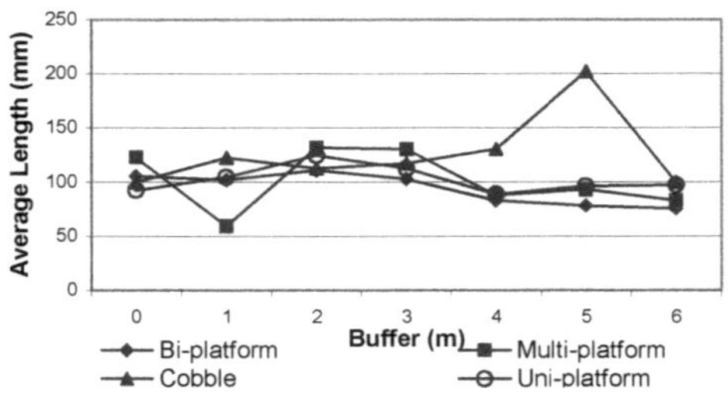

Figure 6.12: Mean Length of the Cores in the Outcrop Buffers

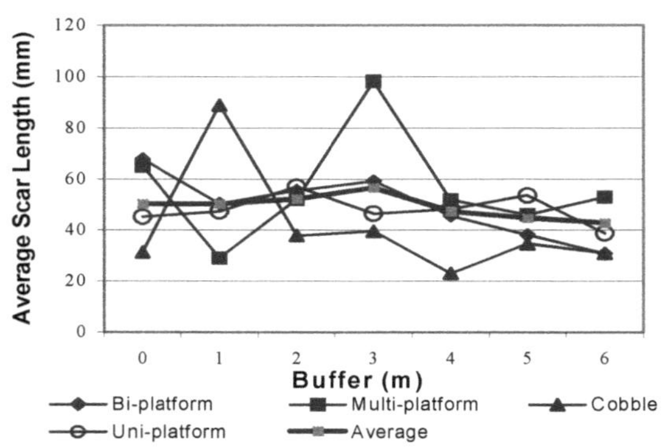

Figure 6.13: The Mean Scar Length on the Cores in the Outcrop Buffers

6.4 SUMMARY

While Quarry 27 is small in size and core reduction was limited, a number of spatial patterns in the distribution of artefacts were found and could be related to the process of core reduction. The results indicate that the core characteristics, which indicate the different amounts of reduction, are spatially congruent with the flake characteristics that reflect the same amount core reduction. Using a density plot analysis in *ArcView* areas of material testing and initial reduction, large flake manufacturing, blade manufacture and tool use were observed in the quarry assemblage. Little variation was seen between the two buffers around the uni-platform and bi-platform cores two assemblages indicating that the flaking debris discarded in the buffers were manufactured during the same stage of core reduction. By buffering all of the cores four concentrations of cores were identified as P49, P19, P18 and P17. Some of the characteristics of the cores and flakes generally indicate that they fall into two main groups. Generally, in P49 and P19 less core reduction occurred in comparison to P17 and P18. That a close spatial relationship is found between the cores and the surrounding flaking debris is highlighted by the assemblage proportion found in the 0.25 m buffer when compared to a 0.25 m buffer of the cobbles.

The distribution and characteristics of the cores can be used to define not only the types of reduction strategies used and the degree of core reduction, but can also be used to gain an understanding of the spatial patterning of artefacts within an entire quarry assemblage. Furthermore, when investigating a larger quarry, with a greater density of artefacts, a more standardised core or a greater core reduction, even stronger spatial patterns of artefact discard may emerge. Consequently, the cores at Quarry 35 will firstly be recorded used to undertake a technological analysis and to search for spatial patterns in the reduction of cores.

A better understanding of the composition and distribution of a quarry assemblage was gained by focusing on the technological aspects and the discard location of the cores. By using this approach many of the problems and biases of other sampling methods can be avoided as the entire quarry assemblage, and not simply a sample of it, is used to make inferences about the types of core reduction strategies employed.
Various landform features have “pulled” the artefact distribution. It is important to account for the “pull” features when undertaking a quarry survey or a sampling bias will occur distorting the accuracy of the information gained. Several “pull” features were found within Quarry 27. The key landscape feature is the presence of a high quality, Group III silcrete source. The outcrop was the obvious focus of the quarry activity and without this feature little concentrated discard of artefacts would have occurred at this location. Furthermore, as the distance from the outcrop increased the characteristics of the discarded cores and flaking debris reflected a greater core reduction. Several other interacting features have influenced the discard of artefacts at Quarry 27. The landform surface and the creek have all contributed to and created patterns of artefact discard.

A spatial redundancy in the patterning of artefact distributions at Quarry 27 has been created by the stages of core reduction and the various landscape features present. The strength of these discard patterns may also be related to the time invested in the reduction of the available material, the time spent at the quarry during each visit, how often the quarry was visited and over what period of time the quarry was used. It is probable that time has created pattern; the more time spent at a place, the stronger the pattern of discard which occurs (Holdaway *et al.* 1998). To test this hypothesis Quarry 35 will be examined in the next chapter and then compared to Quarry 27 in Chapter 9.

Chapter 7

Analysis of Quarry 35

7.1 INTRODUCTION

Quarry 35 is considerably larger, both in extent and artefact density, than Quarry 27. As a result, a targeted sampling method focusing on the distribution of the cores was used to characterise the entire quarry assemblage. This approach incorporated a detailed investigation of the core reduction process, both in terms of the spatial distribution of the cores and as a core reduction strategy. The results of the core analysis highlight locations where different quarrying activities were concentrated and these results are used to justify the placement of sampling locations for the analysis of the flaking debris.

The chapter provides a detailed description of the quarry, the field methodology and the analysis of the assemblage. The analysis is initially divided into two sections focusing on the cores and flaking debris. The cores and flaking debris were analysed separately and then compared in order to determine whether the cores and flaking debris discarded in the same location were manufactured at the same stage of reduction. The tools found in the sampling locations were also described and compared to the manufacturing strategies observed in the analysis of the cores and flaking debris. Lastly, the artefacts from four survey transects were compared to the results of the quarry analysis. The analysis of the survey transects has a two-fold aim; to further define the extent of the quarry and to examine how the characteristics of the artefacts may vary in areas immediately adjoining the quarry.

7.2 DESCRIPTION OF THE QUARRY

Quarry 35 is situated in the upper catchment area of Stud Creek (Figure 3.5) on a bluff approximately 175 m ASL (Figure 7.1). At the base of the bluff, 100 m from the quarried outcrop, a tributary of Stud Creek runs parallel to the quarry. The creek has formed a gorge, intersecting the Mt Wood Ranges and subsequently exposing several silcrete outcrops which occur at irregular intervals along the bluff. Only the outcrop at Quarry 35 has any evidence of quarrying.

The quarry can be divided into two sections based on the topography. The Upper Quarry is a relatively flat area along the edge of the bluff while the Lower Quarry slopes from the outcrops towards the creek (Figures 7.2, 7.3). The gibber pavement forms a somewhat distinctive quarry boundary as few artefacts occur upon this surface (Figure 7.2). Where the gibber pavement ends a vertical drop of 15-20 cm marks the beginning of the sandy, depositional surface on which most of the cores and flaking debris lie (Figure 7.2). The Upper Quarry area continues to the outcrops where the height rises sharply to 176 m ASL. Between the two outcrops the slope gently dips to the south. Directly below the outcrops, over approximately 6 m, an abrupt change in slope of 2 m occurs which then gradually descends towards Stud Creek. At the southern quarry boundary, 30 m from the outcrop, the elevation has dropped to 169 m ASL. This area, between the outcrops and the creek, will be described as the Lower Quarry. The topographic map highlights the degree of slope change across the quarry (Figure 7.3).

Figure 7.1: View of Quarry 35 showing Outcrop 1 and Outcrop 2 (Looking South)

Most of the artefacts were discarded in the Upper Quarry with a maximum density of 9 cores per 50 cm^2. This patterning may represent both a greater artefact visibility, as the area has undergone less disturbance, and a concentration of many knapping activities. The Upper Quarry is 50 m from east to west and 27 m from north to south. The Lower Quarry extends 42 m from east to west and 39 m from north to south and has a maximum density of 10 cores per 1 m^2. The entire surveyed area is 72 m, from north to south, and 50 m, from east to west, covering an area of 3600 m^2 (Figure 7.2).

A total of 3923 cores and core fragments and 14 309 flakes were analysed from the quarry. To extrapolate the number of flakes across the entire quarry, based on the core density and the number of flakes in three sampling locations, a conservative estimate of 700 000 flakes are present at Quarry 35.

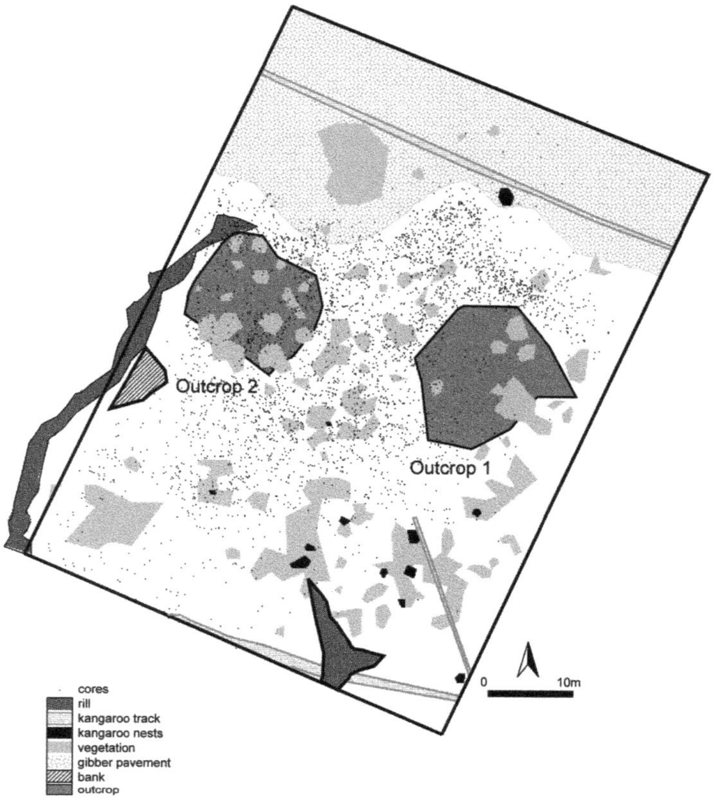

Figure 7.2: Quarry 35

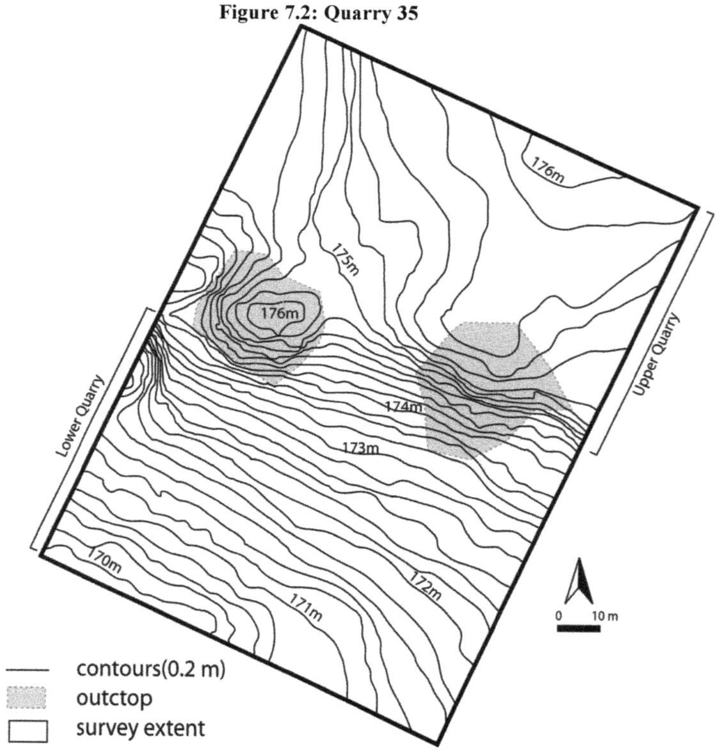

Figure 7.3: Topographic Map of Quarry 35

7.2.1 Geological Characteristics

The raw material available at the quarry varies from a microcrystalline silcrete (Group III) to a medium grained, clast dominated, cryptocrystalline silcrete (Group II). The microcrystalline silcrete has been divided into four sub-groups: amorphous, banded breccia and matrix-dominated (Table 3.2). The clast-dominated materials were divided into three sub-groups: fine, medium and coarse-grained depending on the size of the quartz clasts (Table 3.2). Based on the analysis of the cores (Table 7.1) the raw material available at Quarry 35 is either Group III (28%), medium clast (65%) or fine clast silcrete (7%).

	Material Type	Count	%
Group I	Coarse Clast	2	0.1
Group II	Fine Clast	277	7
	Medium Clast	2547	65
	Sub-total	2824	72
Group III	Amorphous	76	2
	Banded	3	0.1
	Breccia	5	0.1
	Matrix-dominated	1012	26
	Sub-total	1096	28

Table 7.1: The Raw Material Types at Quarry 35 Based on the Number of Cores

The outcrop is a massive silcrete exposure. In profile, the uppermost section has formed columns of concretionary silcrete that has been heavily weathered. Below the columns, mainly outcropping as cobbles, is a massive layer of silicified sandstone with intermittent patches of silicified siltstone. The degree of silicifaction decreases downwards through the profile into the underlying Cretaceous layer until no silification is present in the formation.

The outcrops at Quarry 35 stretch 45 m from east to west and 20 m from north to south. The main area of exposure occurs as two isolated concentrations of cobbles (Figure 7.1, 7.2). Outcrop 1 is 12 m, from north to south, and 15 m, from east to west. Outcrop 2 is larger at 18 m, from north to south, and 18 m, from east to west. Considerably more cores were discarded on Outcrop 2 than Outcrop 1 (Figure 7.1, 7.2). Below both outcrops and towards the creek an overlying jumble of cobbles has formed (Figure 7.2).

On Outcrop 1 the mean cobble length is 232 mm (SD=82, n=189). These cobbles were generally completely covered in cortex (100% cortex, 56%, n=94) or have a highly cortical surface (76-99% cortex, 39%, n=65). Most of the cobbles (n=320) tend to remain wedged within the outcrop with only 20 (13%) cobbles fully exposed at the surface. The mean length of the exposed cobbles (229 mm, SD=81) is not significantly smaller than the unexposed cobbles (244 mm, SD=87; t=0.7509, df=338, p=0.4533).
Most of the cores at Quarry 35 were grey (Table 7.2). A new category, spotty grey, was established which took into account for the presence of distinctive black inclusions within a grey matrix (Table 7.2). A white/grey silcrete and a pink silcrete were also found at the quarry but were not worked. This material was limited in distribution to the lower, central area between the two outcrops. Other colours present in the quarry assemblage include yellow (n=4) and pink (n=6) cores. These cores represent non-local materials discarded at the quarry.

Colour	Count	%
Grey	3516	90
Pink	6	0.2
Spotty Grey	394	9.6
White	2	0.1
Yellow	4	0.1

Table 7.2: The Colours of the Cores at Quarry 35

7.2.2 Identifying Artefact Displacement at the Quarry

The artefacts in the Upper Quarry were not subject to any considerable movement since the area is flat and influenced only to a moderate extent by water run-off and sediment displacement. Since the introduction of European farming practises considerable soil erosion and displacement is thought to have occurred in the region (Fanning 1999; Fanning and Holdaway in press, a). The Upper Quarry is typified by a shallow, sandy deposit of approximately 2-10 cm and overlies an extensive exposure of silcrete. Movement of artefacts has only occurred in localised areas. Kangaroo disturbance has created both tracks and sleeping nests and in conjunction with vegetation growth has either moved artefacts or restricted visibility. These types of disturbances also occur within the outcrops along the edge of the bluff (Figure 7.2).

As the gradient increases away from the outcrop towards the Lower Quarry both gravity and sheet-wash have either transported or buried artefacts. In the area between the two outcrops, sheet-wash from both the outcrops and Upper Quarry has also occurred causing down-slope displacement of the artefacts (Figure 7.2, 7.3). This type of disturbance also appears in the Lower Quarry where the steeper slope gradient immediately below the outcrop is more likely to have created a higher degree of artefact movement (Figure 7.3). It is likely that the cores in this area, due to their shape and larger size, were influenced more by gravity than the flakes (Rick 1976; Reid and Frostick 1985:147-148). However, since the area below the outcrop, where the steepest slope occurs, transects only a two metre section through the quarry then it was presumed that the location of maximum artefact displacement was restricted to this area (Figure 7.3).
It is likely that the large rill, forming the western boundary, has displaced artefacts along its margins (Figure 7.2). This process may also have occurred along the lower rill. Sheet-wash has been observed to move artefacts smaller than 30 mm in similar areas within the study area (Greenwood 1997; Pigdon 1996; Holdaway *et al.* 1998; Fanning and Holdaway 2000). It is likely that small artefacts have been removed from the two rills and redeposited down-slope.

7.3 SPECIFIC FIELDWORK METHODOLOGY FOR THE CORE ANALYSIS

The choice of a targeted sampling strategy, focusing on the core distribution, was based on the results of the spatial analysis of Quarry 27 (Chapter 6). These results indicate a spatial relationship between the flaking debris and the cores reflecting the degree of core reduction. The analysis of all cores from at the quarry was undertaken to identify the reduction strategies employed and search for patterns in the distribution of the cores. A technological and spatial analysis of the cores aided in the selection of sampling areas for the flake analysis.

Three surveying stations for the electronic theodolite were established at Quarry 35. Quarry 35 was incorporated into the Stud Creek grid used to survey Stud 1 and Stud 2 by establishing an intermediate station between the two locations (Figure 3.9). A backsight was located at approximately 200 m from the quarry and used to set the station bearings. Initially an arbitrary quarry boundary was created based on artefact density and landform. The western boundary was a large rill forming a natural boundary. The remaining boundaries were extended beyond the concentrations in artefact densities and onto the gibber pavement. The initial boundary ran 40 m from east to west and 70 m from north to south. However, during the survey the boundary of the Upper Quarry was expanded by 6 m to the east as the core density continued intermittently.

Based on the topography the area was divided into two surveying blocks; the Upper and Lower Quarry. The blocks were pegged at two metre intervals running east to west which allowed the transects to be surveyed for the cores. Each transect was systematically surveyed by two people in order to locate all the cores present at Quarry 35. The cores were tagged, mapped and recorded using the system described in Chapter 4 (Figure 7.4). The second survey block was the Lower Quarry and a similar method was used to record the cores. In the Lower Quarry the boundary was established by following the density of the cores.

The geomorphological surveying, undertaken by Patricia Fanning and students from the Graduate School of the Environment at Macquarie University, recorded the vegetation, watercourses, terraces, outcrop, areas of kangaroo disturbance and landform features such as the gibber pavement (Figure 7.2).

Figure 7.4: The Recording Method used for the Core Analysis

7.4 ANALYSIS OF THE CORES AT QUARRY 35

The core analysis of Quarry 35 has three aims **1.** to examine the manufacturing strategies used at the quarry to reduce the available material, **2.** to define whether any variation in the distribution of the different stages of core reduction were present in the quarry assemblage and finally **3.** to justify a sampling strategy for the flaking debris based on the distribution of cores. In the following section (7.4.1) the characteristics of the different core types is presented. The results are then used to determine the different strategies used in core reduction at Quarry 35.

7.4.1 Description of the Core Assemblage

At Quarry 35 3922 cores (n=3432) and core fragments (n=490) were recorded. The cores are widely dispersed across the Upper and Lower Quarry areas although they mainly concentrate in the Upper Quarry area (Figure 7.2). The largest core has a maximum block length of 840 mm and the smallest core is only 24 mm in block length. The mean length of the cores was 80 mm (SD=31 mm). Most of the cores are uni-platform, bi-platform or microblade cores (Table 7.3).

A considerable amount of reduction and blade manufacturing has occurred at the quarry shown by the size of the cores and the core types (Table 7.3). Core types that were associated with the production of blade manufacture, such as tranchet, bidirectional, microblade, and prismatic, account for 14% of the total population, although other core types were also used to produce blades. The small amount of cortex and the increased number of step and hinge terminations on the flake scars of these core types indicates that they underwent more reduction than the cobble, multi-platform, bi-platform or uni-platform cores (Table 7.4).

The geological characteristics of the Group III silcrete allows for greater flaking control during reduction enabling cores of this material type to be further reduced than cores of Group II silcrete (Doelman *et al.* 2001). Correspondingly, the mean length of the cores differs according to the raw material type (Table 7.3). Cores made from Group III silcrete are smaller in mean length, with the exception of the uni-platform cores than cores made from fine clast or medium clast silcrete (Table 7.3).

A significant difference in the length was found between the cores made from medium clast and Group III silcrete (t=4.551, df=2179, p<0.0001). In contrast, no significant difference in the length was found between the cores made from fine clast and medium clast silcrete (t=0.5080 df=2494, p=0.6612). The relationship between the size of cores and their raw material types is not influenced by the size of the original cobbles as no significant difference occurs in the lengths of the Group II and medium clast cobbles in Area B (t=0.5310, df=91, p=0.5967). Hence, these results suggest that high quality material (Group III silcrete) was selected and reduced further than the poor quality material (Group II silcrete).

The cores with a spotty grey colour are significantly smaller (t=5.243, df=3410, p<0.0001) than the grey cores (Table 7.3). The different frequency of core types by colour also supports the further reduction of spotty grey material (χ^2= 26.860117, df=7, p=0.0004, phi=0.0886; Table 7.4). There are very few cobble cores and considerably more small, multi-platform, tranchet or prismatic cores made from the spotty grey silcrete (Table 7.4). The yellow (n=4) and white (n=2) cores represent material obtained from another source and discarded at Quarry 35 (Table 7.4).

- **The Cobble Cores**

Few cobble cores (n=136) were recorded in the quarry (Table 7.3). The characteristics of these cores, such as their size, indicate that they underwent little reduction. The cobble cores are typically large, heavy and almost completely covered in cortex (Tables 7.3, 7.4). All of these cores have a square cross section, maintaining the original cobble form (Table 7.4). Most of the cobble cores have only one platform with two flakes removed suggesting that these cores were tested and rejected (Tables 7.4).

A selection for smaller, fully exposed cobbles is evident at Quarry 35. All the cobbles (n=262), bar four, were readily available at the surface of the outcrop. This result is consistent with the block length of the cobble cores (142 mm, SD=83 mm; Table 7.3) which is significantly smaller (t=10.104, df=296, p<0.0001) than the length of the unworked cobbles (231 mm, SD=84).

The flake scars on the cobble cores indicate that the flakes removed were generally large, wide, heavily cortical flakes with step terminations (Tables 7.3, 7.4). Most of the flake scars on the cobble cores (n=87, 66%). The presence of step terminations indicates that little care was taken to remove flakes (Whittaker 1994:106). While the largest cobble core (840 mm) is a massive cobble located on Outcrop 1 (Figure 7.1), the mean size of the cobble cores is 142 mm (SD=83 mm), and these range in size between 48 and 840 mm. By comparing the cobble cores to other core types, with the exception of the prismatic and multi-platform cores, the largest flakes were removed from cobble cores (Table 7.3). The size of these flake scars indicates that considerably force was used to remove the flakes (Pelcin 1997; Whittaker 1994:91). However, in comparison to the orientated platform length of the cobble cores size the flake scars are small and wide indicating that the goal of removing flakes at this stage of reduction was simply to test the quality of the material and not to decortify the core (Table 7.3).

Although the cobble cores, when compared to the other core types, have the largest proportion of cortical platforms (73%, n=96) most have no cortex on their platforms. The lack of cortical platforms may be related to a process of natural weathering and spalling removing cortex from the surface of the cobbles. Weathering also creates fractures in the cobbles and weakens their structure (Watts 1974; Bamforth 1992:132) and maybe another reason why these cores were abandoned.

A selection for the further reduction of cobbles made from high quality material, Group III silcrete is evident in the core assemblage at Quarry 35. Most of the cobble cores (n=109, 81%) are made from low quality, Group II, medium clast silcrete (Table 7.4). In contrast, only 15% (n=20) of the cobble cores were made from Group III silcrete and only six cobble cores (4%) were made from fine clast silcrete (Table 7.4). The cobble cores made from medium clast are larger than those made from Group III or fine-clast silcrete indicating that the size of the cobble and the material type were important factors influencing decisions to select and further reduce cores (Tables 7.3). In particular, it seems likely that large cobbles of the high quality, Group III silcrete were usually selected for further reduction.

- **Uni-platform, Bi-platform and Multi-platform Cores**

Most of the cores at the quarry are either bi-platform or uni-platform and made from medium clast, Group II silcrete (Table 7.4). The mean block length of these cores is still large but significantly smaller (f=722.44, df=2, p<0.0001) than the cobble cores (Table 7.3). Few cortical platforms occur on these core types when compared to the cobble cores (Table 7.4). These characteristics indicate that the bi-platform and uni-platform cores underwent greater reduction than the cobble cores (Magne and Pokotylo 1981:38; Andrefsky 1998:12).

Multi-platform cores are larger and heavier than the uni-platform and bi-platform cores even though they have been rotated at least three times to create new platforms and have more flakes removed (Tables 7.3, 7.4). Only the cobble cores are larger than the multi-platform cores at Quarry 35 (Table 7.3). Most of the multi-platform cores still retain a cortical platform and were made from low quality, medium clast silcrete (Table 7.4). These multi-platform cores are significantly larger (t= 2.325, df=131, p=0.0216) than those made from Group III silcrete again indicating that the cores made from this high quality material were reduced further than those made from medium clast silcrete (Table 7.3).

		Bi-platform	Bi-directional	Microblade	Multi-platform	Prismatic	Cobble	Tranchet	Uni-Platform	Total
Total	%	15	3	6	4	2	4	3	63	
	Count	518	89	199	145	73	136	118	2154	3432
Size Dimensions										
Weight (g)	Mean	315	246	101	497	180	1581	150	236	296
	SD	558	519	72	996	93	2014	95	301	613
Length (mm)	Mean	80	73	60	89	72	142	76	78	80
	SD	25	19	14	39	13	83	17	22	27
Width (mm)	Mean	62	58	46	68	55	111	58	59	61
	SD	21	19	12	31	10	64	15	18	23
Thickness (mm)	Mean	44	38	32	48	44	75	31	40	42
	SD	18	17	9	24	9	50	9	14	18
Orientated length(mm)	Mean	62	60	48	68	61	95	61	58	60
	SD	21	18	13	31	13	44	15	18	22
Orientated width(mm)	Mean	60	56	41	69	49	107	54	57	59
	SD	22	19	12	36	9	62	19	19	25
Orientated thickness (mm)	Mean	45	40	34	49	40	86	33	43	44
	SD	22	19	13	27	9	47	15	18	22
Length of the Cores by their Colour (mm)										
Grey	Mean	82	75	60	90	73	140	76	79	81
	SD	26	19	14	40	12	54	18	22	28
Spotty Grey	Mean	71	67	57	82	64	130	75	73	73
	SD	17	18	12	37	16	59	10	17	21
Length of the Cores by the Percentage of Cortex on their Surface (mm)										
0%	Mean	72	67	57	79	73	145	71	71	70
	SD	18	15	14	24	15		15	17	18
1-49%	Mean	79	75	60	86	70	136	77	77	77
	SD	22	20	14	32	12	37	17	20	22
50-99%	Mean	102	102	67	137		134	85	88	99
	SD	38	16	10	78		47	12	27	39
Length of Cores by Material Type (mm)										
Group III	Mean	77	65	58	77	68	123	71	80	78
	SD	25	15	16	29	14	41	14	31	26
Medium Clast	Mean	82	78	61	95	73	136	78	80	83
	SD	26	22	13	44	13	58	18	22	29
Fine clast	Mean	85	69	64	85	74	207	76	78	82
	SD	24	8	21	26	12	22	21	22	30
Flake Scar Characteristics (mm)										
Scar length	Mean	46	46	43	49	55	48	44	44	45
	SD	16	16	11	20	13	26	12	14	15
Scar width	Mean	27	25	16	31	21	43	23	25	26
	SD	13	11	5	18	7	36	8	11	14
Ratio scar length /width										
Cortex	0%	2	2	3	2	3		3	2	
	1-49%	2	2	3	2	2	1	2	2	
	50-99%	1	2	3	1		1	2	2	
	Total	2	2	3	2	3	1	2	2	
Overall Ratio	**1.1**								980 (3[illegible]%)	
	=>2:1								2193 ([illegible]9%)	

Table 7.3: The Number and Measurements of the Cores and their Characteristics at Quarry 35

		Bi-platform	Bi-directional	Microblade	Multi-Platform	Prismatic	Cobble	Tranchet	Uni-Platform	Total
Percentage of Cortex										
0%	Count	110	25	65	32	28		31	306	598
	%	21	28	33	22	38		26	14	17
1-49%	Count	347	62	127	101	45	17	73	1482	2254
	%	67	70	64	70	62	13	62	69	66
50-99%	Count	61	2	7	12		117	14	366	578
	%	12	2	4	8		88	12	17	17
Colour										
Grey	Count	454	75	174	117	63	130	105	1944	3062
	%	88	84	87	81	86	95	89	90	89
Pink	Count	1							4	5
	%									0.1
Spotty Grey	Count	63	14	24	28	10	5	13	202	359
	%	12	16	12	19	14	4	11	9	10
White	Count			1					1	2
	%			1						
Yellow	Count						1		3	4
	%									
Raw Material Types										
Group III	Count	154	28	79	38	17	21	37	560	934
	%	30	32	40	26	23	15	31	26	26
Fine clast	Count	41	9	16	12	9	6	11	138	242
	%	8	10	8	8	12	4	9	8	6
Medium clast	Count	323	51	103	95	47	109	70	1456	2254
	%	62	58	52	66	64	80	59	88	68
Cross-section Type										
Conical	Count	13	1	71	3	65			92	245
	%	3	1	36	2	89			4	7
Square	Count	445	77	92	128		136	7	1622	2507
	%	86	87	46	88		100	6	75	73
Flake	Count	51	8	25	11	5		109	386	595
	%	10	9	13	8	7		92	18	17
Lenticular	Count	9	3	11	3	3		2	54	85
	%	2	3	6	2	4		2	3	2
Platform Characteristics										
Cortical Platform	Mean	0.1	0.1		0.3		0.3			
	SD				1		1			
Platform Number	Mean	2	2	1	3	1	1	1	1	
	SD	1			1		1	1		
Flake Scar Characteristics										
Flake Scars no	Mean	4	5	4	6	6	2	3	3	
	SD	2	2	2	2	2	1	1	2	
Hinges	Mean		1		1					
	SD	1	1	1	1	1		1	1	
Steps	Mean	1	2	1	2	1	1	1	1	
	SD	1	2	1	2	1	1	1	1	

Table 7.4: The Characteristics of the Cores at Quarry 35

Few cores have more than 50% on their surface covered in cortex indicating that once a core was selected it was generally decortified (Table 7.4). Most of the uni-platform, bi-platform and multi-platform cores have between 1-49% of their surfaces covered in cortex (Table 7.4). Cores with more than 50% cortex on their surface tend to be large and do not have elongated flake scars (Tables 7.3). In contrast, cores with no cortex are generally made from amorphous, Group III silcrete, are considerably smaller and have elongated flake scars (Tables 7.3, 7.5). A chi square test shows that significantly more of the amorphous cores have no cortex on their surfaces than the medium clast cores (χ^2= 10.229, *df*=2, *p*=0.006, *phi*=0.63224).

Large flakes were produced from the uni-platform, bi-platform and multi-platform cores in comparison to the other core types (Table 7.3). However, the flake scars are considerably smaller than the orientated platform length on these three core types indicating that flakes did not extend across the entire face of these cores (Table 7.3). No significant difference was found between the flake scars on the uni-platform and bi-platform cores (*t*=1.601, *df*=2288, *p*=0.1094), yet the flake scars on the multi-platform cores (Table 7.3) are much larger than those on the uni-platform or bi-platform cores (*f*=8.895, *df*=2, *p*<0.0001). The consistency in the length of the flake scars on the uni-platform and bi-platform cores indicates a size threshold for the desired flake length manufactured. Most of the flake scars on the uni-platform, bi-platform and multi-platform cores range between 35 to 50 mm in length (Figure 7.5). These results suggest that the goal of this stage of reduction was to decortify the cores and manufacture large flakes from these three core types.

	Total	Group III			Group II	
		Amorphous	Matrix-dominated	Total	Fine clast	Medium clast
Percentage of Cortex						
0%	Count	24	172	196	47	439
	%	32	17	18	17	18
1-49%	Count	44	658	702	186	1631
	%	58	66	66	68	66
50-99%	Count	8	165	173	42	413
	%	11	17	16	15	17
Size Dimensions						
Length (mm)	Mean	67	74		81	83
	SD	18	34		30	29
Width (mm)	Mean	51	56		63	63
	SD	14	26		27	24
Thickness (mm)	Mean	36	38		43	43
	SD	13	20		20	19

Table 7.5: The Influence of Raw Material Type on the Amount of Core Reduction at Quarry 35

- **Cores Used for Blade Manufacture**

Blade manufacture at Quarry 35 is mainly associated with four types of blade producing cores: bidirectional, prismatic, tranchet and microblade cores. These core types are generally small and light, and have little or no cortex or cortical platforms (Tables 7.3, 7.4), characteristics are indicative of greater core reduction (Clarkson and David 1995:31). Evidence that large flakes were used as blade cores was observed at Quarry 35 as many of these cores still retain flake characteristics (Table 7.4). The blade cores have a large number of elongated, parallel flake scars on the core surface (Tables 7.3, 7.4). The flake scars on the uni-platform, bi-platform and multi-platform cores with no cortex are small and have elongated flake scars (Table 7.3). It is likely that these core types were also used for blade manufacture.

More of the blade producing cores are made of Group III silcrete than the cobble, uni-platform or bi-platform cores indicating that high quality material was preferred but not essential for blade manufacture (Table 7.4). Most of the microblade cores (n=79, 40%), representing the smallest type of blade core, were made from Group III silcrete (Tables 7.3, 7.4).

With a greater core reduction the block size and scar length of the cores decreases along with the amount of cortex on the core surface (e.g. Andrefsky 1998:101). This trend is reflected in microblade, bi-directional, tranchet and prismatic cores (Table 7.6). However, although the blade cores are smaller than the other types these cores are generally not exhausted. The limited number of steps on the flake scars on these cores, although more in comparison to the other core types (Table 7.4), and their size indicates that further flake removals were possible (e.g. Whittaker 1994:109).

Multi-platform cores can be divided into two groups: Group 1 are large, cortical, medium clast, grey cores and Group 2 are small, less cortical spotty grey, Group III cores (Tables 7.3, 7.4 and Figure 7.6). Most of the multi-platform cores (n=101) still have cortex (1-49%) on their surfaces. These cores are significantly smaller (*t*= 4.923, *df*=111, *p*<0.0001) than the cores with 50-99% cortex but not significantly larger than the multi-platform cores with no cortex (*t*=1.139, *df*=131, *p*=0.2568). By comparing the length and width of the flake scars of the multi-platform cores (Table 7.3) with the percentage of cortex on the surface it can be observed that the flake scars on the cores with no cortex were twice as long as wide, a characteristic indicative of blade production (e.g. Flenniken and White 1985:135). These results indicate that most of the multi-platform cores were substantially reduced and were probably used to manufacture blades.

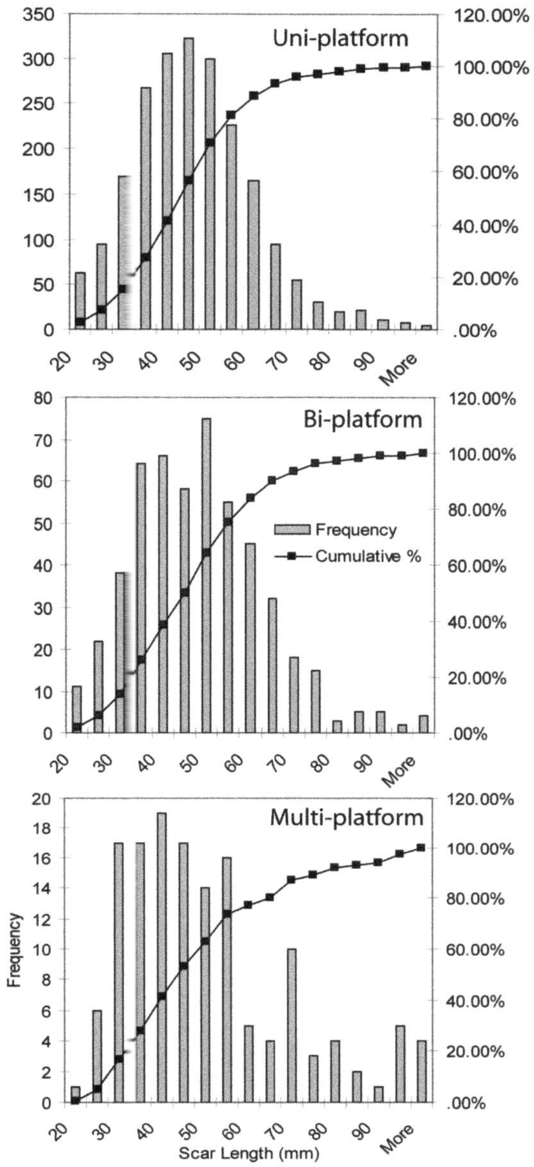

Figure 7.5: Histograms of the Scar Lengths on the Bi-platform, Uni-platform and Multi-platform Cores at Quarry 35

Silicrete type	Length (mm)	Bi-platform	Bi-directional	Microblade	Multi-platform	Prismatic	Cobble	Tranchet	Uni-platform
Fine clast									
0%	Mean	86	69	62	89		81	79	69
	SD	15	8	14	38		14	22	15
1-49%	Mean	83	69	61	84	68	220	79	79
	SD	24	8	21	24	7		12	23
50-99%	Mean	93		109			204	69	82
	SD	34					23	31	21
Medium clast									
0%	Mean	74	73	59	79	74		70	73
	SD	17	14	15	26	14		14	17
1-49%	Mean	80	79	61	92	72	134	80	80
	SD	23	24	12	35	12	32	19	21
50-99%	Mean	115	102	64	143		136	93	90
	SD	42	16	20	83		47	11	27
Group III									
0%	Mean	65	57	54	72	69	145	70	66
	SD	16	12	10	14	16		9	14
1-49%	Mean	77	71	59	74	67	113	71	71
	SD	20	13	14	26	13	15	15	17
50-99%	Mean	82		68	110		124	76	84
	SD	23		6	45		46	7	25

Table 7.6: The Influence of the Material Type on the Type of Core and on the Length of the Cores at Quarry 35

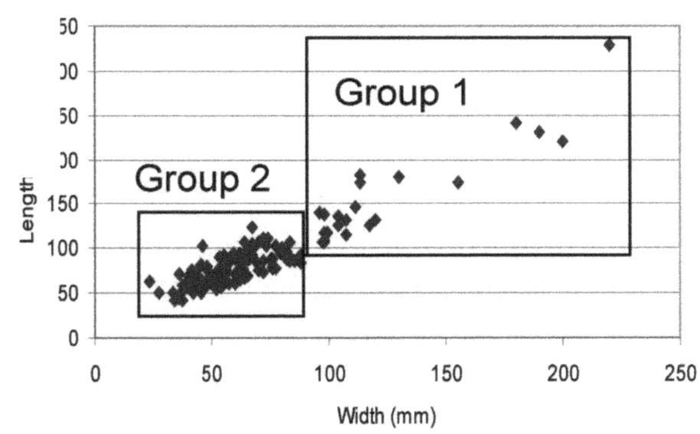

Figure 7.6: Scatter Plot Showing the Length and Width of Multi-platform Cores at Quarry 35

Tranchet cores, with evidence of flake removals across the ventral surface of the flake, were observed at Quarry 35 (Table 7.3). However, after any sign of the initial flake characteristics were removed a core could no longer be defined as being tranchet and was called uni-platform or bi-directional. This process can only be speculated upon as the evidence itself was destroyed by the knapper. These cores typically have a slab shape and elongated flake scars and potentially account for 34% (n=738) of the uni-platform cores. Sometimes the initial flake has a cortical dorsal surface constrained flake removal from anywhere but the ventral surface. This method indicates a need to maximise the use of suitable, high quality stone, even if the dorsal surface of the flake was highly cortical and not generally suitable for further reduction and blade manufacture.

A considerable number of cores still have flake characteristics (Table 7.7; Figure 7.7). These cores are made on large flakes with a mean block length of 79 mm (SD=20) and have smaller flakes removed from their ventral surfaces or their lateral margins. There are also a considerable number of cores with a conical (either microblade or prismatic cores) cross-section (Table 7.7). This cross-section type is a form typical of blade manufacture (e.g. Witter 1992:47; Flenniken and White 1985:135-141). Cores with a conical are significantly smaller in length than other section types (f=49.99, df=3, p<0.0001; Table 7.7). Surprisingly, the section type was not influenced by the quality of the raw material (Table 7.7). No significant difference was found between cores of fine clast, medium clast and Group III silcrete and the section type (χ^2= 7.352, df=6, p=0.2896).

		Conical	Flake	Lenticular	Square
Size Dimensions					
Length (mm)	Mean	67	79	74	82
	SD	17	20	15	34
Width (mm)	Mean	52	58	54	63
	SD	13	16	15	28
Thickness (mm)	Mean	40	37	33	44
	SD	10	12	10	21
Material Type					
Amorphous	Count	1	16		49
	%	2	24		74
Matrix-dominated	Count	64	157	28	598
	%	8	19	3	71
Sub-Total	Count	65	173	28	647
	%	7	19	3	71
Medium clast	Count	156	363	54	1629
	%	7	16	2	74
Fine clast	Count	19	44	2	175
	%	8	18	1	73
Total		245	595	85	2507

Table 7.7: The Characteristics of the Cores with Different Types of Cross Sections at Quarry 35

Figure 7.7: A Blade Core made from a Large Flake at Quarry 35

While most of blade cores were probably produced from large flakes, the exceptions are the prismatic cores. Large thick flakes can be used to produce prismatic cores but these cores are more likely to be fashioned from large cobbles or outcrop fragments (Witter 1992:47). Both varieties of prismatic cores may have occurred at Quarry 35. However, only a small number (n=5) still have flake characteristics (Table 7.4). Furthermore, the mean scar lengths on the prismatic cores are large indicating that most of the prismatic cores were reduced from large blocks or the original cobble (Table 7.3).

7.4.2 Core Reduction Strategies Inferred from the Core Assemblage at Quarry 35

The core assemblage at Quarry 35 represents a complex set of reduction strategies (Figure 7.8). In the quarry assemblage cores with characteristics reflecting material testing, initial decortification and a considerable proportion of blade producing cores were discarded. The large number of blade producing cores indicates that a significant amount of core preparation was undertaken at the quarry. Based on the length/width ratios of the flake scars on the cores at Quarry 35 potentially 69% of the cores were used to manufacture blades as these cores have elongated flake scars (Table 7.3). Material was not simply collected and removed from the quarry but cores were reduced, large flakes and blades were manufactured "on-site" (Figure 7.5).

The results from the core analysis can be divided into four reduction stages: material testing and selection, initial reduction and decortification, large flake manufacture, and blade production. The process of reduction follows a phase of testing and initial reduction, producing large flakes that were often used as cores for blade manufacture (Figure 7.8).

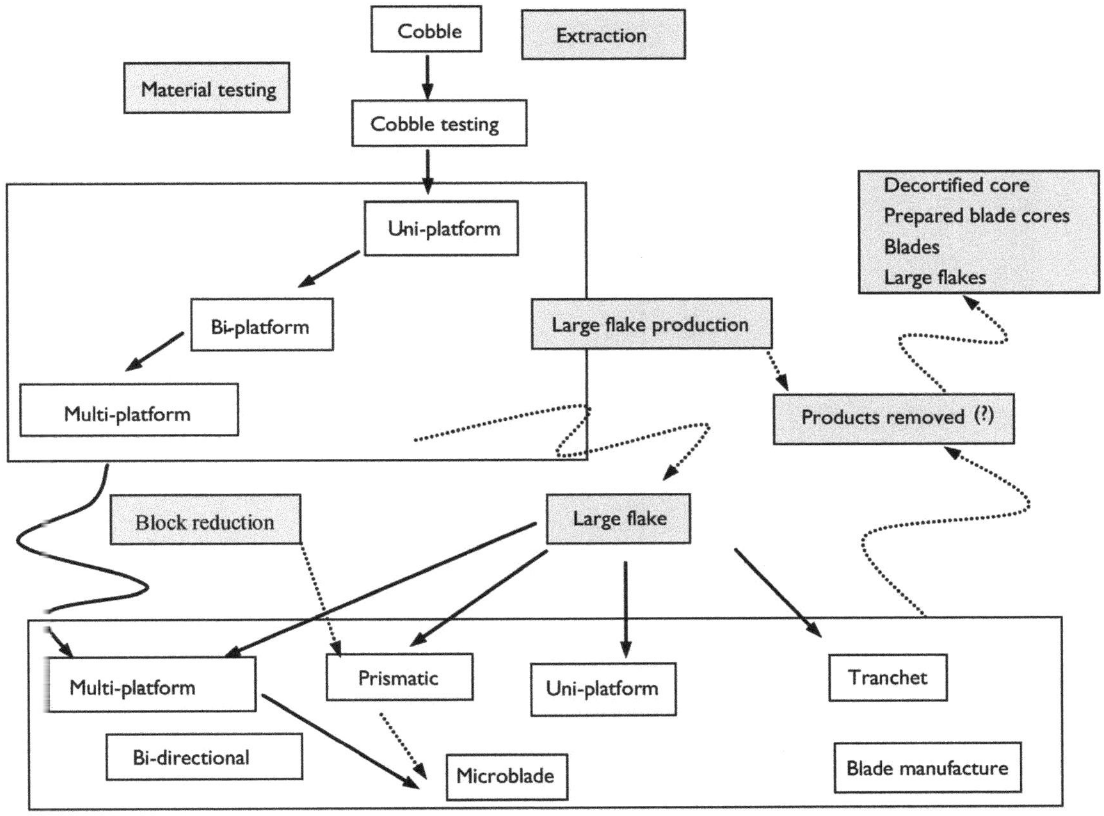

Figure 7.8: A Schematic Diagram Showing the Processes of Core Reduction at Quarry 35

- **Material Testing and Selection**

The testing of material is reflected by the presence of the cobble cores at Quarry 35. 'Testing' may simply have been undertaken to establish the flaking properties of the material as a large number of the cobble cores already had an exposed surface. Decisions about the quality of the fracture and workability of the stone were probably made during this phase of reduction. The discard of a cobble core was an almost instantaneous response to the poor quality of the material. Most of the cobble cores were made from low quality, Group II, medium clast silcrete and this was probably the reason why they were not reduced further. Since few cobble cores of Group III silcrete occur it seems likely that cores of Group III material were preferred for further reduction. However, the small number of cobble cores may also reflect the high material quality of the silcrete at the outcrop since relatively few cores were discarded at this stage.

- **Initial Reduction and Decortification**

The phase of initial reduction is characterised by the decortification of the cores and platform preparation. Once a core was chosen for further reduction, decortification flakes were removed and core rotation occurred. Initial reduction can be observed in the large, highly cortical, uni-platform, bi-platform and multi-platform cores.

Larger cobbles were split on anvils as points of impact were observed on many of the cobbles in the outcrops. However, most of the cobble cores selected have a small size. Cores of this size can be reduced using a free-hand percussion or an anvil-rested method (Flenniken and White 1985:134). The cores were firstly reduced from one platform as uni-platform cores and then rotated to form bi-platform and multi-platform cores. The aim of this strategy was to remove the cortex and prepare a suitable platform from which large flakes could be struck to remove the cortex from the surface of the core.

Relatively few bi-platform, uni-platform or multi-platform cores have more than 50% cortex on the core surface. The limited number of these cores indicates that once a core was selected it was generally reduced and rarely discarded during decortification. These core types represent a point in core discard where either the material quality was deemed inappropriate, too many step or hinge terminations on the flake scars made further flake removals difficult, or alternatively a size threshold was reached in the production of large flakes (e.g. Whittaker 1994:107-109).

- **Large Flake Manufacture**

After the cortex was removed for the cores and a suitable platform was established large flake production could begin. The size of the flake scars of the large uni-platform, bi-platform and multi-platform cores indicates a selection for flakes larger than 44 mm (Table 7.3). The smallest flake scars on these cores types ranges from 35 to 50 mm (Figure 7.5).

- **Blade Manufacture**

Several different strategies were used to manufacture blades at Quarry 35 represented by the characteristics of the different core types. The core types used to produce blades include small multi-platform cores, uni-platform slabs, bidirectional, tranchet and microblade cores.

The selection of higher quality Group III material was evident in the blade producing cores. Material selection is highlighted by the large number of blade producing cores made from high quality, Group III silcrete. Higher quality silcrete generally allows greater reduction and more control during core reduction (Doelman *et al.* 2001).
The blade cores were generally not spent when they were discarded. Most of the blade cores still maintain a relatively large size and have few step and hinge terminations on the flake scars. These characteristics suggest that further core reduction was possible (e.g. Whittaker 1994:107-109). It seems likely that a cut-off threshold in the size of the suitable blades occurred which limited the amount of core reduction undertaken at the quarry.

Blade cores were sometimes made from large flakes, however, another strategy was used to reduce blocks into the prismatic blade cores (Figure 7.8). Large flakes were used as tranchet blade cores, uni-platform slabs, bidirectional and microblade cores as some of these core types still retain their flake characteristics. Small multi-platform cores were also made from large flakes or may have been formed from the reduction of large blocks.

7.5 DEFINING SPATIAL VARIATION IN CORE DISTRIBUTION

The goal of undertaking a spatial analysis of the Quarry 35 core assemblage was to explore whether the different stages of core reduction, already observed in the technological analysis of the cores (Section 7.4), occurred in different locations. The core types and characteristics that reflect the reduction strategies employed at the quarry were plotted in a GIS using three methods already described in Chapter 4: density plots of core attributes, buffering the outcrop and buffering the different core types.

7.5.1 Finding the Location of the Different Stages of Core Reduction at Quarry 35

The following discussion presents the results of the spatial analysis and shows where the different stages of core reduction are located at Quarry 35. The results of the spatial analysis indicate that two distinctive areas occur where different reduction stages were used. The first area occurs on southern slope of Outcrop 1 where a concentration of large, heavily cortical cobble cores were discarded (Figure 7.2). On Outcrop 1 material testing, initial reduction and decortification, and large flake manufacture occurred. In contrast, a concentrated area of blade manufacture was observed on the flat Upper Quarry north of Outcrop 2 (Figure 7.2).

By buffering the cores to a diameter of 1 m overlapping concentrations of different core types were observed (Figure 7.9). Within *ArcView* core buffers with a perimeter of greater then 15 m, representing the densest locations of each core type, were selected and compared. The results indicate that different concentrations of different core types occur in different locations at Quarry 35. The size of the buffers varies between the core types. Core types with the largest area are uni-platform (6599 m^2) and bi-platform (215 m^2) cores. Microblade cores have the third largest concentration at 86 m^2 followed by multi platform cores at 49 m^2. Bidirectional, prismatic and tranchet have smaller concentrations at 21 m^2, 32 m^2 and 30 m^2 respectively. The size of the areas was consistent with the relative frequency of the different core types at Quarry 35 (Table 7.3).

The buffers of the different core types show three overlapping concentrations of the core types used for blade production (Figure 7.9). Two of these concentrations of blade cores occur on the Upper Quarry and another in the central, Lower Quarry. In comparison, the cobble cores tend to occur on or near the outcrops (Figure 7.9).

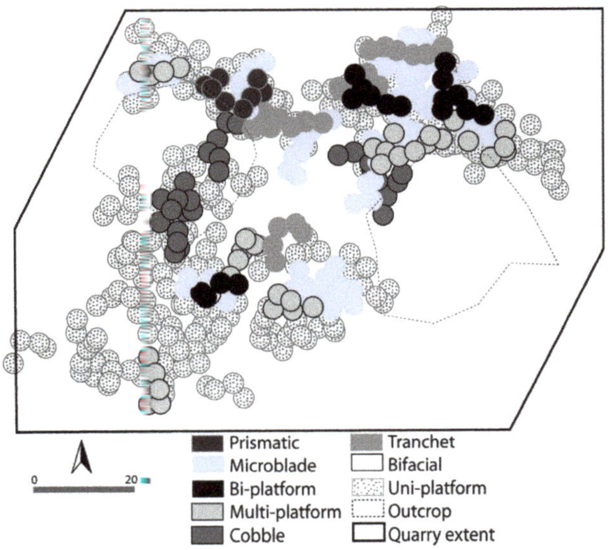

Figure 7.9: 1 m Buffers of the Different Core Types Recorded at Quarry 35

- **Location of Material Testing and Selection**

The overlying concentration of several core characteristics on Outcrop 1 indicates that an area of material testing (Figure 7.10). In this location most of the cobble cores were found. The cores on Outcrop 1 are highly cortical, have long flake scars, are large and heavy and are made from Group II, medium clast silcrete (Figure 7.2). The high number of cores with a single platform and only one flake scar also indicates that limited reduction of the cores on Outcrop 1 (Figure 7.11).

Buffering the cobble cores shows two areas of material testing and selection on Outcrop 1 and one concentration on the west face of Outcrop 2 (Figure 7.9).

- **Location of Initial reduction and Decortification of the Cores**

Initial reduction and the decortification of the cores also occurred on Outcrop 1. Multi-platform, bi-platform and uni-platform cores with their surfaces covered in more than 50% cortex and a weight of greater than 100 g cluster on Outcrop 1 (Figure 7.12). These cores were discarded before most of the cortex was removed from the surface and were made from medium clast silcrete (Table 7.6).

- **Location of Large Flake Manufacturing**

The large multi-platform, bi-platform and uni-platform cores were used to manufacture large flakes. Consequently, these cores have little cortex on the surface, large flake scars and were only discarded when they could no longer produce a large flake. A density plot of the cores with these characteristics indicates that once the core was selected for further reduction it was generally moved away from the outcrop and discarded elsewhere in the quarry (Figure 7.12).

- **Location of Blade Manufacturing Strategies**

The densest concentration of blade producing cores occurs north of Outcrop 2 (Figure 7.1, 713). This "hotspot" converges with several core characteristics that are typical of a greater amount of core reduction. The cores in this area have less cortex, are smaller (<50 mm) and lighter (<100 g), have a greater number of platforms and more flake scars with step terminations (Figure 7.14).

Cores with no cortex occur in three discrete "hotspots" in the Upper Quarry. These three "hotspots" overlie with a concentration of Group III, matrix-dominated silcrete again indicating that cores made from higher quality silcrete were reduced further than those made from medium clast silcrete (Figure 7.11). Cores with a distinctive colour, named spotty grey, also concentrated in the same Upper Quarry areas (Figure 7.15).

Cores made from both medium clast and matrix-dominated silcrete occur as a "hotspot" north of Outcrop 2 indicating the reduction of both material types (Figure 7.11). In contrast, the cores made from fine clast material tend to concentrate in the central, Lower Quarry region and only appear in one isolated area concentrating on the margin of Outcrop 1 in the Upper Quarry (Figure 7.115). This distribution may indicate a series of knapping events where the reduction of fine clast material was preferred.

Cores with more than four scars tend to concentration in two locations; on Outcrop 1 and north of Outcrop 2 (Figure 7.14). The high number of flake scars on the cores at Outcrop 1 may reflect the size of the cores and not the amount of reduction. Large cores can retain a large amount of cortex and still have a high number of flake scars. In contrast, smaller cores have a high number of flake scars and no cortex remaining on the surface (Tables 7.3, 7.4).

In the Upper Quarry concentrations of small multi-platform cores, bidirectional, microblade and tranchet cores can be found (Figure 7.13). These core types are associated blade manufacturing. The Lower Quarry has similar, smaller "hot spots" of microblade, bidirectional, multi-platform and tranchet cores. It is likely that most of the blade manufacture occurred in the Upper Quarry as the prismatic, tranchet and microblade core distributions only overlap in this area.

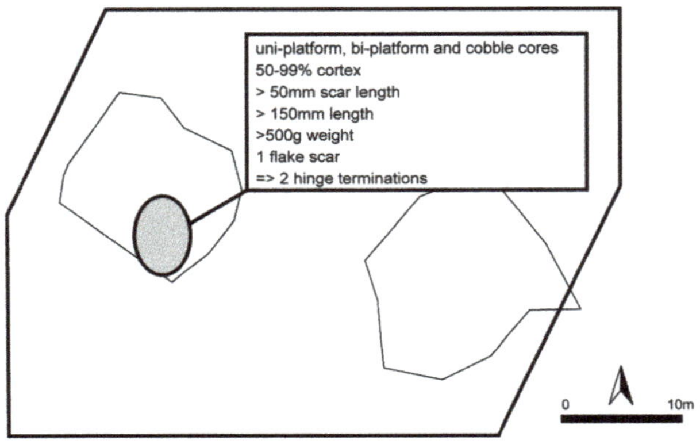

Figure 7.10: Results of the Density Plot Showing the Densest Locations of Cores with Characteristics Indicating Early Core Reduction

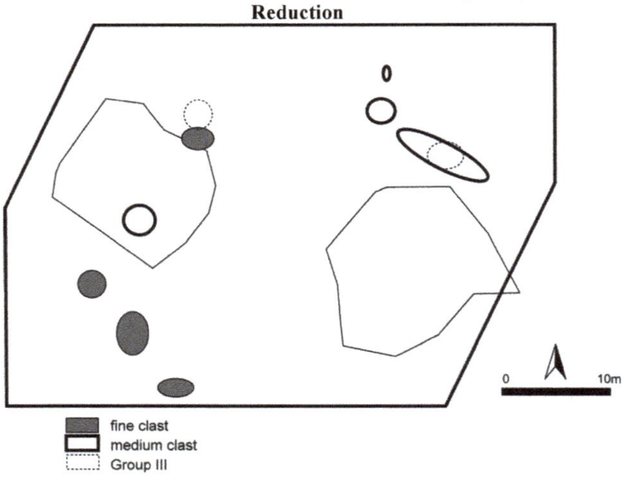

Figure 7.11: Results of the Density Plot Showing the Locations of Cores made from the Different Silcrete Types

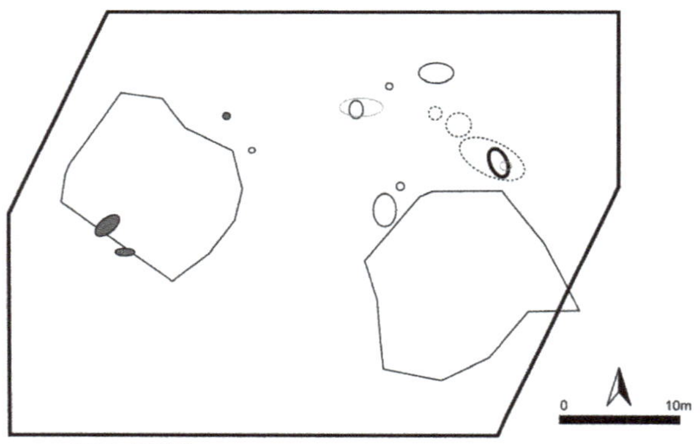

Figure 7.12: Results of the Density Plot Showing the Densest Locations of Multi-platform, Uni-platform and Bi-platform Cores with 50-99% Cortex and a Block Length Greater than 100 mm per 50 cm²at Quarry 35

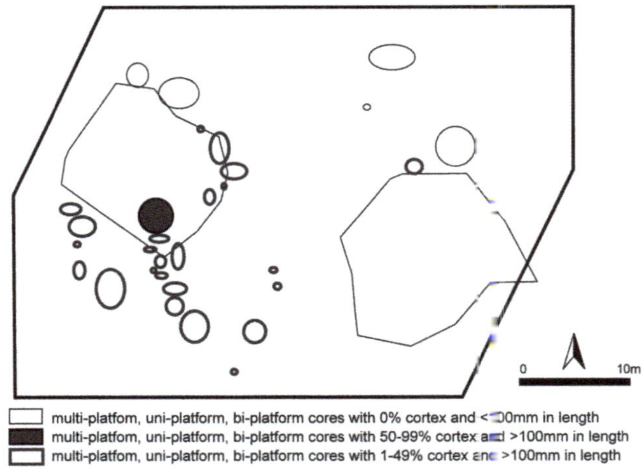

Figure 7.13: Density Plots Showing the Densest Locations of the Blade Producing Cores per 50 cm² at Quarry 35

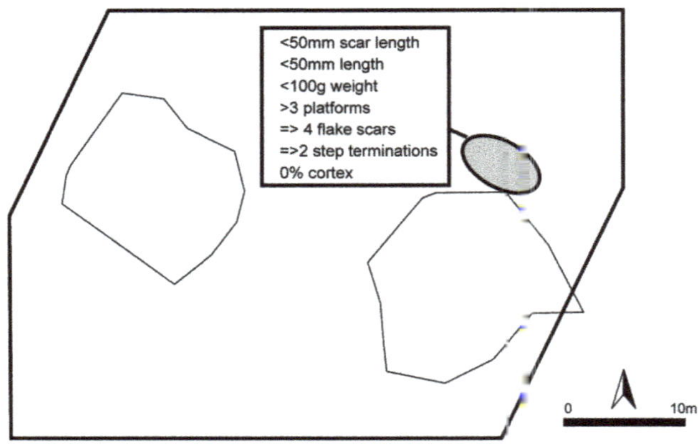

Figure 7.14: Results of the Density Plot Showing the Densest Locations of Cores with Characteristics Indicating Early Core Reduction

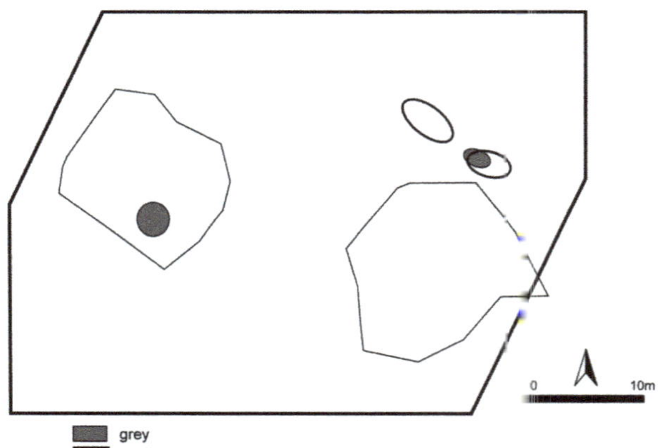

Figure 7.15: Density Plot Showing the Densest Locations of the Cores with Different Colours per 50 cm² at Quarry 35

7.5.2 Buffering the Outcrop

By buffering the outcrop it was possible to identify whether the outcrop influences the spatial distribution of the cores. Comparisons can be made between cores that occur on the outcrop and those that occur at distances away from the outcrop. By contrasting the core populations from the different buffering zones it was possible to determine whether the distance from the outcrop influenced the distribution of the cores and their characteristics (Figure 7.24). Based on the spatial analysis in section 7.5.2 it can be predicted that large, highly cortical cores were discarded at the outcrop, indicative of material testing and initial reduction, while smaller less cortical cores, implying further reduction, were discarded at distances away from the outcrop. It is interesting to note that the numbers of cores within each buffer and on the outcrop are similar although the density of cores decreases away from the outcrop (Table 7.8). By 6 m from the outcrop 63% of the cores were found, in comparison to only 15% of the cores found on the outcrop (Figure 7.24). The even distribution of the cores implies that once cores were tested and selected they were removed from the outcrop and reduced elsewhere.

Buffer	Number
Outcrops 572	
0-2 m	512
2-4 m	512
4-6 m	518

Table 7.8: The Number of Cores in Each of the Buffer Zones

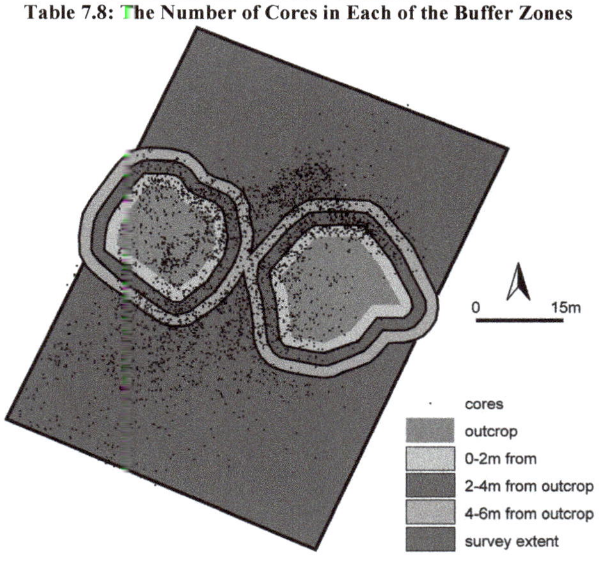

Figure 7.24: The Location of the Different Buffer Zones around the Outcrops at Quarry 35

By comparing the characteristics of the cores in the buffers it can be seen that the distance from the outcrop influences the core types, the amount of cortex, the core section, the size and the length of the flake scars on the cores (Table 7.9). These results indicate that as the distance from the outcrop increases so does the amount of core reduction.

A high number of cobble cores and surprisingly, also a high number of prismatic cores occur on the outcrop and decrease in frequency as the distance from the outcrop increases (Table 7.9). The presence of prismatic cores on the outcrop suggests another core reduction strategy occurring in this location: possible the reduction of large blocks (Figure 7.8). The number of bi-platform and uni-platform cores remains relatively constant across the quarry, despite the distance from the outcrop (Table 7.9). A chi square test comparing the frequency of the different cores types on the outcrop and in the 4-6m buffer gives a significant difference (χ^2=17.153, *df*=7, *p*=0.0164, *phi*=0.125).

The amount of cortex on the surfaces of the cores decreases as the distance from the outcrop increases (Table 7.9). On the outcrop 162 (24%) cores with more than 50% cortex were found. In comparison, only 89 (15%) of the cores with the same amount of cortex were located in the 4-6 m buffer. A significant difference in the amount of cortex on the cores can be found between these two buffers (χ^2=17.638, *df*=2, *p*=0.0001, *phi*=0.118409).

Buffer	Out Crop		0-2m		2-4m		4-6m	
Core Type	**Count**	**%**	**Count**	**%**	**Count**	**%**	**Count**	**%**
Bi-platform	95	17	64	13	95	17	71	14
Bidirectional	7	1	14	3	10	2	17	3
Microblade	33	6	29	6	41	7	29	6
Multi-platform	21	4	25	5	27	5	24	5
Prismatic	19	3	14	3	15	3	6	1
Cobble	35	6	26	5	22	4	19	4
Tranchet	22	4	19	4	13	2	20	4
Uni-platform	340	59	321	63	349	61	332	64
Cortex								
0%	99	15	108	19	107	17	101	17
1-49%	402	61	360	62	439	68	405	68
50-99%	162	24	113	19	102	16	89	15
Section								
Conical	67	13	40	8	37	6	32	6
Flake	87	16	77	15	98	17	93	18
Lenticular	11	2	18	4	17	3	9	2
Square	407	77	377	74	420	73	384	74
Material Type								
Group III	104	28	75	21	58	27	93	24
Fine clast	38	29	24	18	27	21	42	32
Medium clast	430	26	413	25	427	26	383	23
Size Dimensions	**Mean**	**SD**	**Mean**	**SD**	**Mean**	**SD**	**Mean**	**SD**
Length (mm)	89	55	84	35	77	24	78	24
Scar Length (mm)	49	16	48	17	44	14	43	14

Table 7.9: The Characteristics of the Cores in the Different Buffer Zones

A significant difference (χ^2=20.427, *df*=6, *p*=0.0023, *phi*=0.983) in the frequency of the raw material types was found between the buffers (Table 7.9). In the 4-6 m buffer there is an increase in the frequency of cores made from fine clast material. It was expected that more cores made from Group III silcrete would be found as the distance from the outcrop increased, however this was not the case and the frequencies remained relatively equal. This pattern may reflect the proportions of the different material types present in the outcrops: most of the outcrop was composed of medium clast silcrete.

The cross-section of the cores varies in distribution throughout the quarry assemblage (Tables 7.9). On the outcrop there are more cores with square sections (77%, n=407) and surprisingly more cores with a conical section (13%, n=67). The presence of cores with a conical cross section may reflect a different reduction strategy resulting in the production of prismatic cores made from block fragments. The frequency of cores made from large flakes increases slightly as distance from the outcrop increases again suggesting that large flakes were produced, removed from the outcrop and then used as cores. A significant difference was found in the frequency of the different cross sections between the buffers (χ^2=20.472, *df*=9, *p*=0.0152, *phi*=0.097041).

The length of the cores (*f*=12.81, *df*=3, *p*<0.0001) and their scar lengths (*f*=18.98, *df*=3, *p*<0.0001) are significantly difference between the buffer zones (Table 7.9). There was no significant difference in the block lengths (*t*=1.62, *df*=1082, *p*=0.0783) and scar lengths (*t*=0.9974, *df*=1082, *p*=0.3188) of the cores on the outcrop and in the 0-2m buffer. In contrast, significant differences were found between the outcrop and the 2-4m buffer (block length, *t*=4.563, *df*=1082, *p*<0.0001; scar length, *t*=5.447, *df*=1082, *p*<0.0001). Hence, as the distance increases from the outcrop the size of the cores and the length of the core scars decreases. These results show that the outcrop was a feature that influenced the distribution and characteristics of the cores in the quarry.

7.6 SELECTION OF THE SAMPLING AREAS FOR THE ANALYSIS OF THE FLAKING DEBRIS

Spatial variation in the distribution of the various core characteristics was observed throughout the quarry and could be related to different stages of core reduction. The results of the spatial analysis influenced both the size and position of the sampling squares for the analysis of the flaking debris. The location of the sampling squares centred on high-density locations of cores and the characteristics of these cores. In Area B material testing and initial reduction occurred, while in Area A blade manufacture occurred. Area B was located on Outcrop 1 and Area A was located on the flat Upper Quarry area, north of Outcrop 2 (Figures 7.17). For example, in Area A, small, less cortical blade cores occurred, while in Area B large, cobble cores with more cortex were present. This distribution implies that while variation in the characteristics of the cores occurs between each area, within each area there is little variation in the characteristics of the cores. These areas highlight two extremes in the core reduction from material testing and initial reduction to blade manufacture.

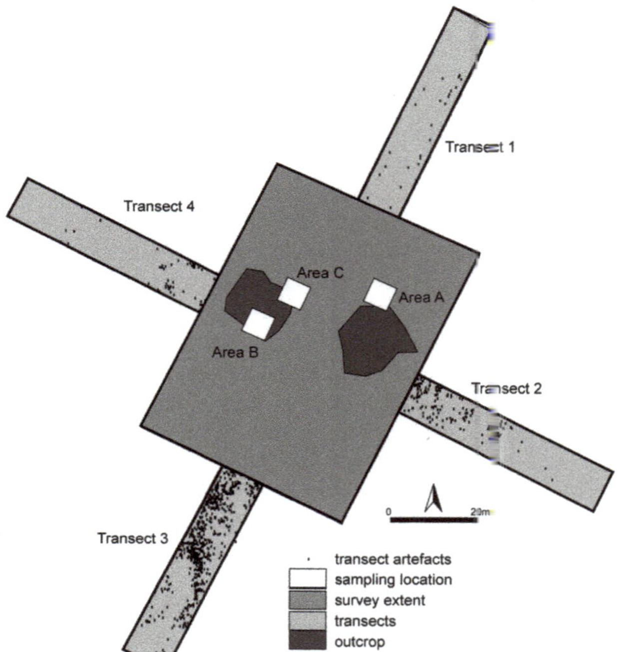

Figure 7.17: The Location of the Sampling Areas for the Analysis of the Flaking Debris and the Survey Transects

The size of the sampling units was based on the results of the core density plots. The sampling units overlaid the "hotspots" observed in the density plots and were measured to a 25 m^2 (5 m by 5 m) area. In the field, the central points of the "hotspots" were relocated using an electronic theodolite with the coordinates generated from *ArcView*. For the easy management of the recording system the area was divided into a grid comprising of 1 m^2 units around the central point of the "hotspot". Each artefact was flagged and labelled with a unique ID number and recorded using the method described in Chapter 4. As the artefact density was high, flagging was replaced with a sticker system, placing the ID number directly upon the artefact in Area B and Area C.

7.6.1 Tests of Diversity

Based on the results of the core analysis, Area A and Area B show locations where different reduction stages were implemented. Tests of diversity were used to measure the degree of variation in core distribution and core characteristics in the quarry to assist in choosing the third sampling area (Table 7.10).

ArcView density plots do not show artefact diversity or variation but simply present areas of high artefact density. By measuring artefact diversity, density as an influencing factor in assessing assemblage variation across space is removed. Diversity (Table 7.10) was measured and compared between each grid unit using *Diversity*, a computer program written by Frankel and Holdaway (1993) based on the diversity formulas explored by Bobrowsky and Ball (1989) and defined in Table 7.10.

Within a GIS the quarry was overlain within a 5 m^2 grid. Grid units with more than 80 cores present were selected and the ID numbers extracted as tables into Microsoft *Access*. Using this relational database, the core attributes (core type, cortex percentage, flake scar number, number of step terminations and material type) were compared between each unit as they reflect the amount of core reduction (Tables 7.11). A comparison of the core characteristics between the units indicates a limited core diversity within Area A and Area B, while two units, unit 84 and unit 54, show the greatest core diversity (Tables 7.11). Diversity comparisons indicate that units 84 and 54 have a low heterogeneity suggesting a mixture of different reduction activities occurred at these locations. Unit 84, (Area C) was chosen because of its greater diversity ('richness') in the characteristics of the cores. Area C is located partly on the outcrop, which may account for the 'mixed' characteristics of the cores. The artefacts in Area C were then used as a comparative sample to Area A and Area B (Figure 7.17).

Richness: indicates the wealth or variety of individuals within a given population. It measures the degree of difference or similarity within a given group. In a population that is similar the species "richness" should be low, closer to 0.1
Evenness: determines the frequency of representation, or proportion, of the attributes present in order to provide an understanding of the relative abundance.
Heterogeneity: measures the relationship between number of species and their individual frequency, encompassing both richness and evenness as a single value. Heterogeneity should be low if the attributes are frequent in more equal proportions.

Table 7.10: Definitions of the Measures of Diversity (Brobrowsky and Ball 1989)

Locations	Area A	Area B	54	106	107	108	84	63	116
Core Type									
Richness	0.61	0.75	0.86	0.62	0.67	0.68	0.86	0.77	0.61
Evenness	0.63	0.70	0.60	0.64	0.65	0.59	0.52	0.70	0.57
Heterogeneity	0.61	0.67	0.57	0.59	0.61	0.57	0.48	0.68	0.48
Percentage of Cortex									
Richness	0.23	0.28	0.31	0.25	0.24	0.25	0.29	0.26	0.29
Evenness	0.83	0.87	0.74	0.78	0.75	0.82	0.56	0.83	0.76
Heterogeneity	0.54	0.57	0.48	0.50	0.47	0.53	0.33	0.54	0.50
Scar Number									
Richness			1.12	0.83	0.89	0.74	1.17	0.87	0.96
Evenness			0.86	0.86	0.85	0.85	0.85	0.89	0.86
Heterogeneity			0.86	0.85	0.86	0.82	0.87	0.86	0.85
Step Terminations									
Richness	0.39	0.47	0.61	0.50	0.65	0.57	0.68	0.61	0.58
Evenness	0.85	0.81	0.75	0.76	0.65	0.73	0.76	0.67	0.73
Heterogeneity	0.72	0.70	0.69	0.71	0.67	0.71	0.72	0.67	0.68
Material Type									
Richness	0.31	0.28	0.41	0.42	0.24	0.33	0.39	0.26	0.48
Evenness	0.46	0.59	0.71	0.42	0.54	0.46	0.52	0.65	0.37
Heterogeneity	0.38	0.38	0.58	0.36	0.38	0.36	0.42	0.42	0.30

Table 7.11 The Results of Diversity Measures for the Core Characteristics at Quarry 35

7.7 ANALYSIS OF THE FLAKING DEBRIS

The analysis of the flaking debris aims to further clarify the reduction strategies used at the quarry and to discover whether variation in the artefact characteristics exist between the three sampling areas selected for flake analysis. To achieve these aims a comparative study of artefacts and unworked cobbles in Area A, Area B and Area C was undertaken.

7.7.1. Characteristics of the Cobbles in the Sampling Locations

Only the characteristics of the cobbles in the sampling locations were recorded as these were thought to influence the composition of the assemblages in each of the sampling areas. Most of the cobbles are found in Area B (n=167) on Outcrop 1 (Figure 7.17). Area C being partly on the outcrop also has a higher number of cobbles (n=77) than Area A (n=38) which is located away from the outcrop (Figure 7.17). The location of these areas explains why the 'richness' values are high (Table 7.11)

Within a silcrete quarry, the natural weathering of cobbles reduces the amount of cortex present on the cobbles. As a result of this process, the amount of cortex on the cores and flakes has been reduced. Consequently, the presence of cortex may not be an accurate indicator of the amount of core reduction. The amount of cortex on the dorsal surfaces of the complete flakes does not vary significantly between the sampling areas (χ^2=3.325, *df*=2, *p*=0.11896). Area B and Area A have the same proportion (14%) of the complete flakes have more than 50% cortex on their dorsal surfaces (Table 7.12). However, the size of the complete flakes in Area B are significantly larger, independent of the amount of cortex on their surfaces, than those in Area A or C (Table 7.12; 0-50%, *f*=57.879, *df*=2, *p*<0.0001; 50-100%, *f*=13.780, *df*=2, *p*<0.0001). These results indicate that the flakes in Area B were manufactured early in core reduction.

Percentage Of Cortex			0-50%	50-100%
Area A	Total	Count	2224	351
		%	87	14
	Length (mm)	Mean	43	51
		SD	16	19
Area B	Total	Count	1408	229
		%	86	14
	Length (mm)	Mean	49	61
		SD	20	27
Area C	Total	Count	2064	287
		%	87	12
	Length (mm)	Mean	43	53
		SD	19	24

Table 7.12: The Amounts of Cortex on the Dorsal Surfaces of Complete Flakes in the Sampling Areas

In Area B only 50 (3%) complete flakes have a cortical termination and a cortical platform. These flakes provide some indication of the initial cobble size and the orientated platform length of the core face (Figure 4.3). The flakes range in length from 21 mm to 152 mm and average at 68 mm (SD=28 mm). In Area A (n=58) and Area C (n=33) only 2% of the complete flakes have these characteristics. The size of these flakes suggests that the cobbles selected for reduction were relatively small. In contrast, the cobbles measured in the sampling areas have a mean width of 164 mm. The difference in size suggests that small cobbles were sometimes selected and reduced. However, if the larger cobbles were split on anvils, as a number of anvils were observed at Quarry 35, (Flenniken and White 1985:133; Gould 1980:123) the amount of cortex on the surface of the cores would have reduced limiting the number of cores with a cortical platform and the amount of cortex on their surfaces. The limited number of flakes with a cortical termination and platform may also be due to the natural spalling of the cortex from the cobbles forming a naturally, decortified platform.

7.7.2 Raw Material Types Found in the Sampling Areas

Both Area A (24%, n= 618) and Area C (30%, n=543) have large numbers of the complete flakes made from Group III silcrete (Table 7.13). In comparison, Area B has only 201 (15%) complete flakes made from Group III silcrete. The results of a chi square show a significant difference between the sampling areas based on the frequency of the complete flakes made from Group III, fine clast and medium clast silcrete (χ^2=236.573, *df*=4, *p*<0.0001, *phi*=0.2844). The proportion of complete flakes made from fine clast silcrete also differs between the sampling locations (Table 7.13). Most of the complete flakes made from fine clast silcrete are in Area C (18%, n=328) and Area B (15%, n=208). In contrast, only 7% (n=180) of the complete flakes in Area A were made from fine clast silcrete.

The difference between the proportion of raw material types in the sampling areas is reflected in the length of the complete flakes. Not only are the complete flakes of Group III silcrete smaller than the other raw material types in each sampling area but are the smallest in Area A and Area C (Table 7.13). A significant size difference was found in the length of the medium clast (*f*=51.99, *df*=2, *p*<0.0001) and Group III, matrix-dominated flakes (*f*= 25.81, *df*=2, *p*<0.0001) between all three sampling areas. Furthermore, significant differences between the complete flakes made from Group III silcrete and medium clast silcrete where found in each of the sampling locations (Area A, *t*=7.891, *df*=2391, *p*<0.0001; Area B, *t*=2.895, *df*=1123, *p*=0.0039; Area C, *t*=6.668, *df*=1808, *p*<0.0001). These size differences indicate that cores made from high quality, Group III silcrete were reduced further in each of the sampling areas and were reduced the most in Area A and C.

Material Type	Area A				Area B				Area C			
	Total		Length (mm)		Total		Length (mm)		Total		Length (mm)	
	Count	%	Mean	SD	Count	%	Mean	SD	Count	%	Mean	SD
Group III												
Amorphous	19	1	38	17	1		20		4		23	8
Banded	2		42	20	1		48				39	14
Matrix-Dominated	597	23	39	15	199	15	48	19	538	30	41	18
Sub-Total	618	24	39	14	201	15	48	18	543	30	41	18
Group II												
Fine Clast	180	7	41	16	208	15	45	19	328	18	44	18
Medium Clast	1775	69	45	17	924	70	53	23	967	53	47	20
Sub-Total	1955	76			1132	85			1295	71		
Group I												
Coarse Clast	2		45	14								
Quartz											26	
Total	2575				1333				1839			

Table 7.13: The Number of Complete Flakes made from the Different Raw Material Types in the Sampling Areas

7.7.3 General Assemblage Characteristics

From the three sampling locations 14 066 artefacts and 313 cobbles were recorded (Table 7.14). The artefacts include 14 143 flaking debris, 52 hammerstones and 51 broken hammerstones. Most of the artefacts are in Area A (n=5969) and Area C (n=4736). Area B has the least number of artefacts (n=3660). Area C (1:29) has the lowest ratio of cores to complete flakes. The ratios indicate that cores were more intensively reduced in these areas than in Area B (1:13) or Area A (1:17). However, it must be remembered that cores were probably removed from the quarry.

The bulk of the assemblage is typical of quarrying debitage (Hiscock 1988:107; Fullagar and Torrence 1991:136; Hiscock and Mitchell 1993:12-17). A high proportion of blocks, complete splits, complete flakes, hammerstones and broken flakes are found in the sampling areas (Table 7.14). However, the high number of retouched flakes (n=1269) in Area A is an anomaly, only 142 tools were found in Area B and 222 retouched flakes in Area C (Table 7.14). Although tools have often been found in quarry assemblages (e.g. Gramly 1980; Torrence 1986; Arnold 1987a, 1987b; Root 1992) large numbers of tools are not characteristic of a quarry assemblage.

	Technological Types	Area A Count	%	Area B Count	%	Area C Count	%
Non-retouched flakes	Complete flake	2569	43	1637	45	2352	50
	Complete split	199	3	202	6	156	3
	Distal flake	762	13	631	17	794	17
	Proximal flake	378	6	218	6	290	6
	Broken flake	541	9	466	13	677	14
	Platform rejuvenation	44	1	45	1	19	
	Sub-total	4493	75	3199	87	4288	91
Retouched Flakes	Complete tool	929	16	88	2	151	3
	Distal tool	113	2	27	1	32	1
	Proximal tool	56	1	3		3	
	Broken tool	97	2	14		24	1
	Core tool	57	1	3		6	
	Core fragment Tool	17		6		4	
	Spall tool			1		2	
	Sub-total	1269	22	142	4	222	5
Other	Block	140	2	79	2	139	3
	Cobble	38	1	198	5	77	2
	Broken Hammerstone	9		30	1	12	
	Hammerstone	34	1	12		6	
	Total	5983		3660		4736	

Table 7.14: The Different Technological Types and Cobbles in the Sampling Areas

Area B has a higher proportion of complete splits, broken hammers, broken flakes and platform rejuvenation flakes than Area A or Area C (Table 7.14). A chi square test gives a significant result between these three areas based on the number of these technological types (χ^2= 65.48, *df*=6, *p*<0.0001, *phi*=0.165182). A large hammerstone force used on a core to remove flakes often leads to broken flakes and hammerstones, or split flakes (Hiscock 1988:365-355). Hence, the difference between the sampling areas indicates that the flaking debris in Area B are more typical of early core reduction, when a large hammerstone force is usually used, than the artefacts in Area A and C.

The flaking debris and tools are considerably larger in Area B than in Area A and Area C (Table 7.15). The complete flakes in Area A, for all raw material types, are consistently smaller than the complete flakes in the other sampling areas (Table 7.15). The size of the complete flakes in Area A are significantly smaller than in Area C (Table 7.15; *t*=5.418, *df*=5231, *p*<0.0001). Flakes with a smaller length indicate greater core reduction (Andrefsky 1998:96). The large size of all the technological types in Area C, when compared to Area A, may be due to the area being located partly on the outcrop where initial reduction of the cores generally occurred (Table 7.15). Only the platform rejuvenation flakes are similar in size in the three sampling areas indicating that this technique was used at the same stage of core reduction (Table 7.15).

It is interesting to note that the size of the broken hammerstones differs considerably between Area A and Area B (Table 7.15). The mean size of the broken hammerstones in Area B (96 mm, SD=32 mm) is larger than in Area A (76 mm, SD=36 mm). Most of the hammerstones are unbroken in Area A (n=34) while in Area B more were broken (n=30). Large gibber cobbles were selected as hammerstones for initial core reduction, these sometimes split on impact when a large amount of force was used on the core. Gibber hammerstones were prone to breaking during initial reduction because they were made from poorly silicified, low quality, medium clast silcrete which has a low fracture strength (Doelman *et al.* 2001). In contrast, smaller cobbles were broken in Area A and may have been used for the finer, more controlled reduction of the cores.

Technological Types	Area A Maximum Dimension (mm) Mean	SD	Area B Maximum Dimension (mm) Mean	SD	Area C Maximum Dimension (mm) Mean	SD
Non-Retouched Flakes						
Complete flake	49	19	59	24	52	21
Complete split	45	14	55	24	45	18
Distal flake	38	13	46	23	40	17
Proximal flake	34	11	39	16	34	13
Broken flake	36	13	38	15	34	14
Platform rejuvenation	68	20	72	24	74	17
Retouched Flakes						
Complete tool	62	16	73	23	64	24
Distal tool	57	15	62	16	57	16
Proximal tool	45	15	32	4	52	25
Broken tool	56	16	58	25	53	18
Spall tool			119		79	47
Other						
Hammerstone	95	29	101	30	96	32
Broken hammerstone	76	36	96	32	89	32
Block	58	17	71	24	51	19

Table 7.15: The Maximum Dimension of the Different Technological Types in the Sampling Areas

7.7.4 Description of the Complete Flakes in the Sampling Areas

The first flakes removed from a core generally have a cortical dorsal surface and a cortical platform (e.g. Andrefsky 1998:112). Only 24 flakes with these characteristics occur in Area B (Table 7.16). These flakes are large, usually have cortical a termination, an indeterminate form and were made from medium clast silcrete (Table 7.16). In Area C 11 complete flakes with these characteristics were found and 15 were found in Area A. There is no significant difference in the length of these flakes was found between the three sampling locations indicating that they were manufactured at the same stage of core reduction but discarded in different locations (*f*=0.4050, *df*=2, *p*=0.6693). These flakes may represent the result of material testing when a flake was removed from a cobble that was completely covered in cortex in order to view the underlying material. This technique was often not necessary as the natural weathering and spalling of the silcrete removed the cortex.

Decortification of the cores is reflected in the large, highly cortical flakes, with cortical or unifacial platforms found in Area B. Complete flakes with more than 50% of their dorsal surfaces covered in cortex account for only 14% (n=241) of the assemblage in Area B. The largest flakes have between 76-99% cortex on their dorsal surfaces or have completely cortical dorsal surfaces (Table 7.17). The complete flakes with between 50-75% cortex on their dorsal surfaces are not significantly smaller than those with 76-99% or 100% cortex (t=1.587, df=2, p=0.2059) and share similar characteristics indicative of early core reduction and decortification. These flakes tend to be made from medium clast silcrete material and most have cortical or abrupt terminations and block or indeterminate forms (Table 7.18).

Attribute		Area	Area	Area
Count		15	24	11
Length	Mean	55	62	56
	SD	15	30	28
Termination	Abrupt	3		1
	Cortical	7	12	7
	Feather	5	11	2
	Hinge		1	
	Plunge			1
Material	Group III	1	9	1
	Fine Clast	1	2	3
	Medium Clast	13	13	7
Form	Expanding		3	5
	Block	3	5	5
	Intermediate	12	16	4

Table 7.16: The Characteristics of the Complete Flakes with a Completely Cortical Dorsal Surface and a Cortical Platform in the Sampling Areas

While almost equal proportions of feather, cortical, platform and plunge terminations were found in the three sampling areas it is the mean size of the complete flakes of these terminations that reflects the amount of core reduction (Tables 7.17, 7.19). In Area B, the complete flakes with these termination types are significantly larger (abrupt, f=15.772, df=2, p<0.0001; cortical, f=9.243, df=2, p<0.0001; platform, f=4.168, df=2, p=0.0246; plunge, f=92.637, df=2, p<0.0001) than in Area A or Area C indicating that they were manufactured earlier in core reduction (Table 7.17).

A significant difference occurs in the frequency of the different termination types in the three sampling locations (χ^2=90.629, df=12, p<0.0001, phi=0.1189). The proportion of the complete flakes with step or hinge terminations demonstrates that further core reduction occurred in Area A and Area C. There are more complete flakes with hinge (7%, n=176) terminations found in Area A than in the other sampling locations. In Area C the highest number of complete flakes with step terminations were found (7%, n=149). Both these termination types are indicative of greater core reduction as they increase in number when a core is reduced (Hiscock 1988:117). The complete flakes with step terminations are consistently smaller than the other termination types within each of the sampling areas but are largest in Area B (Table 7.17). A significant difference in the length of these flakes occurs between the sampling locations (f=3.358, df=2, p=0.036) indicating that the complete flakes with step terminations were removed earlier in core reduction in Area B than in Area A or Area C.

Length (mm)	Area A Mean	Area A SD	Area B Mean	Area B SD	Area C Mean	Area C SD
Termination						
Abrupt	48	18	56	22	51	19
Cortical	51	18	56	24	50	19
Feather	39	14	47	20	41	19
Hinge	38	15	46	23	39	18
Platform	52	19	80	19	77	33
Plunge	55	16	64	19	61	23
Step	37	14	46	21	38	18
Form						
Blade	42	13	50	16	44	13
Block	59	19	68	24	57	23
Expand	40	15	50	25	36	20
Intermediate	42	17	49	9	43	18
Point	40	9	48	7	43	9
Platform Type						
Bifacial	41	16	52	20	43	21
Cortical	46	17	58	26	51	21
Missing	43	18	52	24	45	22
Step	40	18	47	4	39	12
Unifacial	44	17	50	21	44	19
Percentage of Cortex						
0%	38	14	45	19	40	17
1-25%	46	17	53	21	47	20
26-49%	51	17	54	22	53	21
50-75%	53	19	58	24	53	23
76-99%	51	22	63	25	55	23
100%	51	19	63	29	46	25
Mean Width (mm) of the Complete Flakes with Different Flake Forms						
Blade	18	5	20	5	17	5
Block	47	15	55	24	44	18
Expanding	37	16	51	27	44	22
Intermediate	28	12	32	15	29	12
Point	21	5	22	6	23	5
Mean Size Dimensions (mm)						
Length	44	17	51	21	45	20
Width	29	14	36	19	31	15
Thickness	10	7	12	8	10	7
Platform Width	18	13	21	14	17	16
Platform Thickness	7	5	9	6	7	6

Table 7.17: The Size Dimensions of the Complete Flakes Exhibiting Different Characteristics in the Sampling Areas

Attribute	0%		1-25%		26-49%		50-75%		76-99%		100%	
	Count	%	Count	%	Count	%	Count	%	Count	%	Count	%
Material Type												
Fine clast	13[illegible]	18	89	18	23	15	9	12	15	15	7	14
Group III	12[illegible]	16	77	15	21	14	11	14	13	13	18	35
Medium	49[illegible]	66	342	67	106	71	57	74	73	72	26	51
Termination												
Abrupt	13[illegible]	18	75	15	30	20	9	12	24	24	1	2
Cortical			170	33	54	36	30	39	46	46	30	59
Feather	44[illegible]	59	191	38	46	31	22	29	21	21	17	33
Hinge	5[illegible]	8	17	3	6	4	8	10	3	3	3	6
Missing	4[illegible]	6	15	3	1	1						
Platform	7	1	5	1	1	1						
Plunge	2[illegible]	4	16	3	5	3	4	5	1	1		
Step	3[illegible]	4	19	4	7	5	4	5	6	6		
Form												
Blade	3[illegible]	4	16	3								
Block	1[illegible]	2	65	13	42	28	24	31	31	31	12	24
Expand	6[illegible]	8	30	6	17	11	7	9	9	9	6	12
Intermediate	62[illegible]	85	397	78	91	61	46	60	61	60	33	65
Point	1[illegible]	1										
Platform Type												
Bifacial	1[illegible]	1	6	1	1	1	3	4				
Cortical	[illegible]		45	9	24	16	10	13	24	24	24	47
Missing	3[illegible]	4	22	4	10	7	3	4	10	10	4	8
Step	[illegible]		1									
Unifacial	7[illegible]	94	434	85	115	77	61	79	67	66	23	45

Table 7.18: The Characteristics of the Complete Flakes in Area B with Different Amounts of Dorsal Cortex on their Surfaces

Attribute		Area A		Area B		Area C	
		Count	%	Count	%	Count	%
Form	Blade	216	8	50	3	145	6
	Block	210	8	192	12	370	16
	Expand	173	7	129	8	114	5
	Intermediate	1913	74	1256	77	1665	71
	Point	57	2	10	1	52	2
Platform Type	Bifacial	87	3	20	1	51	2
	Cortical	175	7	128	8	121	5
	Missing	21	1	78	5	448	19
	Step	16	1	2		39	2
	Unifacial	2270	88	1409	86	1692	72
Termination Type	Abrupt	544	21	275	17	361	16
	Cortical	512	20	331	20	409	18
	Feather	1129	44	741	45	1103	50
	Hinge	176	7	94	6	85	4
	Platform	10		13	1	12	1
	Plunge	66	3	53	3	99	4
	Step	90	4	68	4	149	7
Percentage of Cortex	0%	1170	46	750	46	1134	48
	1-25%	866	34	508	31	732	31
	26-49%	72	3	150	9	198	8
	50-75%	182	7	77	5	112	5
	76-99%	133	5	101	6	123	5
	100%	146	6	51	3	52	2

Table 7.19: The Characteristics of the Complete Flakes in the Sampling Areas

A significant difference occurs in the frequency of the different termination types in the three sampling locations (χ^2=90.629, *df*=12, *p*<0.0001, *phi*=0.1189). The proportion of the complete flakes with step or hinge terminations demonstrates that further core reduction occurred in Area A and Area C. There are more complete flakes with hinge (7%, n=176) terminations found in Area A than in the other sampling locations. In Area C the highest number of complete flakes with step terminations were found (7%, n=149). Both these termination types are indicative of greater core reduction as they increase in number when a core is reduced (Hiscock 1988:117). The complete flakes with step terminations are consistently smaller than the other termination types within each of the sampling areas but are largest in Area B (Table 7.17). A significant difference in the length of these flakes occurs between the sampling locations (*f*=3.358, *df*=2, *p*=0.036) indicating that the complete flakes with step terminations were removed earlier in core reduction in Area B than in Area A or Area C.

While the size of the platform angle was significantly different between the three areas it seems more likely that the flake form influenced the angle (*f*=66.86, *df*=2, *p*<0.0001). High exterior platform angles are required for blade manufacture (e.g. Flenniken and White 1985:134; Whittaker 1994:99). Blades, points and expanding flakes have a higher platform angle (*f*=19.68, *df*=4, *p*<0.0001) than the block forms (Table 7.20). The type of raw material also influenced the form of the flakes. A considerable number of the flakes with a block form were made from medium clast silcrete (Table 7.20). A chi square test shows a significant difference between the flake form and the type of raw material (χ^2=36.193, *df*=8, *p*<0.0001, *phi*=0.118695). In contrast, blades and points were made from a range of material types including medium clast, fine clast or Group III silcrete (Table 7.20).

Form	Platform Angle (°)		Group III		Medium Clast		Fine clast	
	Mean	SD	Count	%	Count	%	Count	%
Blade	63	9	49	8	158	9	9	5
Block	57	11	25	4	180	10	6	3
Expanding	65	11	44	7	114	6	15	8
Intermediate	62	10	478	78	1290	73	144	80
Point	66	8	19	3	33	2	5	3
Total	62	10	615		1775		179	

Table 7.20: The Influence of the Platform Angle and the Material Type on the Flake Form of the Complete Flakes in the Sampling Areas

Complete flakes with a block form or a cortical platform are considerably larger within each of the sampling areas (Table 7.17). These two characteristics are usually indicative of early core reduction (e.g. Andrefsky 1998:99-102). The largest flakes with these characteristics occur in Area B and it is likely that most of the initial core reduction occurred in this location (Table 7.17). Furthermore, the greatest frequency of complete flakes with cortical platforms occurs in Area B (Table 7.19). However, most of the complete flakes in the three sampling locations have unifacial platforms and an intermediate form (Table 7.19). These flakes may represent the preparation of cores prior to the removal of blades, unlike the flakes with a block form these flakes are elongated (Tables 7.17, 7.21). The manufacture of elongated flakes is consistent with blade technologies (Clarkson and David 1995:31).

There is a high frequency of smashed (missing) platforms in Area B and Area C (Table 7.19). Missing platforms are usually occur during blade manufacture when a small, focalised platform is required (Clarkson and David 1995:31). Chi square tests show a significant difference between the frequencies of the flake forms (χ^2=243.04, *df*=8, *p*<0.0001, *phi*=0.14758) and platform types (χ^2=628.29, *df*=8,*p*<0.0001, *phi*=0.39663) between the three sampling locations. Few complete flakes with bifacial or step platforms were found in the sampling locations. These platform types also tend to occur later in core reduction when core rotation and platform maintenance is required (Andrefsky 1998:88-89). Most of the complete flakes with these platform types are in Area A and Area C (Table 7.19) and are significantly smaller in each of the sampling areas (Area A, *f*=2.982, *df*=4, *p*=0.021; Area B, *f*=4.141, *df*=4, *p*=0.0024; Area C, *f*=5.185, *df*=4, *p*<0.0001) than the other flake forms indicating that they were manufactured later in core reduction (Table 7.17).

Form		Blade	Block	Expand	Intermediate	Point
Area A						
All Flakes	Count	434	491	327	4147	82
	%	8	9	6	76	1
Complete flakes	Length:width	2:01	1:01	1:01	2:01	2:01
Area B						
All Flakes	Count	78	371	188	2550	10
	%	2	12	6	80	
Complete flakes	Length:width	3:01	1:01	1:01	2:01	2:01
Area C						
All Flakes	Count	265	623	131	3355	56
	%	6	14	3	76	1
Complete flakes	Length:width	3:01	1:01	1:01	1:01	2:01

Table 7.21: The Different Flake Forms and the Length:width Ratios of the Complete flakes in the Sampling Areas

Area B has considerably larger artefacts for almost all the technological types (Table 7.15). In particular, the complete flakes from Area B (Table 7.17) are significantly larger than those in Area A and Area C (*f*=77.30, *df*=2, *p*<0.0001). Furthermore, the complete flakes in Area B are also significantly larger, independent of the amount of cortex on their dorsal surfaces (Table 7.17). Complete flakes in Area B with no cortex are larger than in Area A or Area C (*f*=46.78, *df*=2, *p*<0.0001). The same pattern can be found for complete flakes with completely cortical dorsal surfaces (*f*=6.90, *df*=2, *p*=0.001). These characteristics indicate that the greatest amount of core reduction occurred in Area A and Area C.

The dorsal surface of a flake often has scars from previous flake removals. The orientation of the flake scars represents the direction the core was worked and can be used as way of estimating the amount of core reduction (Johnson 1987b:193). However, other researchers argue that the number and direction of the flake scars may be related to the size of the flake (e.g. Shott 1994; Mauldin and Amick 1989:73). Few of the complete flakes in Area B have flake scars originating only from Quadrant 1 compared to Area A and Area B (Table 7.22). The flakes in Area B are larger than those in Area A or Area C, hence this pattern may reflect the size of the flakes (Table 7.15). Another interpretive possibility is that core rotation occurred early in reduction in order to remove cortex from the core. After core preparation flakes were removed from one platform to establish a suitable core face for the removal of elongated flakes

Quadrants	Area A		Area B		Area C	
	Count	%	Count	%	Count	%
1	1924	87	686	67	1814	83
1,2	51	2	55	5	58	3
1,2,3	7		38	4	8	
1,3	107	5	96	9	76	3
1,4	56	3	52	5	48	2
1,3,4	1		19	2	12	1
1,2,3,4	7		20	2		
2	30	1	8	1	46	2
2,3	1		2		6	
2,3,4			1			
2,4	2		2		4	
3	30	1	10	1	46	2
3,4			5		8	
4			23	2	66	3

Table 7.22: The Directions of Flake Scars on the Dorsal Surfaces of the Complete Flakes in the Sampling Areas

.

Most of the complete flakes with feather terminations found at Quarry 35 range from 25 to 55 mm in length (Figure 7.18). In the two areas, Area A and Area C, where cores underwent further reduction a substantial drop in the number of complete flakes with a length of approximately 43 mm are present in both assemblages. It is likely that flakes of this size were removed from the quarry.

C (Table 7.17, 7.19, 7.21) and these are not significantly different in length (*t*=6.90, *df*=359, *p*=0.1[illegible]27). A new flake form was identified to define a point; a tapered blade, leaf-shaped with a triangular cross section already described in Chapter 4. This form has been observed (Campbell and Edwards 1966; McCourt 1975) throughout arid Australia. Campbell and Edwards (1966:195-196) defines a point as being a leaf-shaped, symmetrical blade with a low triangular cross-section mounted as heads on a spear. This type of tapered blade may also have been used for pirri point manufacture (Witter 1992:75). Most of the flakes with a point form also occur in Area A and Area C (Table 7.[illegible]9). It seems likely that away from the outcrop on the flat, Upper Quarry most of the blade and point manufacture occurred. The size of complete flakes with blade or point forms in the sampling areas is similar with a mean length of 45 mm (Table 7.17). It seems likely that suitable flakes with a blade or point form of this size were probably removed from the quarry.

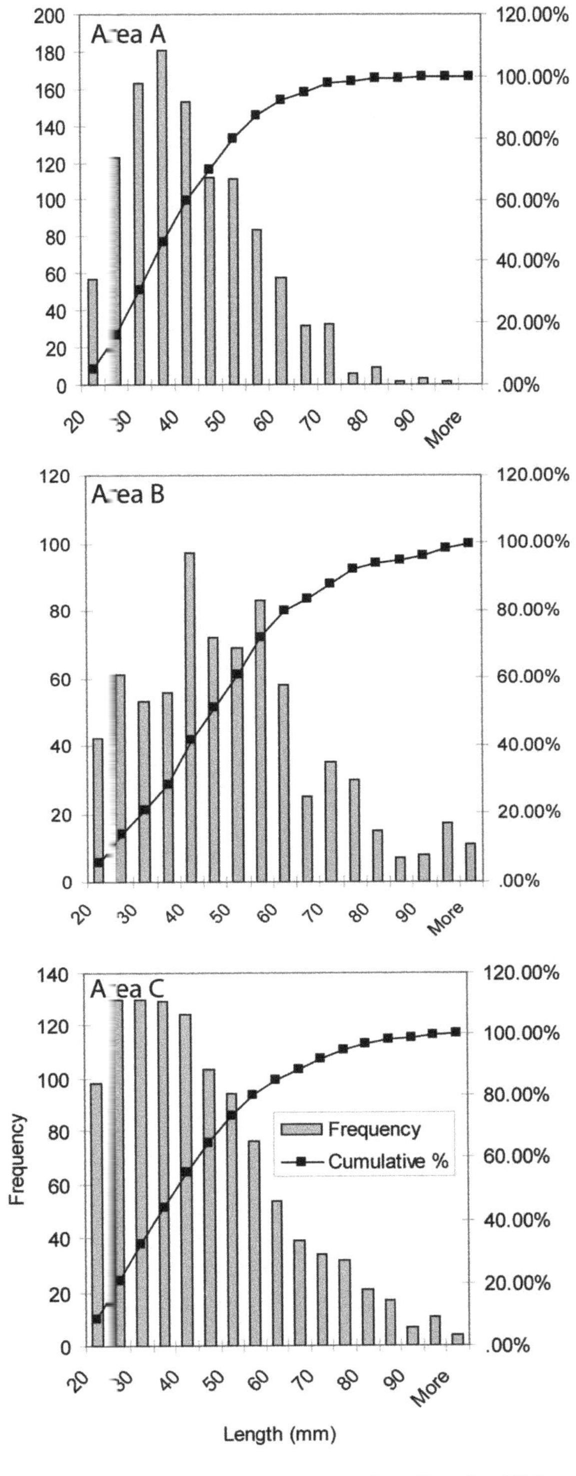

Figure 7.18 Histograms of the Length of the Complete Flakes with Feather Terminations in the Sampling Areas

Most of the flakes with a blade form are in Area A and Area

In Area A the 283 complete flakes with a blade or point form accounts for 10% of the assemblage (Table 7.19). Blade manufacture is generally considered wasteful in the amount of material lost during core preparation prior to the removal of blades (Hayden *et al.* 1996). Experimental studies have shown that between 42 and 60% of the flakes removed from cores occurs during core trimming and preparation and these account for 80% of the assemblage weight (Ammerman and Andrefsky 1982:162-163). Needless-to-say, most of the flakes discarded before or during blade manufacture do not have a blade form but are simply the result of core preparation.

7.8 TOOL TYPES AT QUARRY 35

A high number of discarded tools (n=1272) were found in Area A (Table 7.14). The ratio of complete flakes to complete tools in Area A was 1:3, while in Area B (1:19) and Area C (1:11) low ratios were found. In Area A 79% (n=1272) of all the tools recorded at the quarry were found. In comparison, only 142 tools were recorded within Area B (8%) and 214 tools (13%) in Area C.

The tools identified in the sampling areas can be divided into four groups (after Shott 1989); tools produced as a result of an immediate necessity and used expediently, tools abandoned during manufacturing, tools made from another source and discarded for retooling or tools that were accidentally lost (Figure 7.19). Most of the tools fall into the first group and can be defined as expedient tools (Binford 1973, 1979). These tools (Table 7.23) have either scraper retouch (n=582), notch retouch (n=546) or usewear damage (n=386). The presence of expedient tools is indicative of the need to undertake immediate tasks with minimal effort or time invested in the manufacture of a suitable tool (Nelson 1991). Consequently, large flakes were selected, retouched, used and discarded in the same context. It is probable that 'expedient' tools were used to work wood, possibly for spears, requiring a large size and weight. A significant difference was found in the size of the complete flakes and the complete tools in Area A (Table 7.15) indicating that large flakes were selected for 'expedient' tool use (t=18.838, df=3811, p<0.0001). Interestingly, no significant difference was found between the maximum dimension of the broken tools and the distal tools (Table 7.15) implying that size and not the shape was an important selection criteria for tools (t=0.4820, df=200, p=0.6303). The presence of tools made on weathered spalls in Area C and Area B as supports the opportunistic use of large stone fragments for 'expedient' tool use.

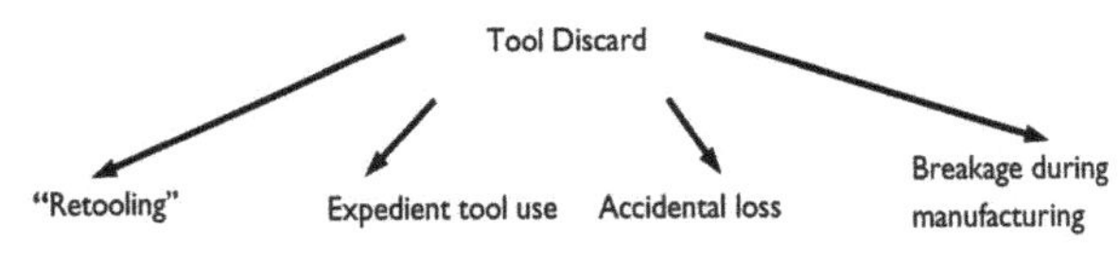

Figure 7.19: The Processes of Tool Discard (after Shott 1989)

Discarded tools, resulting from of a more standardised tool manufacture, were represented by the backed blades, including unfinished geometric microliths and backed points (Table 7.23; Figure 7.20). Very few of these tool types were found, only four backed blades and six backed points were identified in the sampling areas. While large numbers of blade blanks were found in Area A (n=434), none were backed and only a few were retouched (n=27). This result may imply that mostly blade blanks were removed from the quarry and fashioned into tools when required. Symmetrical, untrimmed, hafted spear points (Figure 7.20) have been noted in museum collections and commonly occur among pirri and bondi assemblages (McCarthy 1976:44). The presence of non-retouched points may indicate that a point form was manufactured at Quarry 35.

A wide diversity of tool types were found at Quarry 35, particularly in Area A. Tool types include four burins, two drills and seven nosed-scrapers (Table 7.23). A juan knife was also found during the survey for cores (Figure 7.20). This type of tool typically has an asymmetrical form and a backed ridge along one margin (McCarthy 1976:35-36). All these tools were manufactured from the material present at Quarry 35.

	Area A	Area B	Area C	Total
Backed blade (backed points and geometric microliths)	8		6	14
Denticulate	32	7	3	42
Endnotch	2	2	1	5
Endscraper	21	14	9	44
Nosed Scraper	6	1		7
Notch	445	36	65	546
Scraper	428	54	96	582
Usewear	324	28	34	386
Burin	4			4
Borer	2			2
Total	1272	142	214	

Table 7.23: The Types of Tools Found in the Sampling Areas

A number of tools made from non-local material were also found during the core survey and within the sampling areas. These include a pirri point, made from fine clast, white silcrete material (Figure 7.20), a crystal quartz microlith, two white, opalised quartz geometric microliths in Area C (Figure 7.20) and a number of scrapers, including an end-scraper and several unretouched points made from pink, yellow or white, fine clast silcrete (Figure 7.20). Most of these tool types can be considered more 'curated' in their characteristics (Binford 1979; Nelson 1991). Curated tools have more time invested in the manufacture, hafting and maintenance, the distance transported and use-life of the tool (Torrence 1989b). The discard of 'curated' tools at a quarry may either be indicative of a series of retooling events, resulting in the abandonment of exhausted tools, or simply represent the random loss of tools (Gramly 1980; Shott 1989). Both these processes probably occurred at Quarry 35. Indeed, the presence of two nearly identical geometric microliths, made of the same non-local material, of a similar size may be an example of retooling.

The presence of non-local material can be distinguished by the colour of the tools (Table 7.24). The frequency of yellow non-local material is higher in Area A than in the remaining two sampling areas (1%, n=29). Six of the yellow flakes were utilised as scrapers in Area A. It is likely that this material was derived from a nearby quarry located only 0.5 km from Quarry 35 on an adjoining bluff (Figure 4.1). A large number of pink (n=39) and yellow (n=19) hammerstones made from gibber cobbles were also found in the sampling locations and these were probably obtained from the surrounding gibber pavement.

Most of the tools (Table 7.24) were grey (68%, n=1116) although a large number were spotty grey (31%, n=494). The total number of spotty grey flakes was 1047 of these almost half (n=494, 47%) were tools. In contrast, 13 274 flakes were grey and only 8% (n=798) were tools. These comparisons imply a selection for spotty grey material for tools that was probably associated with the higher quality of the material and the larger size of the flakes. A significant difference in the length of the complete tools (Table 7.24) was found between grey and spotty grey tools (t=9.667, df=1608, p<0.0001). A comparison of the tool types by their colour indicates that most of the spotty grey flakes were used as 'expedient', notched or denticulate tools (33%, n=211). A significant difference occurs in the frequency of notched, denticulate, usewear damage and scraper tools and these two colours (χ^2=13.329, df=3, p=0.093).

Colour	**Grey**	**Spotty Grey**	**Pink**	**Yellow**	**White**
Total count of artefacts	13274	1047	57	46	5
% that are tools	8	47	5	13	60
Count of tools	1116	494	3	6	3
%	68	31			
Mean Length (mm)	52	60			
SD	15	16			
Tool Type					
Backed blade	10	1	1		2
Burin	3	1			
Denticulate	22	20			
Endnotch	5				
Notch	354	191	1		
Borer	1	1			
Nosedscraper	4	3			
Endscraper	34	10			
Scraper	404	160	1	6	1
Usewear	279	107			

Table 7.24: The Tool Types and their Colours in the Sampling Areas

Most of the tools in Area A (Table 7.25) were made from medium clast material (46%, n=589) with smaller proportions of tools made from Group III silcrete (39%, n=491) and fine clast silcrete (15%, n=189). From the total raw material counts in Area A substantially more tools were made from Group III silcrete (29%, n=491) and fine clast silcrete (33%, n=189) than medium clast silcrete (16%, n=589). While more medium clast tools were present in Area A, the number of medium clast flakes selected for tools was low (1:7). It was more likely that a flake made from Group III silcrete (1:3) or fine clast silcrete (1:4) was selected and used as a tool. This result suggests a selection for high quality material (Group III silcrete or fine clast silcrete) to be retouched and used as tools.

Maximum Dimension (mm)		Group III			Group II	
		Amorphous	Banded	Matrix-dominated	Fine Clast	Medium Clast
Broken tool	Mean	50		53	56	59
	SD	13		15	21	14
Complete Tool	Mean	55	53	57	66	64
	SD	12		15	17	17
Distal tool	Mean	26		54	61	61
	SD			14	19	15
Proximal Tool	Mean	39		43	47	47
	SD			11	11	17
Core tool	Mean	59		67	75	68
	SD	8		20	10	16
Core tool fragment	Mean	49		62	55	69
	SD			11		15
Count of tools			491		189	589
Total of complete flakes			1705		569	3694

Table 7.25: The Characteristics of the Tools in Area A

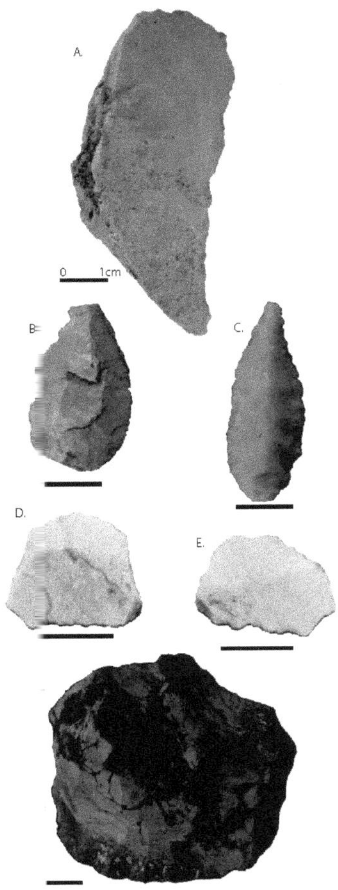

Figure 7.20: Artefacts at Quarry 35. A. juan knife made from Group III silcrete found at Quarry 35. B. Unfinsihed pirri point made from Group III silcrete found at Quarry 35. C. A finished pirri of non-local, white, fine clast silcrete. D/E. Geometric Microliths made from non-local Opalised Quartz. F. A core tool made from Spotty Grey Group III Silcrete

The use of discarded cores and core fragments as tools occurred in all three sampling locations (Table 7.26). Most of the core tools (n=57) were found in Area A (Table 7.26). In total 27 core fragments with retouch were found throughout the sampling areas and most (n=17) of these are also in Area A. It is likely that these tools were firstly used as cores, discarded and then retouched and reused as scrapers (n=38) or notched tools (n=27). Most of the core tools are either uni-platform (n=24) or bi-platform (n=14) cores with smaller numbers of microblade (n=7) and tranchet (n=6) cores. The tools made from these cores (Figure 7.20) and core fragments were made from Group III silcrete (n=43), medium clast (n=37) or fine clast silcrete (n=10). Most of the core tools and fragments have less than 25% (73%, n=66) of their surfaces covered in cortex (Table 7.26) indicating that they underwent considerable reduction before being used as tools (e.g. Hiscock 1988:23; Andrefsky 1998:80).

Attribute	Area A	Area B	Area C	Total
Count of Core tools	57	3	3	63
%	90	5	5	
Count of Core tool fragments	17	6	4	27
%	63	22	15	
Material type				
Group III	37	3	3	43
Medium clast	30	3	4	37
Fine Clast	7	3		10
Percentage of Cortex				
0%	19	5	1	25
1-25%	32	4	5	41
26-49%	16		1	17
50-75%	6			6
Type of retouch				
Denticulate	3	1		4
Notch	25	2		27
Scraper	28	4	6	38
Nosedscraper	1			1
Usewear	15	1	1	17
Core type				
Bi-platform	13		1	14
Microblade	4	3		7
Multi-platform	1			1
Tranchet	6			6
Uni-platform	22		2	24
Length (mm)				
Mean of Core tools	68	58	65	64
SD	17	12	5	11
Mean of Core tool fragments	64	76	55	65
SD	13	23	15	17

Table 7.26: The Characteristics of the Core Tools and Core Tool Fragments in the Sampling Areas

7.9 THE SURVEY TRANSECTS

The survey transects were recorded to further define the extent of the quarry boundary and to assess the whether the artefacts in the transects differed in any way from the artefacts in the quarry. Hence, the artefacts in the transects and those in quarry are compared. Consequently, this comparison helped to gain a better understanding of what was removed from the quarry and discarded nearby.

From the boundary of the core survey four 50 m transects, each 10 metres wide, were surveyed, mapped and the artefacts were recorded. In the survey transects the distribution and density of the artefacts varied considerably. Few artefacts were found in the three transects that extended across the gibber pavement (Figure 7.17). Most of artefacts (70%, n=678) were found in the southern transect (Transect 3) which descends the scarp, below the outcrop, and continues towards Stud Creek (Table 7.27).

Transect	*Count*	*%*
Transect 1 (North)	26	3%
Transect 2 (East)	168	17%
Transect 3 (South)	678	70%
Transect 4 (West)	95	10%

Table 7.27: The Numbers of Artefacts in each Survey Transect

The artefact density decreases at different rates in each transect (Figure 7.17). The northern transect has only a few artefacts, while to the east the flake density continues until approximately 23 m from the edge of the quarry boundary. A similar density pattern occurs in the western transect. In the southern transect most of artefacts concentrate within the first 25 m. It is interesting to note that the transects with the highest numbers of artefacts occur to the south and east. To the south is the Upper Stud Creek landscape sample, a concentration of artefacts along Stud Creek, while to the east is the Lower Stud Creek landscape samples. It is possible that the density of artefacts indicates the main directions of movement between the 'occupation' areas and the quarry.

In the four survey transects 967 artefacts were recorded. Most of these artefacts (Table 7.28) are complete flakes (56%, n=539). Few broken (8%, n=75), distal (12%, n=114) or proximal flakes (6%, n=54) were recorded in the transects (Table 7.28). The small number of broken flakes suggests either a limited amount of treadage or a lack of core reduction occurring in these transects (e.g. Andrefsky 1998:81-83). The density of cores (8%, n=73) in the transects is very low with 0.004 cores per m^2 in an area of 20 000 m^2. In contrast, the quarry has a core density of over 1 per m^2. The number of platform rejuvenation flakes (1%, n=7), complete splits (2%, n=19) and blocks (2%, n=15) is also low in the transects compared to the sampling areas in the quarry (Table 7.14, 7.27). Chi square tests comparing the frequency of complete flakes, broken flakes and blocks from each the sampling location and the transects give significant results (Area A, χ^2=15.183, *df*=2, *p*=0.0005 *phi*=0.0626; Area B, χ^2=31.982, *df*=2, *p*<0.0001, *phi*=0.1067; Area C, χ^2= 37.871, *df*=2, *p*<0.0001, *phi*=0.0999). The frequency of the different technological types found within the transects indicates that the assemblage is not "typical" of a quarry but reflects the selection and transportation of certain artefact types with particular characteristics away from the quarry.

A surprising amount of artefacts made from fine clast silcrete was found in the survey transects (30%, n=289; Table 7.29). Unlike the sampling areas (Table 7.13), more artefacts of fine clast material were found than Group III silcrete (26%, n=253). There is also a low proportion of medium clast silcrete (43%, n=416) when compared to the sampling areas indicating that artefacts of medium clast silcrete were generally not removed from the quarry (Tables 7.13, 7.29). The frequency of the different raw material types indicates that artefacts made from both matrix-dominated and fine clast materials were selected and removed from the quarry. The size of the Group III, complete flakes in the transects are smaller than those in Area B (*t*=4.514, *df*=319, *p*>0.0001) and the same size (*t*=0.000, *df*=736, *p*>0.9999) as those in Area A (Tables 7.17, 7.30). The medium clast flakes are large, akin in size to those in Area B (Table 7.17, 7.30), with no significant difference found between the lengths of these flakes in the transects and Area B (*t*=1.854, *df*=1164, *p*=0.06). In contrast, a highly significant difference was found between the complete flakes of medium clast material in Area A and the transects (*t*=4.197, *df*=2015, *p*<0.0001). These results indicate that large flakes of medium clast material were removed early in core reduction from Area B, on the outcrop, while smaller, further reduced flakes of Group III silcrete were removed from Area A during a later of stage reduction.

	Technological Types	Count	%
Core	Core	73	8
	Core Fragment	20	2
	Sub-total	**93**	**10**
Non-retouched flakes	Complete Flake	539	56
	Complete Split	19	2
	Distal Flake	114	12
	Proximal Flake	54	6
	Broken Flake	75	8
	Platform Rejuvenation	7	1
	Sub-total	**808**	**84**
Retouched flakes	Complete Tool	35	4
	Broken Tool	5	1
	Core Fragment Tool	1	
	Distal Tool	8	1
	Sub-total	**49**	**5**
Spall	Spall Tool	**2**	
Other	Block	**15**	**2**
	Total	**967**	

Table 7.28: The Technological Types in the Survey Transects

	Grey	Spotty Grey	Pink	Yellow	White	Total
Amorphous	6					6
Matrix-dominated	252	1				253
Sub-total	258	1				259
Fine clast	280	7			2	289
Medium clast	404	8	3	1		416
Sub-total	684	15	3	1		703
Coarse Clast	1					1
Quartz					2	2
Total	943	16	3	1	4	967

Table 7.29: The Colour and Material Types of the Flaking Debris found in the Survey Transects

	Material Type	Total Count	Total %	Length (mm) Mean	Length (mm) SD
	Amorphous	6	1	28	11
Group II	Matrix-dominated	114	21	40	16
	Sub-total	**120**	**22**	**39**	**16**
	Fine clast	166	31	45	18
Group I	Medium clast	242	47	50	20
	Sub-total	**408**	**78**		
Other	Quartz	1		18	
	Total	**529**		**46**	**19**

Table 7.30: The Material Type and Size of the Complete Flakes in the Survey Transects

Most of the complete flakes discarded in the transects are relatively large, highly decortified flakes, with feather terminations, unifacial platforms and an intermediate form (Table 7.31). These characteristics indicate that most of the flakes were removed from the quarry after initial core reduction and decortification occurred, indicated also by the flake size. Large, spotty grey flakes (n=16) were also selected and removed, these flakes were considerably larger then those found in Area A (Tables 7.17, 7.31).

Most of the flakes in the survey transects have feather terminations (54%, n=376). The frequency of flakes with feather terminations in the transects is higher than in the sampling areas (Table 7.19, 7.31). In contrast, there were proportionally less abrupt (15%, n=82) and cortical terminations (13%, n=90) than in the sampling areas. These results imply that flakes with a feather termination were removed for the quarry. A significant difference was found in the frequency of the termination types on the complete flakes found in Area A (χ^2= 32.892, *df*=6, *p*<0.0001, *phi*=0.104096) and Area B (χ^2= 16.96, *df*=6, *p*=0.0094, *phi*=0.90211) and the transects but no difference was found between Area C and the transects (χ^2= 10.596, *df*=6, *p*=0.1017). The complete flakes with a feather termination (Table 7.31) have a mean length of 44 mm (SD=18 mm). These flakes are significantly smaller than the flakes found in Area B (*t*=2.189, *df*=1015, *p*=0.0288; Table 7.17) but were larger than those flakes with feather terminations in Area A (*t*=5.022, *df*=1407, *p*<0.0001; Table 7.17).

Attribute	Total %	Total Count	Length (mm) Mean	Length (mm) SD
Colour				
Grey	528	98	46	19
Pink	3	1	49	19
Spotty grey	5	1	67	11
White	2		27	12
Yellow	1		33	
Percentage of Cortex				
0%	224	42	42	17
1-25%	181	34	48	20
26-49%	62	12	50	16
50-75%	33	6	55	25
76-99%	32	6	45	19
100%	7	1	54	24
Termination Type				
Abrupt	82	16	52	18
Cortical	71	14	51	21
Feather	278	55	44	18
Hinge	27	5	37	12
Platform	3	1	60	7
Plunge	22	4	61	27
Step	26	5	46	18
Platform Type				
Bifacial	5	1	35	14
Cortical	41	8	45	20
Missing	77	14	47	24
Step	6	1	38	14
Unifacial	410	76	47	18
Form				
Blade	21	4	39	13
Block	71	13	63	20
Expand	15	3	42	17
Intermediate	425	79	44	18
Point	7	1	41	7

Table 7.31: The Characteristics of the Complete Flakes in the Survey Transects

The low proportion of flakes with more than 50% of their dorsal surfaces covered in cortex (13%, n=72) indicates that complete flakes were removed from the quarry after initial core reduction (Table 7.31). Most of the complete flakes in the transects either have no cortex (42%, n=224) or less than 25% cortex (34%, n=181). This trend reflects the amount of core reduction; as distance increases from the outcrop the amount of cortex on the complete flakes also reduces. A chi square test between Area B (Table 7.19) and the transects (Table 7.31) shows a slightly significant difference between the complete flakes with varying amounts of cortex of their dorsal surfaces (χ^2=11.423, *df*=5, *p*=0.0436, *phi*=0.072455). This result suggests that a substantial distance-decay in the amount of dorsal cortex occurred as a selection and transportation of non-cortical flakes becomes more evident. The size of the complete flakes with different amounts of cortex has a similar profile to Area A with the exception of the flakes with completely cortical dorsal surfaces (1%, n=7) which were considerably larger and were more likely to be obtained from Area B (Tables 7.17, 7.31).

The vast majority of the flake forms were intermediate (79%, n=425; Table 7.31). Few points (n=7) or blades (n=21) were found in the transects (Table 7.31). In comparison to the sampling areas a high frequency of flakes with a block form (13%, n=71) were also found in the transects (Tables 7.19, 7.31). These flakes are large and similar in length to those in Area B (Tables 7.31,

7.17) with no significant difference between the lengths of the complete flakes (t=1.093, df=250, p=0.276). Flakes with a blade, expanding, intermediate or point form in the transects were similar in size to Area A and Area C (Tables 7.17, 7.31). These results suggest that larger flakes with a block form were removed from the quarry during initial reduction, while smaller flakes were removed during blade manufacture. Nearly all of the complete flakes have a unifacial platform (76%, n=410). Few stepped (1%, n=6) or bifacial (1%, n=5) platforms were present in the survey transects (Table 7.31). A relatively high frequency of the flakes in the transects have a cortical platform (8%, n=41) compared to Area A and Area C (Tables 7.19, 7.31). The frequency of cortical platforms is similar to Area B, situated on the outcrop, however, these flakes are smaller than those found in Area B and similar in size to those in Area A (Tables 7.19, 7.31). Indeed, a significant difference in the length of the complete flakes with cortical platforms occurs between Area B and the transects (t=2.1812, df=157, p=0.0056), while no difference occurs between Area A and the transects (t=0.3266, df=200, p=0.7443).

Few tools (n=51) were found in the survey transects (Table 7.32). The tools are large (Table 7.32) and mainly have scraper (n=27) or notched retouch (n=14). Only two backed blades were found in the transects. Both of these tools were made on complete flakes which have a point form and may represent the accidental loss of the more curated tool types produced in the quarry. The backed blades are similar in size to the flakes with a point form found in the quarry (Tables 7.17, 7.32). Only eight flakes of non-local material were found in the survey transects. These were either quartz flakes (n=2) or yellow or pink silcrete from another source.

Tool Type	Count	Maximum Dimension (mm)	SD (mm)
Backed Blade	2	37	0.7
Denticulate	1	68	
Notched	14	68	25
Scraper	27	54	17
Usewear	7	51	16
Total	51		

Table 7.32: TheTool Types found in the Survey Transects

The quarry boundary was based on the density of the cores and natural creek or land-surface boundaries. Consequently, the transects have a very low number of cores with only 73 cores identified (Table 7.33). The transects which cross the gibber pavement have the least number of cores (T1, n=3, T4, n= 12, T2, n=11) while the transect (T3) that extends towards the creek has considerably more cores (n=46). The higher density of cores within this transect may be indicative of an access-way between the creek, the Upper Stud Creek sampling unit and the quarry (Figure 3.5). Most of the cores have very little cortex with 53% (n=42) having less than 25% of their entire surfaces covered in cortex (Table 7.33). Cores with either no cortex or less than 25% cortex are similar in size to those found in the quarry with the same amount of cortex and suggest a large amount of core reduction (Tables 7.3, 7.33).

		Bi-platform	Microblade	Multi-platform	Cobble	Tranchet	Uni-platform
Count		10	6	6	1	3	47
%		14	8	8	1	4	64
Length (mm)	Mean	70	57	70	115	66	70
	SD	19	7	14		6	24
Scar length (mm)	Mean	45	42	38	67	49	38
	SD	17	8	14		7	13
Percentage of Cortex							
0%	Mean		44				59
	SD						10
	Total		1				6
1-25%	Mean	69		59		63	62
	SD	22		5		4	20
	Total	6		3		2	21

Table 7.33: The Characteristics of the Cores in the Survey Transects

The types of raw materials used to make the cores in the survey transects best reflects what was removed from the quarry. Surprisingly, most of the cores were made from fine clast silcrete (Table 7.34). In the survey transects a considerably larger proportion of the cores (43%) were made from this material from fine clast silcrete compared to the quarry (6%; Table 7.4, Table 7.34). The frequency of Group III cores in the quarry and the survey transects remains the same while the number of cores made from medium clast silcrete drops substantially from 68% to 31% (Tables 7.4, 7.34). These results indicate that cores of poor quality, medium clast silcrete were generally not removed from the quarry.

Core type	Medium clast	Fine clast	Group III
Bi-platform	3	2	5
Microblade	1	2	3
Multi-platform	2	2	2
Tranchet	1		1
Uni-platform	14	25	8
Cobble	1		
Total	22	31	19
Mean Length (mm)	77	67	59
SD (mm)	24	21	10

Table 7.34: The Types of Raw Materials used to Manufacture the Cores in the Survey Transects

7.10 CORE REDUCTION STRATEGIES INFERRED FROM THE QUARRY 35 ASSEMBLAGE

In the following section the results from analysis of the cores and flaking debris are drawn together to make further inferences about the reduction strategies already identified from the analysis of the cores. The artefacts at Quarry 35 represent a complex set of reduction strategies. Four different stages of reduction were observed at Quarry 35; material testing and selection, initial reduction and core preparation, large flake manufacture and blade production. A number of different techniques were used to reduce the cobbles and manufacture blades.

7.10.1 Material Testing and Selection

The outcrops at Quarry 35 represent an abundant, high quality raw material source. While numerous outcrops occur along the bluff above Stud Creek only the outcrop at Quarry 35 has any evidence of quarrying. It is a large, massive outcrop with three types of silcrete available: fine clast, medium clast and Group III silcrete. The outcrop was probably targeted for quarrying because of its large size, material quality and location near to the large artefact scatters found in the Stud Creek catchment. Most of the cores and flaking debris are associated with Outcrop 1 and it is likely that most of the worked material was extracted from this outcrop.

Most of the cobbles are made of low quality, Group II, medium clast silcrete. This proportion may reflect the selection and reduction of Group III cobbles. Similarly, most of the cobble cores are medium clast silcrete and only a few were made from Group III silcrete. Cobble cores made from low quality, medium clast silcrete reflect the testing of cobbles. One or two large flakes were removed from these cores to assess the type and quality of the material. Root (1992:146) identified large numbers of "tested cobbles", made from poor quality material with obvious flaws that were discarded at the Knife River flint quarries of western North Dakota.

Testing was a technique often not required as most of the cobbles already have exposed surfaces due to the weathering and spalling of the cortex from the surface of the cobbles. This process explains why so few of the complete flakes have a cortical platform and dorsal surface. Another reason why cobble cores were discarded may be due to their unsuitable size and shape (Dibble and Rolland 1992:12-13; Hiscock 1986b). Smaller cobbles were selected that were fully exposed and easily available at the surface of the outcrop. The mean size of the flakes with a cortical platform and a cortical termination indicates that these cores have a mean orientated platform length of 68 mm (SD=28mm). Cores of this size could be easily reduced using freehand percussion (Flenniken and White 1985:133).

7.10.2 Initial Reduction and Core Preparation

Although small cobbles were generally selected and reduced evidence for the reduction of large cobbles was also identified at Quarry 35. Larger cobbles were split on anvils as points of impact were observed on numerous cobbles. Splitting cobbles is a technique often used at quarries by Aboriginal people in the Western Desert of Australia (Gould 1980:123). A split cobble provides a suitable platform for unidirectional flake removals and may explain why so many uni-platform cores and flakes with unifacial platforms are present in the quarry assemblage.

Initial reduction was observed in the large, highly cortical, uni-platform, bi-platform and multi-platform cores. Cobbles were reduced from one platform (uni-platform cores), to remove the cortex from the surface of the cores, and sometimes rotated to form bi-platform or multi-platform cores. The aim of this strategy was not only to remove the cortex but also to produce a suitable platform from which large flakes could be struck using the entire face of the core. That core rotation generally occurred early in reduction is supported by the flake scars on the complete flakes and the size of the multi-platform cores. More of the flakes in Area B, an area of initial core reduction and decortification on Outcrop 1, have flake scars that originate from two for more quadrants than in Area A or Area C. The cores in Area B were also considerably larger than the other sampling areas, another characteristic of early core reduction, and more were made from medium clast silcrete.

Few bi-platform, uni-platform or multi-platform cores have more than 50% cortex on the core surface. The limited number of these cores indicates that once a core was selected it was generally reduced and rarely discarded during decortification.

7.10.3 Large Flake Manufacture

Large flakes could be removed from the cores after the cortex was removed and a suitable platform was prepared. Most of the cores in the quarry have one or two platforms and less than 50% of their surfaces covered in cortex. The manufacture of large flakes has a two-fold aim; to produce large flakes that were used as blade cores and to provide suitable flakes for tool blanks. Large flakes made from medium clast silcrete were removed from the quarry and discarded in the survey transects. These flakes are the same size as those found in Area B. Large flakes were also used as expedient tools in Area A. In contrast, flakes made from Group III silcrete are the same size in Area A and the survey transects.

The cores with less than 50% of their surfaces covered in cortex are dispersed throughout the quarry indicating that once a core was tested and selected at the outcrop it was transported away from the outcrop for further reduction.

Cores with very little cortex have flake scars that are elongated, unlike the cores with more than 50% of their surfaces covered in cortex, indicating that the faces of these cores were being prepared for blade manufacture. Sometimes platform rejuvenation was required to undertake further reduction of the cores. The large size of the platform rejuvenation flakes indicates that this technique was used early in core reduction.

A number of reasons led to the discard of cores at Quarry 35, either the material quality was deemed inappropriate, too many step or hinge terminations on the flake scars made further flake removals difficult, or a size threshold was reached in the production of large flakes (*cf* Whittaker 1994:107-109; Flenniken and White 1985:140).

7.10.4 Blade Manufacture

A considerable focus on blade production and blade core preparation is evident in Areas A and C of Quarry 35. Several different reduction strategies were used to manufacture blades represented by the characteristics of the different core types. The core types used to produce blades include small multi-platform cores, uni-platform slabs, bidirectional, prismatic, tranchet and microblade cores. Blade cores represent an efficient form of transporting material but require sufficient material, skill and effort to prepare (Hayden *et al.* 1996).

Selection for high quality, Group III silcrete was reflected in the cores, tools and flaking debris. Cores of this material type were reduced further than those made from Group II silcrete. A large number of small, blade producing cores were made from high quality, Group III silcrete.

The face of the blade cores was prepared to allow for continued blade removals. Overhang removal, indicated by the increased number of flakes with a step platform, was more common in Area A and Area C.

The size and characteristics of the blade cores indicates that while the blade producing cores result in the greatest amount of core reduction they were generally not exhausted and more flakes could be removed. It seems likely that a cut-off threshold in the size of the suitable blades limited the amount of core reduction undertaken at the quarry. The consistent size of the blades and points in Area A and Area C indicates a size threshold of approximately 40 mm.

Blade cores were often made from large flakes. Large flakes were used as tranchet blade cores, small multi-platform, prismatic, uni-platform slabs, bidirectional and microblade cores. In contrast, small multi-platform cores and prismatic cores were made from block fragments. The highest density of prismatic cores is found on Outcrop 1 indicating that blocks were reduced near to the source. In contrast, the large flakes that were used as blade cores were mainly reduced on the flat, Upper Quarry area. The flake scars on the blade cores are elongated and are associated with similar sized flakes in Area A and Area C.

The use of large flakes as tranchet cores has been associated with the manufacture of blades with a triangular cross section (Flenniken and White 1985:136). These flakes have been defined as having a point form at Quarry 35. The ventral surfaces of the large flakes were also used as a platform to manufacture blades producing cores with a conical cross section. If these cores were rotated and underwent further reduction small, multi-platform cores were produced. Once the ridges for blade manufacture were present on the face of the core but the platform was not longer useable cores were sometimes rotated 180° to create a bidirectional core.

Although blade production did occur at Quarry 35 it seems likely that blade cores were also prepared and removed from the quarry as an efficient way of transporting material for later use.

7.10.5 Tool Manufacture and Use in the Quarry

Four processes of tool discard were identified at Quarry 35. The most common process, resulting in the bulk of the discarded tools, was from the manufacture of expedient tools in Area A. Other processes include the abandonment of spent tools for retooling (Gramly 1980), accidental loss and the discard of tools broken during manufacturing (Shott 1989). Broken blades that were partially backed were discarded in Area A and Area C. Two near identical geometric microliths made from non-local stone were found in Area C and may be examples of retooling.

Evidence of extensive heavy-duty tool use observed in the reuse of cores as tools, spall tools and the selection of large flakes, was observed at Quarry 35. Most of these tools are large and have notch or scraper retouch. It is possible that these heavy-duty tools were used expediently for woodworking activities, such as the shaping of spears, at Quarry 35.

7.10.6 The Form Material was Removed From Quarry 35

Stone was removed from the quarry in several forms: prepared blade cores, large flakes, blade blanks and fashioned, blade tools. Another definable product, which has never been systematically detected in an Australian quarry, is the point form which may have been used as an unretouched spear point (Thomson 1983:55; Campbell and Edwards 1966:195-196) or as blanks for pirri point manufacture. Two pirri points were found at the quarry, one with evidence of trimming along only one lateral margin was made from stone available at Quarry 35 and the other a perfectly finished pirri made from non-local white, fine clast silcrete. It is probable that other blade tools such as geometric microliths were also manufactured and removed from the quarry, however only 14 flakes were found with evidence of backing. It is likely that blade blanks and other elongated flakes were also removed from the quarry and could be made into tools when required. The similar size of the blades and points in Area A and Area C indicates that suitable flakes of this size were removed from the quarry.

Blade manufacture generally occurs at a quarry where large amounts of high quality material are available to balance the high rate of failure (Hayden *et al.* 1996:37). Most material is discarded during the preparation of a blade core but the blades themselves provide an efficient way of transporting and utilising material (Sheets and Muto 1972; Nelson 1991:68; Hayden *et al.* 1996:37). Blades have a long cutting edge to unit weight, can be fashioned into many tool types, such as backed points and geometric microliths, and are suitable for the demands of a highly mobile hunter-gatherers (Hayden *et al.* 1996:37). Prepared blade cores can also be viewed as an efficient way of transporting material. If the core shape, platform and face is already prepared then the likelihood of failure to remove a suitable blade is lowered if the core is hit with the correct force, at the right position and angle on the platform (Whittaker 1994:95-98).

Large flakes were used as expedient tools in Area A and discarded in the survey transects. Large flakes are a convenient form to transport material as they can be continually resharpened or used as cores.

7.10.7 The Spatial Distribution of the Core Reduction Stages

Variation was observed in the distribution of the cores, the sampling areas and the survey transects. There was a strong technological and spatial relationship between core and flaking debris at Quarry 35. This relationship is reflected in the raw material type, the amount of cortex on the cores and flakes, the core type and flake form, platform type and the size of the artefacts found in different areas of the quarry. For example, cobble cores concentrate in Area B with the large, broken hammerstones, complete splits, block fragments and highly cortical, large flakes. In this area material testing and initial core reduction took place. In contrast, blade producing cores concentrate near the flakes with blade or point forms in Area A and Area C.

Most of the tools recorded (n=1272) were discarded in the in Area A. In contrast, only 134 tools were found in Area B. The concentration of tools in Area A was associated with blade production and the manufacture of curated tool types.

A number of 'pull factors' were observed within the quarrying that influenced the distribution of the artefacts. The 'pull factors' fell into four groups; the landform, topography, distance from outcrop and the location of the creek. The landform "pull factors" included the type of surface. The gibber pavement formed a natural boundary to the distribution of artefacts as few artefacts were found on this surface. Most of the artefacts concentrated on the flat, sandy depositional surface in the Upper Quarry. This pattern of artefact distribution was also related to topography. The flat area of the Upper Quarry was preferred and used for the bulk of the core reduction, blade manufacturing and tool manufacture. Once a core was selected and decortified at the outcrop it was further reduced at the Upper Quarry. Hence, the distance from the outcrop influenced the characteristics and size of the cores and flakes. The creek and the presence of a nearby artefact scatter, the Upper Stud Creek sampling unit, may also have influenced the patterns of discard in areas immediately surrounding the quarry. A large number of artefacts only occurred in Transect 3 which ran towards the creek and may reflect the movement of people between the quarry, the gorge and the Upper Stud Creek sampling unit.

Area C had a series of mid-range values, indicating a mixture of different reduction stages. The area is partly located on Outcrop 1 and this may have influenced the characteristics of discarded artefacts.

7.11 SUMMARY

Quarry 35 exhibits a number of characteristics that are not a typical of a quarry assemblage (*cf* Hiscock and Mitchell 1993:12-17). Evidence of extraction, initial reduction, core preparation, blade production, non-local tool discard, tool manufacture and a considerable amount of expedient tool manufacture was observed within the quarry assemblage. The presence of these activities has created a complex pattern of core reduction indicating the organisation of both the stone and space. As a stone source it was probably used at many times, potentially in a multitude of different ways, over a long period of time. However, the degree of spatial segregation of the different reduction strategies indicates that the quarry was used in a similar fashion during multiple visits (Torrence 1986; Arnold 1987a).

Of fundamental importance when creating a good field methodology is to gain understanding of the spatial variation of artefact types and characteristics throughout the quarry. The danger of not taking this approach creates a trap that many quarry researchers have fallen into by using a sampling strategy that does not define artefact variation and creates a bias in the results. By carefully justifying the position of the sampling areas at Quarry 35 a better understanding of the quarry assemblage was gained.

While cores were discarded at all stages of reduction the cores that were reduced the most were those used for blade manufacture. Considerable core preparation and working occurred at the quarry. It is likely that blade and point blanks, prepared blade cores, large flakes and finished blade tools were removed from the quarry.
Various landform features have constrained or "pulled" the discard of artefacts over time. An understanding of the "pull factors" was of fundamental importance to understanding the formation of the assemblage at Quarry 35. Of importance is the distance-decay profile observed in the artefacts at Quarry 35; as the distance increased from the outcrop the characteristics of the artefacts indicated greater core reduction. This trend has significant implications as it indicates that the material removed from the quarry had little or no cortex, a characteristic that has greater meaning in the following chapter.

It is likely that Quarry 35 was used at many different times over a long period of time. Consequently, overprinting of many different knapping events has occurred, however, patterns in the reduction of stone can be observed in the distribution and technological characteristics of the artefacts.

Chapter 8

The Landscape Sampling Units:

Upper & Lower Stud Creek

8.1 INTRODUCTION

The focus of this chapter is on the reduction of stone obtained from quarried and non-quarried stone sources at the three landscape sampling units in the Stud Creek Catchment. Three types of stone resources are present in the catchment, outcrops, creek cobbles and gibber cobbles. This chapter largely concentrates on the stone obtained from the gibber pavement and the quarried outcrops that were discarded in the landscape sampling units. The cores, flaking debris and tools recorded in the units are described and then compared to gain a better understanding of the ways stone from these sources were acquired, reduced and used in the study area. As a result, it is possible to identify some of the factors that influenced the composition of the assemblages at each of the landscape sampling units.

This chapter has four aims:

- To identify the sources of the stone available at each landscape sample
- To determine how the quarried and gibber stone sources were reduced and used in each landscape sampling unit
- To examine whether a correlation occurs between the quality of the raw material, the availability of material, the reduction strategies employed and the tool types manufactured at the landscape sampling units
- To identify the factors that influenced the composition of the assemblage at each landscape sampling unit

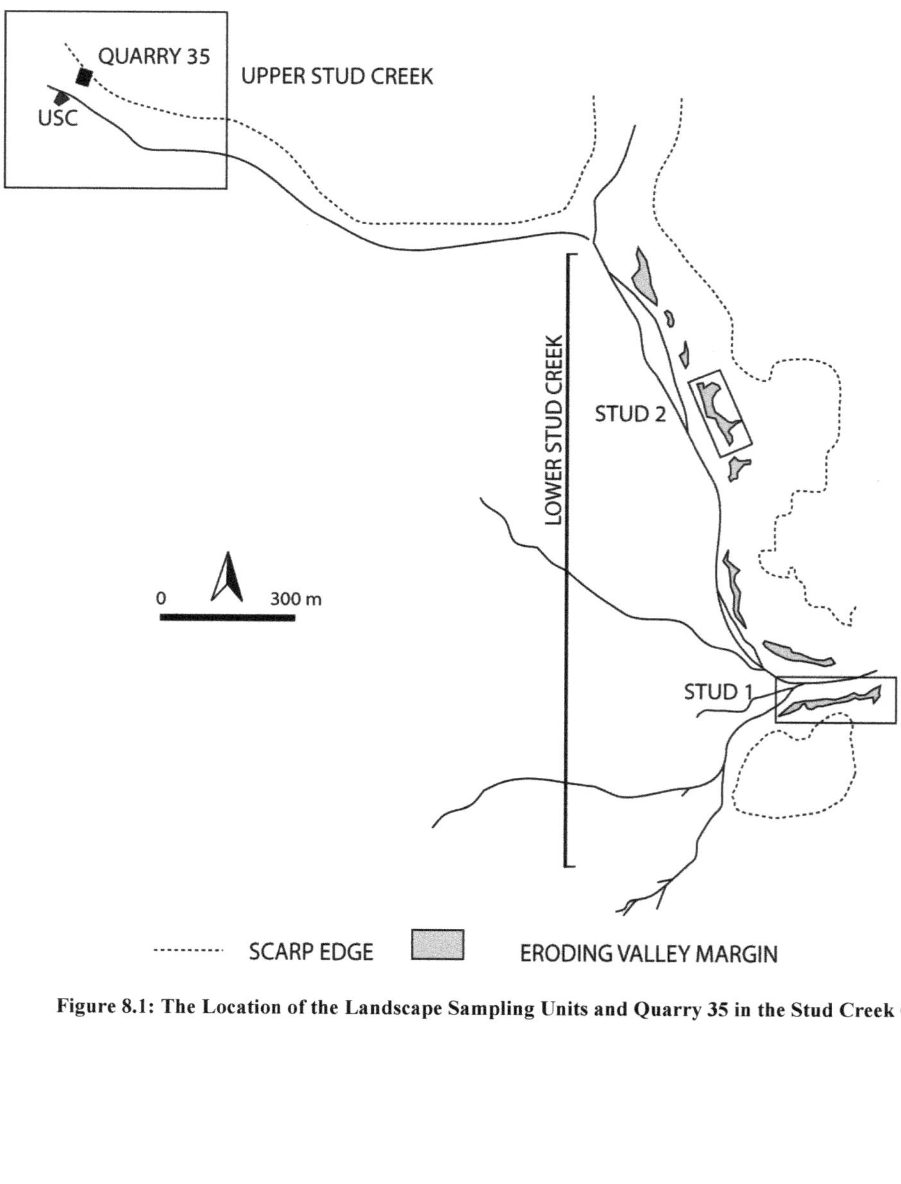

Figure 8.1: The Location of the Landscape Sampling Units and Quarry 35 in the Stud Creek Catchment

The landscape sampling units are divided into two major landform units (Figure 8.1) based on the topography of the catchment: Upper Stud Creek (USC) and Lower Stud Creek (LSC) Within the Lower Stud Creek unit two locations were recorded by WNSWAP; Stud 1 and Stud 2 (Figures 3.9, 8.1). The composition of the assemblages at each landscape sampling units were analysed to search for similarities and differences in the ways stone from the different sources was worked, transported and in what form it was discarded in the Stud Creek catchment.

8.2 DESCRIPTION OF THE LANDSCAPE UNITS

The Upper Stud Creek (USC) unit comprises of a small scatter of 3774 artefacts. The unit is located along Stud Creek, within a narrow gorge, and is approximately 200 m from Quarry 35 (Figure 8.1). On the northern bluff of the gorge a series of silcrete outcrops occur. The outcrop along the southern ridge, directly above the scatter, comprises of ferricrete and laterite (Figure 8.2). The surveyed area, not including the transects, stretches 75 m (from north to south) and 58 m (from east to west) to encompass an area of 4524 m^2 (Figure 8.3). The distance from the ferricrete outcrop to the stream is only 55 m. This area is the only flat, lagged surface within the gorge and was the focus of the survey. To the east and west of the lagged surface are gibber pavements (Figure 8.2, 8.3). On the northern side of Stud Creek another gibber pavement occurs between a silcrete outcrop at the edge of the bluff and the creek. Within the creek, alongside the scatter is a pool creating an ephemeral water source (Figure 8.3) lined with large mulga (*Acacia aneura*) and gidgee (*Acacia camagei*) (Figure 8.2).

Figure 8.2: View Across the Upper Stud Creek Landscape Sampling Unit

Approximately 2 km from the Upper Stud Creek unit on the catchment floor are the two Lower Stud Creek units (Stud 1 and Stud 2) located along the margins of Stud Creek (Figure 3.9, 8.1). Stud 1 is situated south of Stud 2, between the creek and a large mesa that forms the southern boundary of the catchment (Figure 8.4). Two small channels to the east and west delineate the boundaries of the surveyed area while Stud Creek forms the northern boundary of Stud 1. Upstream of Stud 1 lies Stud 2 situated at the mouth of a small valley, surrounded by a series of mesas, and stretching along the eastern margin of Stud Creek (Figure 8.5). In this location the field survey focused on the Eroding Valley Margin (Figure 8.1). During the two field seasons (1996 and 1997) 50 339 artefacts were mapped and recorded by WNSWAP using the methodology described in Chapter 4. At Stud 1 24 554 artefacts were recorded over an area of 5321 m^2 giving an artefact density of 5 per squared metre. At Stud 2 the artefact density is much higher (9 per m^2) with 25 785 artefacts recorded from a smaller area (2987 m^2).

8.2.1 Geological Description

- **Sources of Stone Available in the Stud Creek Catchment**

Three sources of stone are present in the immediate vicinity of the USC sampling unit. These sources include creek cobbles, gibber stones and quarried stone. In order to identify the use of the stone sources a new attribute category was recorded - the type of cortex. A distinctive difference in the cortex occurs on the outcropping stone and the gibber cobbles. Consequently, the type of cortex was used as a way of distinguishing the source of the stone. The cortex was divided into two types: gibber having a smooth, heavily tarnished, red-brown surface and weathered cortex, with a rough, white/grey patina resulting from exposure as part a silcrete outcrop. Two quarries occur within a 500 m radius of the USC unit, Quarry 35 is only 200 m away while Quarry 34 is 450 m away on an adjacent bluff (Figure 4.1)

In the Lower Stud Creek sampling units three sources of stone occur locally. These include five quarries, within a 2 km radius, gibber cobbles and creek cobbles immediately available around and within the units. Cochrane's (1998) analysis of the artefacts at Stud 1 indicates that the gibber pavement influenced the spatial distribution of the artefacts. This influence resulted from the use of the gibber pavement as a source of stone to manufacture expedient tools (Doelman *et al.* 2001).

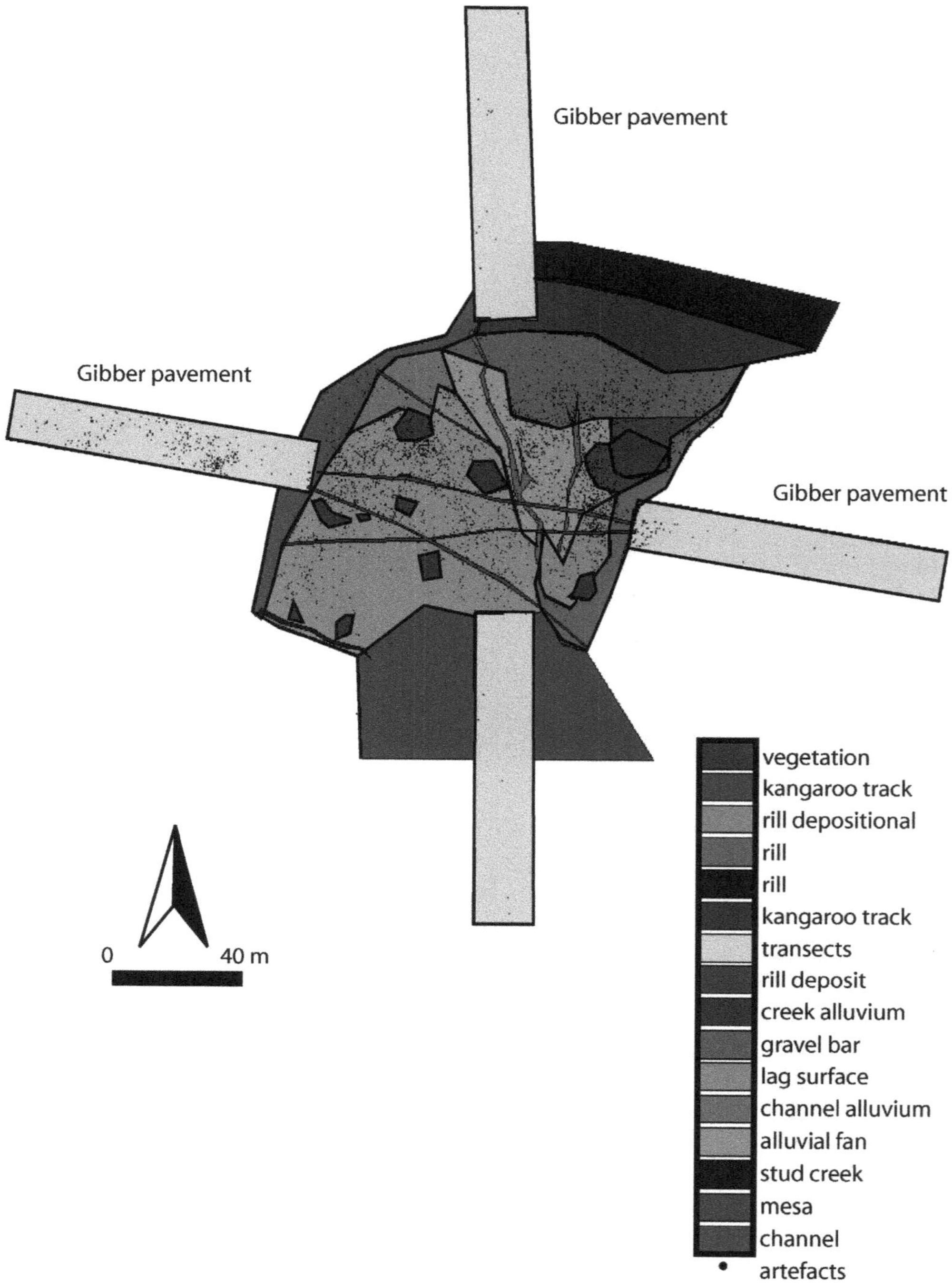

Figure 8.3: The Upper Stud Creek Landscape Sampling Unit

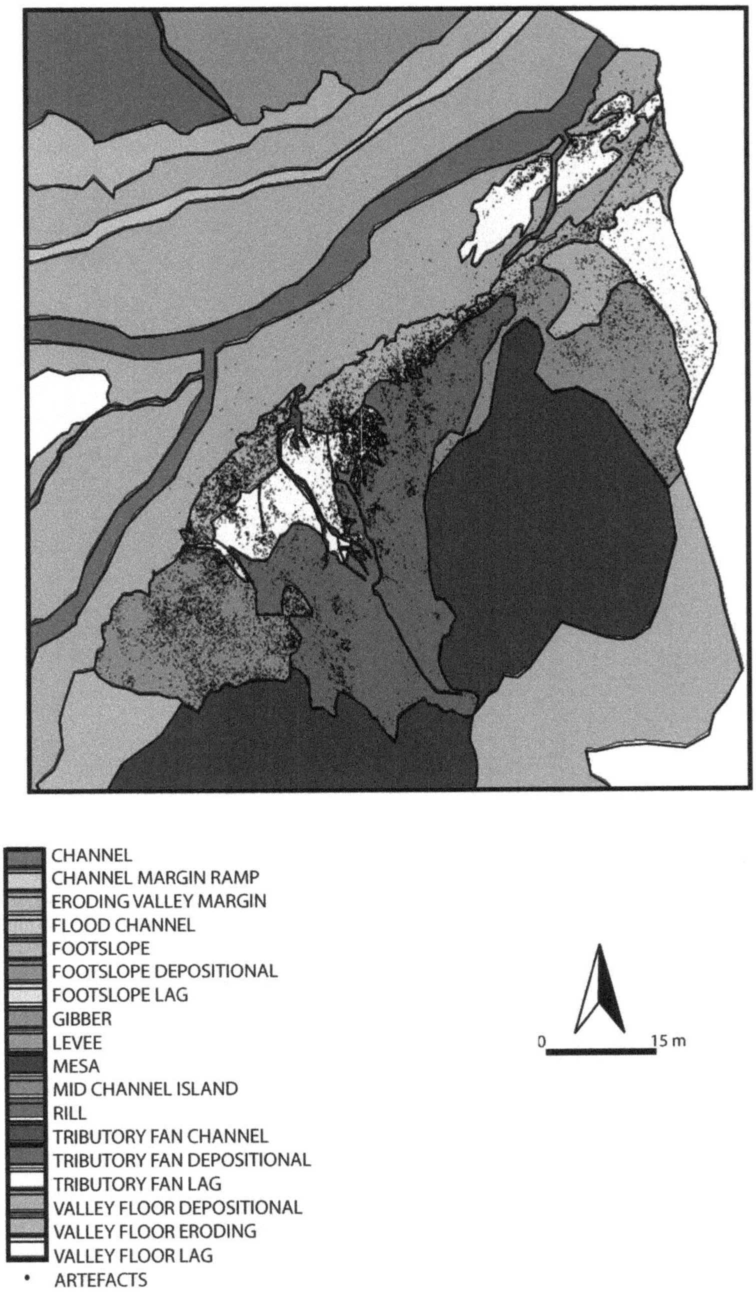

Figure 8.4: Stud 1 in the Lower Stud Creek Catchment Area

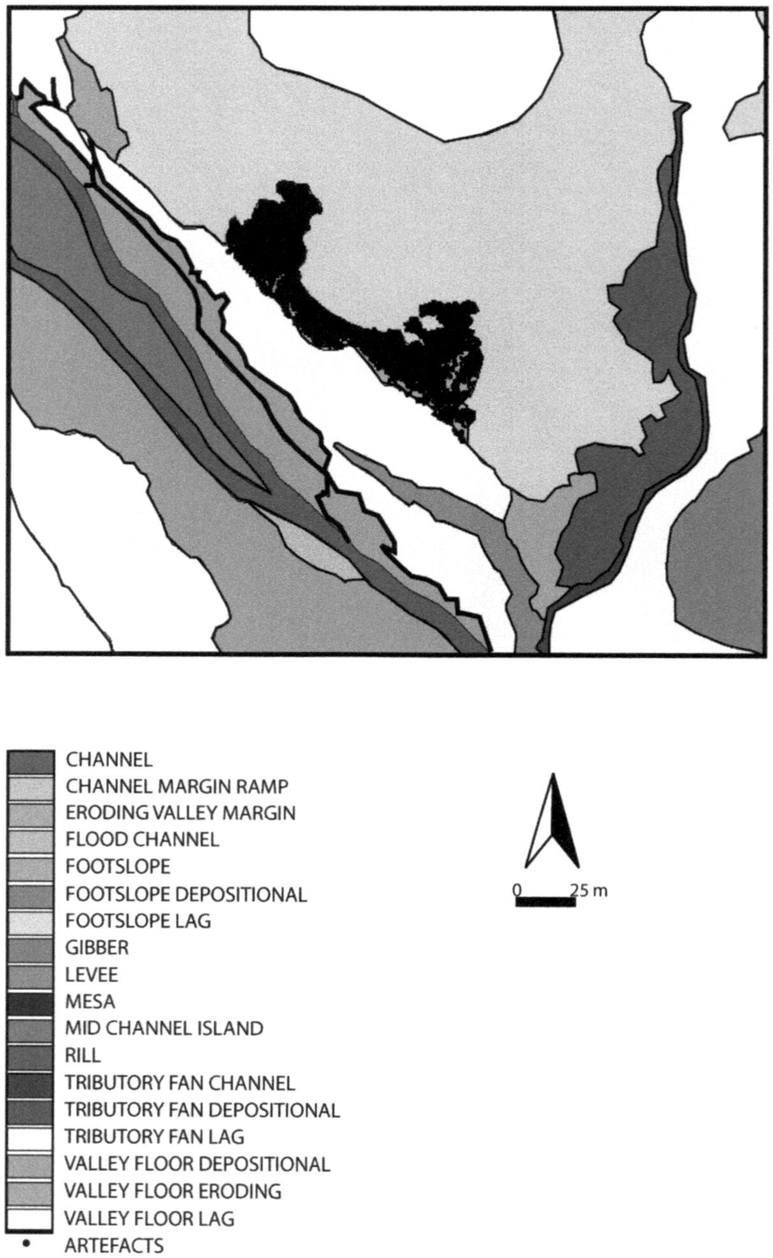

Figure 8.5: Stud 2 in the Lower Stud Creek Catchment Area

- **Raw Material Types Recorded in the Landscape Sampling Units**

A significant difference in the frequency of the raw material types occurs between the units for the Group III, fine clast and medium silcrete, quartz and quartzite (χ^2=2819.5, *df*=8, *p*<0.0001, *phi*=0.230087). Most of the artefacts found at the Upper Stud Creek unit were made from medium clast silcrete (70%, n=2652; Table 8.1). The remaining raw material types include Group III silcrete (20%, n=731) and fine clast silcrete (10%, n=366). A limited number of other material types also occurs including quartzite (n=10), quartz (n=10), hornfels (n=1), laterite (n=1) and sandstone (n=2).

Unlike the Upper Stud Creek unit Stud 1 and Stud 2 have considerably more artefacts made from Group III silcrete (Table 8.1). Almost half of the assemblage at Stud 2 is composed of Group III silcrete (42%, n=10414) while only 33% (n=8444) of the artefacts at Stud 1 were made from this material. Eight hornfels artefacts were found in the Lower Stud Creek units. The presence of hornfels in the catchment is indicative of the transportation of material from the contact metamorphic zone in the Tibooburra Ranges (Figure 3.4).

	Material Type	Stud 1		Stud 2		USC	
		Count	%	Count	%	Count	%
Group I	Coarse Clast	516	2	269	1	2	
	Sub-total	516	2	269	1	2	
Group II	Fine Clast	7469	30	6537	25	366	10
	Medium Clast	7312	30	7747	30	2652	70
	Sub-total	14781	60	14284	55	3018	80
Group III	Amorphous	2358	10	2591	10	59	2
	Banded	650	3	490	2	33	1
	Breccia	1731	7	494	2	5	
	Matrix-Dominated	3665	15	6839	27	631	17
	Porcellanite					3	
	Sub-total	8404	35	10414	41	731	20
Other	Hornfels	3		5		1	
	Laterite					1	
	Chalcedony	2					
	Ochre			22			
	Quartz	204	1	259	1	10	
	Quartzite	626	3	518	2	10	
	Petrified wood	7		2			
	Sandstone	11		9		2	
	Sub-total	853	4	815	3	24	
	Total	24554		25782		3775	

Table 8.1: The Different Raw Material Types in the Landscape Sampling Units

The complete flakes made from fine clast silcrete (*f*=38.58, *df*=2, *p*<0.0001), medium clast silcrete (*f*=160.87, *df*=2, *p*<0.0001) and Group III silcrete (*f*=87.78, *df*=2, *p*<0.0001) are significantly larger at the USC unit than the equivalent flakes in Stud 1 and Stud 2 (Table 8.2). However, within each landscape sampling unit the complete flakes made from the higher quality Group III silcretes are significantly smaller than those made from medium clast or fine clast silcrete (USC, *f*=22.31, *df*=2, *p*<0.0001; Stud 1, *f*=262.41, *df*=2, *p*<0.0001; Stud 2, *f*=189.01, *df*=2, *p*<0.0001) indicating that cores made from Group III silcrete underwent greater reduction. Interestingly, there is no significant difference in the size of the complete flakes made from fine clast silcrete and Group III silcrete in the USC unit (*t*=0.000, *df*=569, *p*>0.9999). In contrast, at the LSC units there is a significant difference between the complete flakes made from fine clast and Group III silcretes (Stud 1, *t*=3.162, *df*=5256, *p*=0.0016; Stud 2, *t*=8.302, *df*=2023, *p*<0.0001).

Material Type	Stud 1			Stud 2			USC		
	Count	Length (mm)		Count	Length (mm)		Count	Length (mm)	
		Mean	SD		Mean	SD		Mean	SD
Group II									
Amorphous	1019	25	9	1074	24	9	23	30	9
Banded	296	26	9	198	24	9	18	33	12
Breccia	802	27	10	217	26	10	3	37	12
Matrix-dominated	1679	28	10	2850	27	10	318	34	13
Sub-total	3796	28	10	4339	26	9	362	34	13
Group II									
Medium clast	3578	33	12	3243	31	13	1414	38	15
Fine clast	3579	29	11	2686	28	11	209	34	13
Group I									
Coarse clast	256	34	12	113	30	11	2	32	6
Other									
Hornfels				3	26	12	1	51	
Petrified wood	2	18		1	18				
Quartz	82	28	9	101	25	8	3	20	7
Quartzite	318	32	11	215	31	13	1	25	

Table 8.2: The Material Type and Mean Lengths of the Complete Flakes in the Landscape Sampling Units

8.2.2 Identifying Artefact Displacement in the Landscape Sampling Units

- **Upper Stud Creek**

Below the ferricrete outcrop, along Stud Creek, is the only lagged surface within the gorge. The lag comprises mainly of gertite and laterite and has an area of 1806 m^2 covering 40% of the surveyed area (Figure 8.3). On this surface most of the artefacts are located (n=1367). Forming the boundary of the surveyed area, are rills originating from the ferricrete outcrop. Along the rills several alluvial deposits and fans are present (Figure 8.3). The artefacts found on these surfaces have been redeposited along the watercourses by sheet-wash. This process has created an alluvial deposit along Stud Creek, while a gravel bar, with large gibber cobbles, occurs beyond the alluvial deposit (Figure 8.3). The artefacts on these surfaces have undergone substantial movement and redeposition. The remaining landscape surfaces, particularly the lagged surface, are relatively stable. A distinctive area of conjoining cores and flakes made from white, fine clast silcrete suggests that limited movement has occurred on the lagged surface (Figure 8.6).

The mean maximum dimension of the complete flakes at the USC unit is 44 mm (SD=16mm). The large size of the artefacts has hindered substantial post-exposure movement. Analysis of the artefacts at Stud 1 indicates that artefacts with a maximum dimension of greater than 30 mm were not influenced by slope or waterwash (Fanning 1999; Fanning and Holdaway in press, a). In the Upper Stud Creek unit 82% (n=1643) of the complete flakes are larger than 30 mm suggesting that limited artefact movement occurred in the unit.

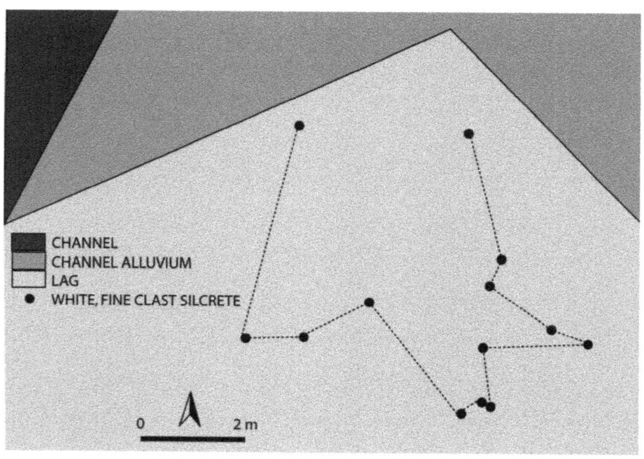

Figure 8.6: The Location of the Conjoining White, Fine Clast Artefacts in the USC Unit

- **Stud 1**

The geomorphology of Stud Creek 1 is very complex. Due to the degradation of the landscape, post-European arrival and pastoral practices, severe erosion has caused the stripping of the topsoil exposing artefact concentrations intermittently (refer to 3.3.3). Nine landform units were surveyed for artefacts along the catchment floor (Figure 8.4). These landforms include the gibber pavement, areas of erosion (e.g. Eroding Valley Margin), deposition (e.g. Footslope Depositional) and lagging (e.g. Tributary Fan Lag) defined by Patricia Fanning, a geomorphologist from Macquarie University (Fanning 1999). Most of the artefacts (n=7028) were located on the Eroding Valley Margin, while 5874 artefacts were found on the Tributary Fan Lag (Cochrane 1998:53). A meandering system of interconnecting rills concentrates the water run-off from the mesas towards the Eroding Valley Margin at Stud 1. As this process continues the Tributary Fan Depositional has eroded, creating the Tributary Fan Lag and exposing hearths and artefacts. Most of the artefacts on this surface concentrate along the eastern margin of the lag and within the rills (Figure 8.4). It appears that sheet-wash (*cf* Shackley 1978) has removed the smaller artefacts and redeposited them within the Eroding Valley Margin where they were subsequently moved by water run-off concentrating in areas of low topography. Analysis of the artefacts at Stud 1 indicates that artefacts under 30 mm in maximum dimension were subject to movement (Fanning and Holdaway in press, a). At Stud 1 41% (n=4814) of the complete flakes have a maximum dimension between 20 and 30 mm indicating that considerable artefact movement has occurred within the assemblage.

From the landforms mapped within Stud 1 it seems likely that only the gibber pavement maintained a high degree of spatial integrity. The gibber pavement remains unaffected by any fluvial action, due to the absence of sheet-wash or rills and the higher elevation of the pavement, creating greater artefact visibility. An ANOVA comparing the length of complete flakes on the Tributary Fan Lag, the gibber pavement and the Eroding Valley Margin shows a significant size difference (f=45.132, df=2, p<0.0001) between these three landforms (Table 8.3). This result indicates that the smaller flakes were removed from the Eroding Valley Margin (EVM). The Eroding Valley Margin has the highest intensity of concentrated sheet-wash as it channels water towards Stud Creek. This process of surface degradation and artefact movement was reflected in the micro-scale, geomorphological mapping which shows where the artefacts have concentrated in the sand, mud, gravel and vegetation patches (Cochrane 1998:55). Tests of artefact size and the calculation of the mean distance between the size categories of different artefact types (Cochrane 1998:53) support the argument that artefact movement has occurred. His results show that as size increases the mean distance also increases indicating that small artefacts clustered together at Stud 1. Other studies in the area have suggest that only limited artefact movement occurred, dependent on landform surface, artefact size and slope, with a size threshold of 24 mm restricting artefact movement (Pigdon 1996, Greenwood 1997).

Landscape Surface	Count	Mean (mm)	SD (mm)
EVM	3307	29	11
Gibber pavement	2044	31	11
Lag	2913	28	11

Table 8.3: The Maximum Dimension of the Complete Flakes in the Different Landform Units at Stud 1

- **Stud 2**

The survey of Stud 2 focused on the Eroding Valley Margin as artefacts concentrated on this landform unit at Stud 1 (Figure 8.4). This unit forms a narrow, eroding band only separated from the creek at a distance of 40 m by the post-European alluvium (Valley Floor Depositional) and the Valley Floor Lag (Figure 8.4). The same process of erosion and lagging occurring at Stud 1 has also occurred at Stud 2. In this assemblage 46% (n=4940) of the complete flakes are 30 mm or smaller indicating that considerable artefact movement may have occurred at this location particularly on this landform surface.

8.3 GENERAL ASSEMBLAGE COMPARISONS

Obvious differences in the frequency of the technological types occur between the sampling units. Large numbers of complete splits, blocks and broken flakes, including distal and proximal flakes, were found in Stud 1 and Stud 2 reflecting repeated occupations or a greater amount of core reduction and discard than at the USC unit (Tables 8.4, 8.5). These artefact types are produced as by-products of core reduction or were broken as a result of trampling (Andrefsky 1998:81-82; Hiscock 1988:365-356). The USC unit has a greater frequency of cores and complete flakes than Stud 1 or Stud 2 (Tables 8.4, 8.5). A chi square test gives a significant result (χ^2=703.457, *df*=10, *p*<0.0001, *phi*=0.13115) comparing the number of complete and broken flakes and tools, hammerstones, blocks and cores in the landscape sampling units.

More tools (Tables 8.4, 8.5) were found in Stud 1 (13%, n= 3152) than in Stud 2 (7%, n=1980) or the USC unit (8%, n=344). Grinding equipment was identified in all three sampling units (Table 8.5). The presence of grinding equipment and hearths has been considered indicative of a "residential base camp" (Binford 1980:9).

Ratio	USC	Stud 1	Stud 2
Complete tool/complete flake	1:9	1:5	1:8
Complete split/complete flake	1:14	1:8	1:4
Complete flake/block	1:52	1:13	1:12
Complete flake/ broken flake	1:9	1:4	1:3
Core/complete flake	1:6	1:8	1:10

Table 8.4: The Ratios of the Different Technological Types in the Landscape Sampling Units

	Technological Types	Stud 1 Count	%	Stud 2 Count	%	USC Count	%
Non-retouched Flakes	Complete Flake	11611	47	10660	42	1993	53
	Complete Split	1377	6	2755	11	145	4
	Platform Rejuvenation	*		*		35	1
	Distal Flake	1709	7	2603	10	416	11
	Proximal Flake	1490	6	1690	7	180	5
	Broken Flake	2811	11	3467	14	223	6
	Broken Split			419	2		
	Sub-total	18998	77	21594	86	2992	79
Retouched Flakes	Complete Tool	2154	9	1365	5	232	6
	Distal Tool	278	1	167	1	49	1
	Proximal Tool	151	1	49		10	
	Broken Tool	552	2	399	1	29	1
	Coretool	*		*		18	
	Core Fragment Tool	*		*		3	
	Spall Tool	17				3	
	Sub-total	3152	13	1980	7	344	8
Hammerstone	Broken Hammerstone	11		*		1	
	Hammerstone	12		16		12	
	Sub-total	23		16		13	
Cores	Core	1477	6	1079	5	320	8
	Core Fragment					62	2
	Sub-total	1477	6	1079	5	382	10
Grinding Equipment	Mortar					1	
	Muller	15		9		2	
	Broken Grinder					4	
	Sub-total	15		31		7	
Other	Cylcon	4					
	Ochre			22			
	Anvil			2			
	Axe	1					
	Block	884	4	862	3	38	1

Table 8.5: The Different Technological Types in the Landscape Sampling Units (* not recorded)

8.4 DESCRIPTION OF THE CORES IN THE LANDSCAPE SAMPLING LOCATIONS

The following section aims to describe the cores in each of the landscape sampling units and establish where material was obtained and how it was reduced. Unfortunately, different core characteristics were recorded at the localities making any comprehensive comparison between the landscape units difficult. Only four core types were recorded in the Lower Stud Creek units by WNSWAP, whereas eight core types were recorded in the Upper Stud Creek unit. However, after considering these differences the frequency of the different core types still varies considerably between the Upper and Lower Stud Creek landscape units (Table 8.6). In the Upper Stud Creek unit there are less bi-platform cores and more blade cores than the Lower Stud Creek units. High numbers of cobble cores occur in all of three units (Table 8.6). By comparing the number of cobble, uni-platform, bi-platform and microblade cores found in all three sampling locations a significant difference was found (χ^2=106.47, *df*=6, *p*<0.0001, *phi*=0.19345).

The low ratio of cores to complete flakes in the Lower Stud Creek units indicates that the cores in these two locations were reduced further than the cores in the USC unit. At Stud 1 a ratio of 1:10 cores to complete flakes occurs while a ratio of 1:8 was found at Stud 2. In contrast, the USC unit has a ratio of 1:6 (Table 8.4). These ratios indicate that the Lower Stud Creek units have similar proportions of cores while more cores are present in the USC assemblage (Table 8.5).

Most of the cores (Table 8.7) were made from medium clast silcrete (78%, n=250). Few cores were made from Group III silcrete (14%, n=47) or fine clast silcrete (5%, n=18). The size of the cores varies greatly between the different raw material types (Table 8.7). The medium clast cores are significantly larger, probably due to the size of the original cobble and the limited amount of reduction they underwent, than those made Group III or fine silcrete (*f*=6.347, *df*=2, *p*=0.0020; Table 8.7). Furthermore, the size of the cobble cores indicates that the gibber cobbles selected for reduction have a mean length of 105 mm (SD=66), smaller than the cobbles recorded in Area B at Quarry 35 (231 mm, SD=61). Consequently, the size of the cores reflects not only the amount of reduction, the flaking properties and quality of the different types of silcrete but also in the type of stone source; cores made from gibber cobbles at the USC unit are large when discarded and mostly made from medium clast silcrete.

Core Type	Stud 1				Stud 2				USC			
	Total		Length (mm)		Total		Length (mm)		Total		Length (mm)	
	Count	%	Mean	SD	Count	%	Mean	SD	Count	%	Mean	SD
Bi-platform	548	37	48	15	371	34	43	18	65	19	72	25
Horsehoof*									1		91	
Microblade	53	4	37	10	43	4	27	7	50	15	46	12
Multi-platform*									24	7	90	78
Prismatic*									2	1	49	18
Cobble	169	12	63	18	136	13	53	20	37	11	106	33
Tranchet*									13	4	55	15
Uni-platform	697	48	49	16	530	49	39	16	146	43	71	25

Table 8.6: The Number and Size of the Different Core Types in the Landscape Sampling Units

8.4.1 The Cores in the Upper Stud Creek Unit

Of the 338 cores present in the USC sampling unit over half (53%, n=191) were derived from the gibber pavement (Table 8.7). The remaining cores have either no cortex (21%, n=57) or a weathered cortex indicating that their source was probably a quarried outcrop (23%, n=72). It seems likely that cores with no cortex were also derived from a quarried outcrop based on the results of Quarry 35 (Chapter 7). Nearly all the cobble (94%, n=35) and uni-platform (65%, n=93) cores in the Upper Stud Creek unit have a gibber cortex. In contrast, the microblade, prismatic and tranchet cores have either no cortex or a weathered cortex (Table 8.7).

The cores with a gibber cortex were probably obtained from the immediate vicinity of the USC sampling unit as gibber or creek cobbles. The large block length (85 mm, SD=38 mm) of these cores indicates that they were not heavily reduced (Table 8.7). The size of the flake scars (Table 8.7) on the gibber cores are also large with a mean length of 39 mm (SD=15 mm), but small in comparison to the overall size of the cores (Figure 8.3). While a significant difference was found between the block length (Table 8.7) of cores with a gibber and weathered cortex (*t*=7.588, *df*=261, *p*<0.0001) no significant difference was found in the length of the cores with no cortex or a weathered cortex (*t*=1.891, *df*=127, *p*=0.0609). These results indicate that the cores with no cortex or weathered cortex underwent greater reduction then the cores with a gibber cortex and suggest that cores with no cortex did not derive from the gibber pavement but were also obtained from a quarried outcrop.

Core Type		Bi-platform	Horsehoof	Microblade	Multi-platform	Cobble	Tranchet	Uni-platform	Prismatic	Total
Cortex Type										
Gibber	Count	40	1	4	16	35	2	93		191
	%	21	1	2	8	18	1	49		60
	Mean Length (mm)	81	91	62	107	105	66	77		85
	SD (mm)	27		5	91	33	25	25		38
	Mean Scar Length (mm)	41	58	33	47	40	30	37		39
	SD (mm)	15		11	26	15	4	13		15
Weathered	Count	11		25	6	2	5	23		72
	%	15		35	8	3	7	32		23
	Mean Length (mm)	57		46	54	119	61	61		50
	SD (mm)	12		13	15	42	15	24		15
	Mean Scar Length (mm)	31		31	28	25	32	33		29
	SD (mm)	10		11	6	6	11	15		8
No Cortex	Count	11		18	1		6	19	2	57
	%	19		32	2		11	33	4	21
	Mean Length (mm)	61		42	73	49	47	52		56
	SD (mm)	17		9		18	7	17		22
	Mean Scar Length (mm)	36		28	40	34	24	27		31
	SD (mm)	7		8		5	5	8		12
Material Type										
Fine clast	Count	4		4	2			1	9	18
	Mean Length (mm)	66		48	58			77	51	55
	SD (mm)	18		17	20				16	19
Group III	Count	10		17	5	2		2	11	47
	Mean Length (mm)	62		44	54	123		64	54	60
	SD (mm)	21		10	15	100		27	21	31
	Count	48	1	27	16	33	2	10	122	250
Medium clast	Mean Length (mm)	75	91	47	107	104	49	51	75	76
	SD (mm)	27		12	91	29	18	11	25	37
Percentage of Cortex										
0%	Count	11		20	1		2	6	20	60
	%	18		33	2		3	10	33	18
1-25%	Count	28	1	22	10			5	31	97
	%	29	1	23	10			5	32	27
26-49%	Count	9		7	9			1	36	62
	%	15		11	15			2	58	18
50-75%	Count	13		1	4	8			37	63
	%	21		2	6	13			59	19
76-99%	Count	4				29		1	22	56
	%	7				52		2	39	17

Table 8.7: The Characteristics of the Cores in the USC Landscape Sampling Unit

The colour of the cores gives some idea about where it was acquired. If a core is grey, pink or yellow, has a large amount of cortex and was made from medium clast silcrete it was more likely to be manufactured from a gibber cobble. Most of the cores at the USC unit were grey (Table 8.8) and obtained from three sources; creek or gibber cobbles (n=171) and quarried outcrops (n=116). The grey cores with a gibber cortex are mostly made from medium clast silcrete (n=153, 89%). The cores with the distinctive spotty grey colour were from Quarry 35 (n=3). The pink cores are either creek cobbles of quartzite (n=1) or gibber cobbles (n=10) of medium clast silcrete. The yellow cores were probably obtained from a quarry, such as Quarry 34 (n=4), gibber cobbles (n=7) or the nearby laterite outcrop (n=1). The white cores are either fine clast silcrete (n=2) or quartz (n=2). These results indicate that the most of the cores were not made from quarried material even though Quarry 35 is located only 200 m away from the USC unit

Colour	Quarry Source	Gibber Cobble	Creek Cobble
Grey	46 (Group III) 6 (fine clast) 75 (medium clast) 1 (quartzite)	153 (medium clast) 10 (Group III) 7 (fine clast)	
Sub-total	116	171	
Spotty grey	2 (medium clast) 1 (Group III)		
Pink	1 (fine clast) 1 (medium clast)	10 (medium clast)	1 (Quartzite)
Yellow	1 (fine clast) 1 (laterite) 2 (Group III) 1 (medium clast)	1 (fine clast) 6 (medium clast)	
White	2 (fine clast)	1 (Quartz)	
Total	128	189	1

Table 8.8: The Raw Material Types and Colours of the Cores in the USC Landscape Sampling Unit

The material type and the amount of cortex on the cores in the USC unit were used to establish the sources of the stone exploited. Only a few cores of Group III silcrete (n=47) were found at the USC unit and these have little or no cortex (Table 8.9). The remaining cores are mainly medium clast with large amounts of cortex on their surfaces (Table 8.9). A chi square test comparing the amount of cortex on the cores made from Group III and medium clast silcrete gives a significant result (χ^2=21.324, *df*=3, *p*<0.0001, *phi*=0.267949). Generally, cores of poor quality material, such as the medium clast silcrete, with a high percentage of cortex were made from gibber cobbles, while cores of high quality material, such as the Group III silcrete, have little or no cortex and were probably obtained from a quarried outcrop. Furthermore, if cortex was present on the Group III cores it was usually weathered (Table 8.9). The cores with a gibber cortex were mostly made from medium clast silcrete (n=169; Table 8.9). In contrast, only 10 cores made from Group III silcrete have a gibber cortex (Table 8.9). These results presumably reflect the composition of the different types of silcrete in the gibber pavement. Hence, only 5% of the gibber cobbles were made from Group III silcrete and 4% are fine clast silcrete, most (88%) of the cores were made from medium clast silcrete.

	Medium Clast				Fine Clast				Group III			
	Total		Length (mm)		Total		Length (mm)		Total		Length (mm)	
	Count	%	Mean	SD	Count	%	Mean	SD	Count	%	Mean	SD
Type of Cortex												
Gibber	169	68	87	39	8	47	54	18	10	21	85	41
No cortex	38	15	53	17	4	24	48	9	5	32	43	8
Weathered	43	17	55	19	5	29	63	23	22	47	51	17
Percentage of Cortex												
0%	38	15	53	17	4	24	48	9	5	32	43	9
1-25%	66	26	57	16	6	35	51	18	20	43	55	20
26-49%	44	18	70	18	5	29	62	19	8	17	63	21
50-75%	54	22	97	53					3	6	108	77
76-99%	48	19	103	31	2	12	66	34	1	2	52	

Table 8.9: The Amount and Type of Cortex on the Cores made from Different Raw Material Types in the USC unit

The type of core made from the different stone sources reflects not only the amount of reduction but also the goal of core reduction. Most of the cores (Table 8.7) made from the gibber cobbles are uni-platform (49%, n=93), bi-platform (21%, n=40) or cobble cores (18%, n=35) and have over 50% of their surfaces covered in cortex indicating that they underwent little reduction (Table 8.7). A chi square test gives a significant difference between the type of cortex and the number of bi-platform, uni-platform and microblade cores (Table, 8.7; χ^2=55.496, *df*=4, *p*<0.0001, *phi*=0.475936). It seems likely that only a few flakes were removed from the bi-platform, uni-platform and multi-platform cores before they were discarded. Multi-platform cores made from gibber cobbles (8%, n=16) represent further rotation and flake removals but were still large when discarded (Table 8.7). An ANOVA comparing the length of the flake scars on the bi-platform, uni-platform, cobble and multi-platform cores made from gibber cobbles gives no significant difference (*f*=2.242, *df*=3, *p*=0.0850). This result indicates that large flakes, with a minimum flake size of approximately 40 mm were removed from the gibber cores before they were discarded (Table 8.7).

The use of the different stone sources can be gained by comparing the material type, size, the type and amount of cortex with the core type. The blade producing cores, such as microblade, prismatic and tranchet, are considerably smaller, have less cortex, which if present was weathered, and were usually made from Group III silcrete (Table 8.7). Most of the cores with a weathered cortex have less than 25% cortex (67%, n=48), are smaller in length and have smaller flake scars than the cores with gibber cortex (Figure 8.7). These characteristics indicate that the material used to manufacture blade cores was obtained from quarried outcrops and these were reduced further than the cores made from gibber cobbles.

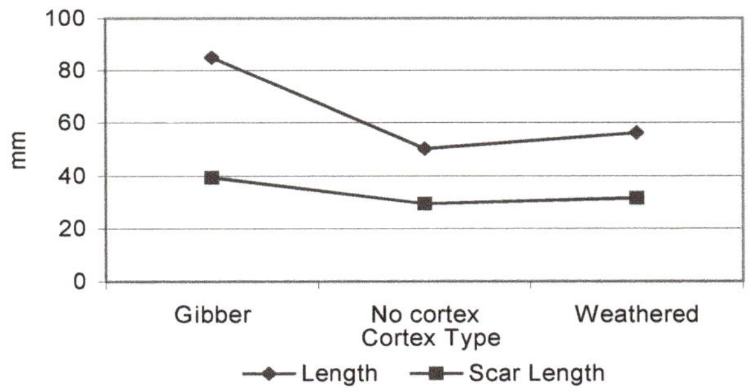

Figure 8.7: The Mean Length of the Flake Scars and Mean Length of the Cores with Different Types of Cortex on their Surfaces at the USC Unit

It is possible to gain an idea of the amount of core reduction for each type of stone source by comparing the ratio of cores to complete flakes with the same type of cortex. The gibber cobbles did not undergo substantially reduction as they have a low flaking ratio (1:3). In contrast, the cores and complete flakes with no cortex (1:20) or a weathered cortex (1:11) underwent a greater amount of reduction as more flakes were removed from these cores.

A more controlled and standardised reduction of the cores obtained from a quarry source was observed in the reduction of tranchet or prismatic cores (*cf* Witter 1992:47-53). Indeed, these two types of blade cores reflect two very different reduction strategies. While tranchet cores reflect the reduction of a large flake, similar to the Redbank A technique described by Hiscock (1993), the prismatic cores are usually made from large blocks. Both these core reduction strategies were also observed at Quarry 35 (refer to Chapter 7).

The results of the core analysis indicates that two reduction strategies were used in the USC sampling unit; one strategy for quarried stone and another for gibber cobbles. The core type, the percentage of cortex, cortex type, core size and raw material type reflects the source of the raw material. Based on the results of the Upper Stud Creek unit several predictions can be made regarding the characteristics of the cores at Stud 1 and Stud 2 which identify the source of the stone (Table 8.10). Since the type of cortex was not recorded at either Stud 1 or Stud 2 these predictions were used to define the source of the material (Table 8.10). In sections 8.4.2.and 8.4.3 the characteristics of the cores at Stud 1 and Stud 2 are decribed and used to determine where stone was acquired and how it was reduced at these locations.

Non-quarried Stone Sources

- Less reduced, large block size and larger flake scars
- Generally poor quality material, such as medium clast silcrete
- High percentage of surface cortex
- Gibber cortex
- Either cobble, uni-platform or bi-platform core types

Quarried Stone Sources

- Further reduction (small block size, smaller flake scars)
- Generally made from high quality material, such as Group III silcrete
- Less or no surface cortex
- Weathered cortex or no cortex
- Blade producing cores

Table 8.10: The Characteristics of the Cores that Indicate the Type of Stone Source

8.4.2 The Cores at Stud 1

The analysis of the cores at Stud 1 indicates that approximately one third of the cores were obtained from a non-quarried source such as the gibber pavement. At Stud 1 most of the cores (Table 8.11) are medium clast (37%, n=527), fine clast (31%, n=416) or Group III silcrete (31%, n=420). The cores made from fine clast and Group III silcrete probably originated from a quarried source, while medium clast silcrete cores were more likely to have been obtained as cobbles from a gibber pavement.

	Fine clast				Medium clast				Group III			
	Total		Length (mm)		Total		Length (mm)		Total		Length (mm)	
Cortex	Count	%	Mean	SD	Count	%	Mean	SD	Count	%	Mean	SD
0%	122	29	43	13	116	22	46	12	200	48	40	10
1-49%	160	38	49	15	205	39	51	15	158	38	43	11
50-99%	135	32	61	19	206	39	62	16	62	15	56	18
Total	416	31	52	18	527	37	54	16	420	31	43	13
Core Type												
Bi-platform	159	38	50	17	162	31	54	16	183	44	42	12
Microblade	12	3	39	12	13	2	41	8	27	6	35	10
Cobble	41	10	64	22	104	20	63	15	22	5	62	19
Uni-platform	208	50	51	17	251	47	52	16	188	45	43	12

Table 8.11: The Amount of Cortex and the Core Type of the Cores made from Different Silcrete Types at Stud 1

The proportions of the raw material types indicates that approximately two thirds of the grey silcrete came from quarries - presuming that mainly medium clast cobbles of gibber occur and that cores of medium clast material were not removed from the quarries. Almost equal proportions of fine clast (n=303), medium clast (n=385) and Group III silcrete (n=313) grey cores were found (Table 8.12). It is likely that based on the quality of the material the fine clast and Group III cores were obtained from a quarried outcrop. Most of these cores were not used to manufacture blades as only 53 microblade cores are present in the assemblage (Table 8.12). The blade cores were obtained from at least three sources based on their colour (Table 8.12). Of the white cores, 29 are fine clast silcrete, suitable for blade manufacture, and 26 are quartz. As already described in Chapter 3, quartz is found in the gibber pavement while fine clast, white silcrete is only found as outcrops.

Core Type	Grey	Yellow	White	Pink
Microblade	40	11	2	
Bi-platform	390	90	31	36
Cobble	106	38	1	34
Uni-platform	506	111	27	50
Total	1042	250	61	120
Material Type				
Medium clast	385	94	3	48
Fine clast	303	59	29	29
Group III	313	81	2	23
Quartz			26	

Table 8.12: The Colours of the Different Core Types and their Material Types at Stud 1

medium clast silcrete and have no cortex. These differences indicate that cores of Group III silcrete were reduced further than the medium clast cores and were obtained from a quarried outcrop.

At Stud 1 the medium clast cores with more than 50% (Table 8.14) of their surfaces covered in cortex represent cobbles obtained from the nearby gibber pavement (Figure 8.4). Most of these cores were discarded as cobble cores (n=94, 46%). In comparison to the block length, relatively small flakes were removed from the cobble, uni-platform and bi-platform cores, although these scars are much larger than those on the microblade cores (Table 8.13, 8.14). Furthermore, the length of the scars (f=1.398, df=2, p=0.2494) and the block length of these three core types (f=0.09338, df=2, p=0.9109) are not significantly different (Table 8.13). It seems likely that only one or two large flakes of a similar size were removed from these core types, as their block length remains large, indicating that the original gibber cobble was not substantially reduced.

Core Type	Bi-platform		Microblade		Cobble		Uni-platform	
Percentage of Cortex	**Count**	**%**	**Count**	**%**	**Count**	**%**	**Count**	**%**
0%	220	40	30	57			216	31
1-49%	229	42	21	40	16	9	302	43
50-99%	97	18	2	4	158	91	178	26
Size Dimensions (mm)	**Mean**	**SD**	**Mean**	**SD**	**Mean**	**SD**	**Mean**	**SD**
Scar Length	29	10	26	7	30	12	28	10
Length	48	15	37	10	63	18	49	16

Table 8.13: The Amount of Cortex and the Size Characteristics of the Different Core Types at Stud 1

The size of the cores made from medium clast material are significantly larger (Table 8.11) than the cores of Group III silcrete (t=11.405, df=945, p<0.0001) and have considerably more cortex on their surfaces (χ^2=94.910, df=2, p<0.0001, phi=0.316577; Table 8.11). A comparison of the core types and the same material types (χ^2=56.606, df=3, p<0.0001, phi=0.244102) indicates that most of the cobble cores were made from gibber cobbles of medium clast silcrete (Table 8.11). The amount of cortex remaining of the different core types also varies (Table 8.13). Most of the microblade cores have no cortex (57%, n=30) and were probably derived from a quarried outcrop. In contrast, more of the cobble (91%, n=158) and uni-platform cores (26%, n=178) have more than 50% of their surface covered in cortex and were probably made from gibber cobbles (Table 8.13). Furthermore, 45% (n=85) of the uni-platform cores made from Group III silcrete have no cortex. In comparison, only 24% (n=62) of the uni-platform cores are made from

Core Type	Bi-platform	Microblade	Cobble	Uni-platform
Stud 1				
Count	35	1	94	76
%	17		46	37
Mean Length (mm)	61	41	62	61
SD (mm)	17		15	18
Mean Scar Length (mm)	33	33	30	32
SD (mm)	11		11	9
Stud 2				
Count	21		61	57
%	15		44	41
Mean Length (mm)	66		58	49
SD (mm)	16		19	15
Mean Scar Length (mm)	42		34	34
SD (mm)	18		15	10

Table 8.14: The Characteristics of the Medium Clast Cores at Stud 1 and Stud 2 with more than 50% of their Surfaces Covered in Cortex

Only 53 (4%) microblade cores were found at Stud 1 and it is likely that they were obtained from a quarried outcrop (Table 8.6). Unlike the cores derived from the gibber pavement this core type was heavily reduced. Most of the microblade cores have no cortex, are small and have small flake scars (Table 8.13). The length of theflake scars and the length of the core are very similar (Table 8.13; Figure 8.8). This resemblance in size suggests not only that the largest possible flakes were removed from the cores but also that their use-life was maximised until they were exhausted (Holdaway and Stern in press). This intensity of core reduction may be an economising strategy in response to the quality of the material or simply reflects the distance from the source (Hiscock 1986b, 1988).

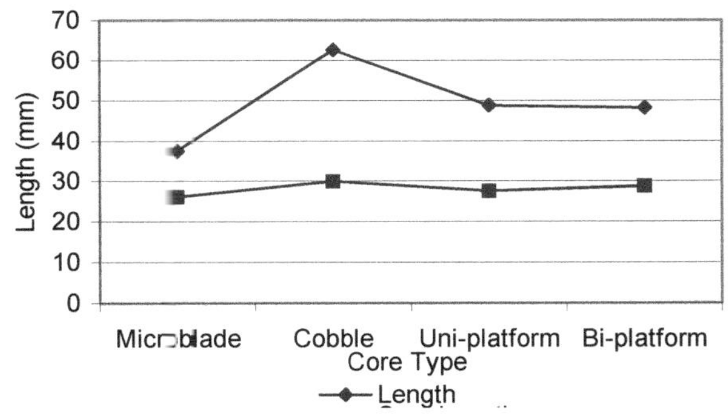

Figure 8.8: The Lengths of the Flake Scars and Lengths of the Different Types of Cores at Stud 1

Most of the microblade cores were made from Group III silcrete or fine clast silcrete (Table 8.11). These types of silcrete allow for better control and versatility in core reduction (Doelman et al. 2001). Based on the colour and the type of material used to manufacture the microblade cores at least three quarry sources were identified. The sources include grey, yellow or fine clast white outcrops. All these sources are available within the Stud Catchment. However, it is likely that at least some of the cores were transported and curated over considerable distances before being discarded.

8.4.3 The Cores at Stud 2

A similar trend in the use of stone sources was seen in Stud 1 and Stud 2. Most of the cores in Stud 2 were made from medium clast silcrete (39%, n=399), however, a substantially larger proportion of the cores were made from fine clast (22%, n=234) and Group III silcrete (38%, n=389) than at Stud 1 (Table 8.11, 8.15). A chi square test gives a significant difference in the number of cores made from fine clast, medium clast and Group III silcrete between Stud 1 and Stud 2 ($\chi 2=21.526$, df=2, $p<0.0001$, phi=0.095003). Based on the results of the USC sample it can be predicted that the medium clast cores were derived from the gibber pavement indicating that at least one third of the cores at Stud 2 were made from gibber cobbles.

Material Type	Fine clast				Medium clast				Group III			
	Total		Length (mm)		Total		Length (mm)		Total		Length (mm)	
% of Cortex	Count	%	Mean	SD	Count	%	Mean	SD	Count	%	Mean	SD
0%	78	33	34	12	94	24	38	13	206	53	33	12
1-49%	94	40	41	14	166	42	42	14	129	33	38	19
50-99%	62	27	49	19	139	35	55	22	54	14	53	20
Total	234	22	41	16	399	39	46	19	389	38	38	17
Core Type												
Bi-platform	85	36	41	17	119	30	47	19	153	40	43	18
Microblade	10	4	30	9	3	1	28	6	29	5	25	6
Cobble	26	11	51	18	75	19	55	21	27	8	50	20
Uni-platform	113	48	39	14	206	51	42	16	184	47	38	19

Table 8.15: The Number and Size of the Cores at Stud 2 made from Different Types of Silcrete in Relationship to the Amount of Cortex on the Core Surface and their Core Type

It is difficult to assess the type of source exploited based on the different colours of the cores. The pink material represents either quartzite (n=9) or cores made from gibber cobbles (Table 8.16). Most of the pink cores are medium clast silcrete (n=46), however, a large number of pink cores, made from Group III silcrete (n=31) were also found. These cores were probably obtained from quarried sources. The yellow cores represent stone acquired from quarried outcrops and gibber cobbles (Table 8.16). Most of the yellow cores were made from Group III silcrete (n=108) while only 69 medium clast and 41 fine clast yellow cores were found. These numbers suggest that most of the yellow cores were obtained from quarried outcrops. The white cores are mainly quartz cobbles (n=16), found in the gibber pavement, or fine clast cores probably obtained from quarried outcrops (n=14). Almost equal numbers of grey, Group III silcrete (n=286) and medium clast silcrete (n=254) were present at Stud 2. In contrast, only 150 fine clast, grey cores are present in this assemblage. These results suggest that almost two thirds of the cores were obtained from a quarried outcrop.

Core Type	Grey	Pink	White	Yellow
Bi-platform	242	32	12	86
Microblade	33	1		9
Cobble	71	31		34
Uni-platform	363	53	19	96
Material Type				
Medium clast	286	46	1	69
Fine clast	150	29	14	41
Group III	254	31		108
Quartzite		9		
Quartz			16	1

Table 8.16: The Colours of the Different Core Types and the Material Types at Stud 2

A comparison of the length of the cores made from medium clast and Group III silcrete gives a significant result (t=6.224, df=786, p<0.0001) indicating that cores of Group III silcrete are considerably smaller than the medium clast cores (Table 8.15). This result shows that

Group III cores were reduced further than the medium clast cores. The large size of the medium clast cores may also suggest that their source was the gibber pavement while the cores made from Group III silcrete were probably obtained from quarried outcrops.

The amount of cortex on the cores made from medium clast and Group III silcrete also reflects the source of the material as cores obtained from the gibber have greater amounts of cortex on their surfaces. A chi square test gives a significant difference between the amount of cortex on the cores made from medium clast and Group III silcrete (χ^2=83.776, *df*=2, *p*<0.0001, *phi*=0.326059). The cores made from Group III have considerably less cortex than those made from medium clast silcrete indicating fthat they underwent greater reduction (Table 8.15).

The quality of the material influenced the type of core manufactured. Most of the cobble, uni-platform and bi-platform cores were made from the poor quality, medium clast silcrete (Table 8.15). However, a significant difference in the lengths of cobble, bi-platform and the uni-platform cores (*f*=36.103, *df*=2, *p*<0.0001) was found which shows that the cobble cores are considerably larger than the uni-platform or bi-platform cores and were discarded early in reduction (Table 8.15). No difference was found in the size of the bi-platform and uni-platform cores indicating that they were discarded at the same stage of reduction (*t*=3.40, *df*=839, *p*=0.0009). In contrast, only 1% (n=3) of the microblade cores were made from medium clast silcrete (Table 8.15). Most of the microblade cores were made from high quality Group III or fine clast silcrete (Table 8.15).

Most of the cores at Stud 2 with more than 50% of their surfaces covered from cortex and made from medium clast silcrete are cobble cores (Table 8.14). A large number of uni-platform cores have the same characteristics as the cobble cores (Table 8.14). The cores with these characteristics were probably made from a gibber cobbles. Surprisingly, the bi-platform cores with these characteristics are significantly larger (*f*=8.815, *df*=2, *p*=0.0003) than the cobble and uni-platform cores with the same characteristics indicating that they were obtained from another source – possibly as creek cobbles (Table 8.17). The size of the cobble cores probably indicates the size of the gibber cobbles selected, tested and rejected (Table 8.17).

At Stud 2 only 43 microblade cores were found and these are significantly smaller than the other core types at Stud 2 (*f*=35.025, *df*=3, *p*<0.0001). These cores are also significantly smaller (*t*=5.545, *df*=94, *p*<0.0001; Table 8.6) and have smaller flake scars (*t*=2.089, *df*=94, *p*=0.00395) than the microblade cores at Stud 1 indicating that they underwent further reduction (Tables 8.13, 8.17). The microblade cores at Stud 2 have a similar size and the scar length indicating that the cores were probably exhausted (Table 8.17; Figure 8.9). Based on the colour of the microblade cores at least three quarry sources were used including grey, pink and yellow outcrops (Table 8.16).

Core Type	Bi-platform		Microblade		Cobble		Uni-platform	
Percentage of Cortex	Count	%	Count	%	Count	%	Count	%
0%	163	44	28	65	6	5	196	37
1-49%	161	43	14	33	17	13	220	42
50-99%	47	13	1	2	107	82	111	21
Total	371	35	43	4	130	12	527	49
Size Dimensions (mm)	**Mean**	**SD**	**Mean**	**SD**	**Mean**	**SD**	**Mean**	**SD**
Scar Length	30	11	23	7	34	14	29	16
Length	43	18	27	7	53	20	39	16

Table 8.17: The Characteristics of the Different Core Types at Stud 2

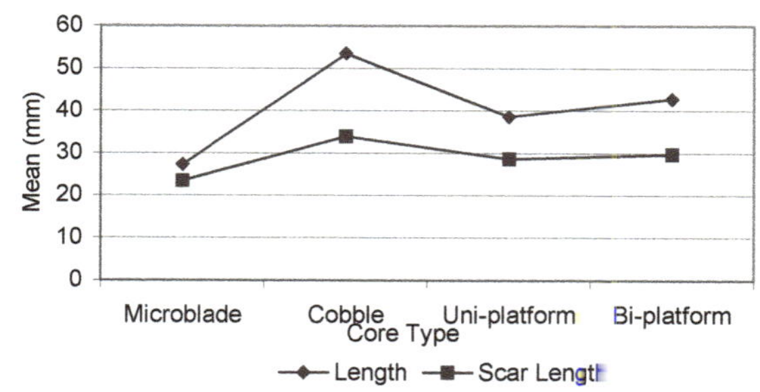

Figure 8.9: The Mean Lengths of the Flake Scars and Mean Lengths of the Cores at Stud 2

8.4.4 Comparing the Cores in the Landscape Sampling Units

Based on the analysis of the cores in the landscape sampling units some understanding can be gained about the use of the available stone sources. The proportion of the core types varies between the sampling units (Table 8.6). At the Lower Stud Creek units more bi-platform cores and smaller blade cores are present in the assemblages hinting at a greater core rotation and reduction occurring in these areas. However, proportionately fewer microblade cores were found in the LSC units than in the USC unit. The high number of microblade cores in the USC unit is probably due to the presence of a nearby quarry where blade manufacture took place (Quarry 35).

In the USC unit the cores made from Group III silcrete have little cortex and if present is generally weathered. In contrast, the medium clast cores have a greater amount of cortex and this is usually gibber cortex. The amount of cortex and the type of cortex generally indicates that poor quality material was derived from the gibber pavement while higher quality material was obtained a quarried outcrop. In the LSC units a similar pattern in the use of stone sources to the USC unit was found. It seems likely that if a core was medium clast it was procured from the gibber pavement, although this was not always the case.

Cores in the Lower Stud Creek units are significantly smaller then those in the Upper Stud creek unit. This result may be due to the distance from the surrounding quarries and the absence of large creek cobbles at Stud 1 and Stud 2, although the bi-platform at Stud 2 are large and may have been obtained from the creek bed.

In each of the three units, medium clast cores are larger than the cores of Group III silcrete indicating the selection and further reduction of high quality material. The frequency of cores made from Group III silcrete in the LSC units is higher than in the USC unit. This pattern may occur for two reasons; the high number of medium clast cobbles being flaked at the USC unit or the selection and transportation of cores made from Group III silcretes away from the quarries for further reduction at the LSC units.

There are a high number of cobble cores in each of the sampling units again suggesting the use of non-quarried stone sources. In the USC unit the cobble cores were made from large creek or gibber cobbles and their characteristics indicate that cobbles were tested and either reduced further or discarded immediately. Half the cores present in this unit were made from gibber cobbles. The remaining cores were obtained from a quarried outcrop due to presence of weathered cortex or the absence of cortex on the cores.

To summarise these findings, potentially two thirds of the cores in the Lower Stud Creek units were derived from the gibber pavement based on the amount of cortex and the raw material types. These cores were made from medium clast silcrete and are considerably larger than the cores made from Group III. The similarity in core size and characteristics between Stud 1 and Stud 2 may reflect their similar location at the catchment floor, at a greater distance from any quarry than the USC unit. Furthermore, the presence of a gibber pavement within the survey area at Stud 1 has increased the numbers of medium clast, cobble cores present in the unit. In contrast, the Upper Stud Creek unit is close to a quarry, thereby influencing the composition of the core assemblage. However, even at this location a considerable number of gibber and creek cobbles were reduced. Hence, the distance from a quarried outcrop influenced the composition of each assemblage, highlighted by the high number of microblade cores in the USC unit compared to the LSC units.

8.5 DESCRIPTION OF THE FLAKING DEBRIS IN THE LANDSCAPE SAMPLING LOCATIONS

The complete flakes in the landscape sampling units are described to further determine the types of source reduced, what was manufactured from each source and how intensively the cores were reduced at each of the landscape sampling units.

8.5.1 The Complete Flakes in the Upper Stud Creek Unit

Most of the complete flakes (n=1414, 71%) in the USC unit were made from medium clast silcrete (Table 8.2). In contrast, few complete flakes were made from Group III silcrete (n=362, 18%) or fine clast silcrete (n=209, 10%). The reduction of cobbles is reflected in the high number of medium clast flakes with gibber cortex. This type of cortex suggests that approximately one third of the complete flakes were manufactured from gibber cobbles (Table 8.18). It is likely that the cores used to manufacture the flakes with no cortex or a weathered cortex were obtained from a quarry based on the results of Quarry 35 (Chapter 7) and the analysis of the cores in the USC unit (Section 8.4.1)

Cortex Type	Total		Length (mm)		Width (mm)	
	Count	%	Mean	SD	Mean	SD
Gibber	636	32	41	16	34	13
No cortex	944	47	33	12	24	11
Weathered	410	21	40	15	30	13

Table 8.18: The Number and Size of Complete Flakes at the USC Unit with Different Types of Cortex on their Surfaces

Most of the yellow (n=17) and pink (n=18) complete flakes were made from gibber cobbles of medium clast silcrete (Table 8.19). In contrast, only one complete flake of yellow, Group III silcrete and no complete flakes of pink, Group III silcrete were made from gibber cobbles (Table 8.19). The white complete flakes were either quartz (n=3), also from the gibber pavement, or fine clast silcrete (n=30) from a quarried outcrop (Table 8.19). The white, fine clast flakes generally have either no cortex or a weathered cortex (Table 8.19). Most of the medium clast complete flakes were derived from a quarried outcrop (62%) but a greater proportion (38%), in comparison to the other material types, were made from gibber cobbles (Table 8.19).

Type	Colour	Medium	Fine	Group III	Quartz
No Cortex	Grey	613	105	159	
	Pink	4			
	Spotty Grey	3		1	
	White	4	20		
	Yellow	8	5	1	
Weathered	Grey	236	25	127	
	Pink	3			
	Spotty Grey	2		2	
	White	1	10		
	Yellow	2	3	16	
	Total	876 (62%)	164 (80%)	306 (85%)	
Gibber	Grey	500	35	49	
	Pink	18	2		
	Spotty Grey				
	White			1	3
	Yellow	17		6	
	Total	535 (38%)	41 (20%)	56 (15%)	3

Table 8.19: The Raw Material Types and Colours of the Complete Flakes at the USC Unit

The length of the complete flakes has a direct relationship to the type of cortex on their dorsal surfaces (Table 8.18). The complete flakes with gibber or weathered cortex are not significantly different in length (Table 8.18; t=1.399, df=2015, p=0.1619). This similarity in size suggests that large, cortical flakes were manufactured from both gibber cobbles and quarried outcrops. However, a significant difference in the width of flakes with a gibber cortex or weathered cortex was found at the USC unit (Table 8.18; t=6.726, df=2015, p<0.0001). This result suggests that the flakes from a quarried source are elongated while the flakes manufactured from the gibber cores have the same length but are much wider.

Most of complete flakes with a completely cortical dorsal surface have a gibber cortex (n=44) and only 11 have a weathered cortex (Table 8.20). Of the 44 complete flakes with their dorsal surfaces completely covered in gibber cortex 38 (86%) were made from medium clast silcrete. Indeed, 84% (n=535) of all the complete flakes with a gibber cortex were made from medium clast silcrete (Table 8.19).

Most of the flakes have either no cortex or less than 50% of their dorsal surfaces covered in cortex (Table 8.20). Two scenarios explain why so few flakes have large amounts of cortex on their dorsal surfaces, cores were generally completely decortified at the quarry or only a few flakes were removed from each of the gibber cobbles. Both these scenarios probably influenced the composition of the assemblage at the USC unit.

The aim of reducing gibber cobbles was simply to produce a few large, wide flakes. The complete flakes with a gibber cortex have a higher proportion of cortical platforms and more cortex on their dorsal surfaces (χ^2=33.553, df=4, p<0.0001, phi=0.183365) than the complete flakes with a weathered cortex (Table 8.20).

Based on the characteristics of the complete flakes with no cortex or a weathered cortex two flake forms were manufactured from quarried material and discarded in the USC unit. These two forms include large flakes with a weathered cortex, a block form and cortical terminations and smaller, non-cortical flakes with a blade form and a feather termination (Table 8.20). The presence of cortex on the termination indicates that large flakes were removed from the quarry early in reduction, while the increase in the number of steps and hinges on flakes with no cortex indicates a later stage of core reduction (Table 8.20). Most of the blades in the USC unit were made from quarried stone (n=97) with only one blade having a gibber cortex (Table 8.20). The same pattern was observed in the complete flakes with a point form (Table 8.20). These results indicate that the blades and other flake forms with no cortex were made after considerable core reduction. The flakes with a weathered cortex mostly have less than 25% cortex on their dorsal surfaces (66%, n=269). The size of these flakes reflects a selection for large, elongated flakes, probably manufactured at a quarry, and subsequently removed and discarded in the USC unit. Flakes with these characteristics may also be the result of core preparation before the removal of blades (Flenniken and White 1985:136).

Attribute	Gibber		Weathered		No Cortex		Total	
	Count	%	Count	%	Count	%	Count	%
Termination								
Abrupt	79	12	52	13	94	10	225	11
Cortical	81	13	92	22			173	9
Feather	371	58	215	52	633	67	1219	61
Hinge	11	2	6	1	32	3	49	2
Missing	31	5	10	2	57	6	98	5
Platform	3		3	1	12	1	18	1
Plunge	13	2	13	3	34	4	60	3
Step	47	7	19	5	85	9	151	8
Form								
Blade	1	1	13	13	84	86	98	5
Block	66	46	35	25	41	29	142	7
Expanding	31	36	13	15	41	48	85	4
Intermediate	536	33	347	21	764	46	1647	83
Point			1	8	12	92	13	1
Platform Type								
Bifacial	12	2	13	3	27	3	52	3
Cortical	163	26	37	9			200	10
Missing	77	12	52	13	157	17	286	14
Step	1		3	1	11	1	15	1
Unifacial	383	60	305	74	751	79	1439	72
Percentage of Cortex								
0%					988		988	50
1-25%	316	50	269	66			585	29
26-49%	111	17	64	16			175	9
50-75%	75	12	24	6			99	5
76-99%	60	9	24	6			84	4
100%	44	7	11	3			55	3
Size Dimensions (mm)								
	Mean	**SD**	**Mean**	**SD**	**Mean**	**SD**	**Mean**	**SD**
Length	41	16	40	15	33	12	37	15
Width	34	13	30	13	24	11	29	13
Thickness	11	6	10	5	7	4	9	5

Table 8.20: The Characteristics of the Complete Flakes at the USC Unit

8.5.2 The Complete Flakes at Stud 1

Based on the results of the USC unit it is likely that most of the grey, yellow and pink complete flakes at Stud 1 made from medium clast silcrete were manufactured from gibber cobbles. The medium clast material accounts for 29% (n=4576) of the complete flakes found at Stud 1 (Table 8.21). Gibber cobbles of petrified wood and quartz acquired from the gibber pavement were also used to manufacture flakes. Cobbles of quartzite can be found in the nearby stream bed and account of 3% (n=318) of the complete flakes at Stud 1 (Table 8.21). In contrast to the USC unit, more of the complete flakes at Stud 1 were made from Group III (30%, n=3796) or fine clast silcrete (29%, n=3576). It is likely that these two material types were obtained from quarried outcrops. An easily distinguishable material type, derived from an outcrop, was the fine clast white silcrete. Based on the frequency of the raw material types, Group III and fine clast silcrete, and the results of the USC unit approximately 65% (n=7385) of the complete flakes were manufactured from quarried material (Table 8.21). These results are only approximate values as flakes of medium clast silcrete were also manufactured at the quarries, such as Quarry 35 (Chapter 7), and transported away.

A further assessment of the stone sources used at Stud 1 was made by comparing the amount of cortex on the dorsal surfaces of the complete flake with the length of these flakes. In the Stud 1 assemblage 61% of the complete flakes have no cortex on their dorsal surfaces (Table 8.22) and most of these flakes were made from Group III silcrete (71%, n=2697) or fine clast silcrete (62%, n=2232). In contrast, few complete flakes made from medium clast silcrete have no cortex on their dorsal surfaces (49%, n=1758) and it is possible that these flakes may also have derived from a quarried outcrop. A chi square shows a significant result when comparing the amount of cortex on the complete flakes made from Group III, fine clast and medium clast silcrete (χ^2=429.66, *df*=6, *p*<0.0001, *phi*=0.198079).

A significant difference was found in the length of the complete flakes with no cortex (*t*=13.254, *df*=4453, *p*<0.0001) and made from medium clast, fine clast or Group III silcrete (Table 8.22). This result indicates that the cores made from Group III and fine clast silcrete were reduced further and may have been obtained from a quarried outcrop. The remaining complete flakes were probably made from gibber cobbles, as the gibber cobbles are small, with a mean length of 63 mm (SD=16 mm), consequently the likelihood of cortex being present on the flakes is high

Material Type	Material Colour	Material Source		
		Gibber Pavement	Quarried outcrop	Stream Cobble
Group III	Grey		2812	
	White		14	
	Yellow		837	
	Pink		131	
	Sub-total		3796	
Group II				
Fine clast	White		266	
	Grey		2715	
	Pink		126	
	Yellow		469	
	Sub-total		3576	
Medium clast	Grey	3578		
	White	28		
	Pink	222		
	Yellow	748		
	Sub-total	4576		
Petrified wood		2		
Quartzite	Pink			318
Quartz	White	82		
	Total	**4660**	**7372**	**318**

Table 8.21: The Classification of Stone Sources used at Stud 1Based on the Number of Complete Flakes in the Assemblage

Percentage of Cortex	Length (mm)	0%	1-49%	50-99%	100%	Total
Group III						
Amorphous	Mean	24	26	30	28	25
	SD	8	9	10	6	9
Banded	Mean	25	29	37	31	26
	SD	8	10	20	16	9
Breccia	Mean	25	29	42	44	27
	SD	9	11	16	17	10
Matrix-dominated	Mean	27	30	34	33	28
	SD	10	11	12	13	10
Sub-total	Count	2697	872	189	36	3796
	%	71	23	5	1	32
	Mean	26	29	34	34	27
	SD	9	10	13	13	10
Group II						
Fine clast	Count	2235	955	324	65	3576
	%	62	27	9	2	30
	Mean	27	32	34	31	29
	SD	10	11	12	13	11
Medium clast	Count	1758	1247	432	141	4576
	%	49	35	12	4	38
	Mean	30	34	38	35	33
	SD	11	12	13	14	12
Total	Count	6690	3074	945	242	30
	%	61	28	9	2	11

Table 8.22: The Amount of Cortex and the Size of the Complete Flakes made from Different Raw Material Types at Stud 1

Complete flakes with no cortex generally have feather terminations, however, more flakes with hinge and step terminations are present in the Stud 1 assemblage in comparison to the USC unit (Table 8.20, 8.23). These termination types are indicative of a greater amount of core reduction (e.g. Whittaker 1994:109). A chi square test shows a significant difference in the amount of cortex on the complete flakes and the type of termination on the complete flakes at Stud 1 (χ^2=226.37, *df*=12, *p*<0.0001, *phi*=0.1399).

Few of the complete flakes made from Group III silcrete have a cortical platform in comparison to those made from medium clast silcrete (Table 8.24). Cortical platforms usually occur early in core reduction (e.g. Andrefsky 1998:93-94). The presence of a cortical platform on a complete flake at Stud 1 probably reflects the reduction of a gibber cobble. A significant difference (χ^2=349.02, *df*=6, *p*<0.0001, *phi*=0.178575) was found between the complete flakes made from Group III, fine clast and medium clast silcrete and the type of platform (Table 8.24).

% of Cortex	0%		1-49%		50-99%		100%		Total		Length (mm)	
	Count	%	Count	%	Count	%	Count	%	Count	%	Mean	SD
Form												
Blade	675	10	167	5	13	1			856	7	35	11
Block	211	3	154	5	79	8	13	5	457	4	36	13
Expanding	1372	20	663	20	213	21	65	25	2315	20	27	10
Intermediate	4761	68	2294	70	703	70	181	70	7939	69	30	11
Termination Type												
Abrupt	1938	28	1251	38	347	34	96	37	3632	31	32	11
Feather	3381	48	1357	41	466	46	119	45	5325	46	29	11
Hinge	828	12	214	7	54	5	11	4	1107	10	26	10
Plunge	371	5	238	7	80	8	20	8	709	6	35	13
Step	492	7	212	6	58	6	15	6	777	7	28	10

Table 8.23 The Different Termination Types and Flake Forms on the Complete Flakes at Stud 1

Material Type	Total		Group III		Medium Clast		Fine Clast	
Form	Count	%	Count	%	Count	%	Count	%
Blade	856	7	218	6	265	7	339	9
Block	457	4	137	4	155	4	139	4
Expanding	2315	20	719	19	727	20	703	20
Intermediate	7939	69	2701	72	2422	68	2389	67
Platform Type								
Bifacial	918	8	349	9	252	7	270	8
Cortical	1469	13	214	6	712	20	432	12
Missing	1227	11	432	11	357	10	377	11
Unifacial	7989	69	2799	74	2255	63	2496	70

Table 8.24: The Platform Types and the Flake Forms of the Complete Flakes made from Different Raw Material Type at Stud 1

Of the complete flakes with more than 50% cortex on their dorsal surface most were made from medium clast silcrete (Table 8.22) and were significantly larger than those made from Group III or fine clast silcrete (f=11.694, df=2, p<0.0001). It is likely that these complete flakes were manufactured from gibber cobbles. Flakes with large amounts of dorsal cortex have more abrupt and plunge terminations, and less step or hinge terminations than the complete flakes with no cortex (Table 8.23). A chi square test of the termination types shows a significant difference between the flakes with no cortex and those with between 50 and 99% cortex on their dorsal surfaces (χ^2=61.458, df=4, p<0.0001, phi=0.087566).

Complete flakes with a blade form account for 7% of the assemblage at Stud 1 (Table 8.24). Most of these flakes (n=339) were made from fine clast silcrete (Table 8.24) and have no cortex on their dorsal surfaces (Table 8.23).

8.5.3 The Complete Flakes at Stud 2

At Stud 2 more of the complete flakes were made from Group III silcrete (42%, n=4339) and less were made from fine clast silcrete (26%, n=2686) or medium clast silcrete (32%, n=3242) than at Stud 1 (Table 8.22, 8.25). A chi square test gives a significant result when comparing the frequency of complete flakes made from Group III, fine clast and medium clast silcrete in Stud 1 and Stud 2 (χ^2=264.39, df=2, p<0.0001, phi=0.109153). The large number of flakes made from Group III silcrete suggests a concentrated use of stone from quarries and less reduction of gibber cobbles. However, the Stud 2 survey focused on the EVM and did not incorporate an area of gibber pavement like Stud 1, hence the location of Stud 2 may have influenced the composition of the assemblage (Figure 8.5).

The complete flakes made from Group III or fine clast silcrete have more characteristics indicative of further core reduction in comparison to the complete flakes made from medium clast silcrete. The complete flakes made of Group III silcrete are smaller (t=13.254, df=4453, p<0.0001) and have less cortex (χ^2=4[illegible]7.01, df=3, p<0.0001, phi=0.234611) than the flakes of medium clast silcrete (Table 8.25). A similar pattern emerges for the complete flakes made from fine clast silcrete. These flakes are also smaller (t=9.797, df=5926, p<0.0001) and have less cortex (χ^2=135.57, df=3, p<0.0001, phi=0.151227) than the medium clast flakes.

	Group III				Fine clast				Medium clast			
	Total		Length (mm)		Total		Length (mm)		Total		Length (mm)	
Form	Count	%	Mean	SD	Count	%	Mean	SD	Count	%	Mean	SD
Blade	107	3	31	9	93	4	32	9	100	3	36	10
Block	99	2	31	11	53	2	31	11	75	2	40	15
Expanding	1221	29	25	10	688	26	27	11	8[illegible]3	28	31	11
Intermediate	2810	66	27	10	1805	68	28	10	2124	67	31	13
Percentage of Cortex												
0%	3303	76	25	9	1850	69	27	10	1770	55	29	11
1-49%	830	19	30	11	647	24	30	11	1049	32	33	15
50-99%	171	4	33	12	143	5	32	12	321	10	36	15
100%	35	1	29	12	46	2	34	13	102	3	34	11
Total	4339	42	26	10	2686	26	28	10	3242	32	31	13

Table 8.25: The Length of the Complete Flakes at Stud 2 in Relationship to the Type of Silcrete, the Amount of Cortex and their Flake Forms

Flakes with more than 50% cortex on their dorsal surfaces and/or a cortical platform were probably manufactured from gibber cobbles. These flakes are larger (t=19.476, df=7850, p<0.0001) and have more abrupt and plunge terminations (χ^2=91.435, df=4, p<0.0001, phi=0.106808) than the flakes with no cortex (Table 8.26). The presence of more plunge and abrupt terminations generally indicates the beginnings of a reduction, when a large force was used on the core to remove a flake (e.g. Andrefsky 1998:86; Cotterell and Kamminga 1987:699). The low number of cortical platforms indicates that only 12% (n=1227) of the complete flakes were made from gibber cobbles – although this may simply reflect the reduction strategy (Table 8.26). Correspondingly, the number of flakes with more than 50% cortex on their dorsal surfaces is also low, again suggesting a limited reduction and use of the gibber cobbles (Table 8.26).

Platform Type	Total Count	%	0% Count	%	1-49% Count	%	50-99% Count	%	100% Count	%
Bifacial	412	4	310	4	78	3	20	3	4	2
Cortical	1227	12	63	1	850	32	193	28	121	63
Missing	1508	14	1057	15	316	12	120	17	15	8
Unifacial	7509	70	5696	80	1405	53	356	52	52	27
Form										
Blade	341	3	271	4	43	2	27	4		
Block	450	4	132	2	78	2	236	35	4	2
Expanding	3095	29	1876	27	756	3	409	60	54	28
Intermediate	6613	63	4735	68	1735	29	11	2	132	69
Termination										
Abrupt	2783	26	1618	23	901	34	215	31	49	26
Feather	5280	50	3620	51	1202	45	343	50	115	60
Hinge	1247	12	1021	14	189	7	29	4	8	4
Plunge	678	6	405	6	200	8	62	9	11	6
Step	671	6	475	7	152	6	36	5	8	4
Length (mm)	Mean	SD	Mean	SD	Mean	SD	Mean	SD	Mean	SD
	28	11	26	10	31	13	34	13	33	12
Total (%)	10698		7163	67	2654	25	689	6	192	2

Table 8.26: The Characteristics of the Complete Flakes at Stud 2 in Relationship to the Amount of Cortex on their Dorsal Surfaces

At Stud 2 most of the complete flakes (67%, n=7163) have no dorsal cortex and were probably manufactured from quarried material. A greater number of steps and hinge terminations and a decrease in the size of the flakes with no cortex indicate a more intensive reduction of the stone acquired from a quarried outcrop (Table 8.26).

Most of the blades have no cortex (86%, n=271) and were made Group III silcrete (Table 8.25). In comparison to Stud 1 (7%, n=856) few complete flakes at Stud 2 have a blade form (3%, n=314). This result may suggest that less emphasis was placed on blade manufacture at Stud 2.

8.5.4 Comparison of the Complete Flakes in the Landscape Sampling Units

Variation in the frequency of the technological types and characteristics occurs between the three landscape sampling units. In the Lower Stud Creek units more evidence of "on-site" manufacturing occurs shown by the greater number of complete splits, blocks and broken flakes than in the USC unit. The high proportion of broken flakes, including proximal flakes and distal flakes, may also be due to the repeated or long-term use of the area causing trampling and the subsequent breakage of the artefacts (Boot 1987; Hall and Love 1985). However, while proportionally more cores occur in the USC assemblage than in the LSC assemblages these were not substantially reduced.

Raw material types occur in different proportions in each of the landscape sampling units. In the Lower Stud Creek units more flakes of Group III (amorphous, breccia, matrix dominated and banded) silcrete occurred than flakes of medium clast silcrete. In contrast, the opposite raw material proportions occurred in the USC unit, with more flakes of medium clast silcrete present in the assemblage. This difference may be due to two reasons; a preference for the transport and use of Group III silcrete to the LSC units or the focus on the flaking of the gibber cobbles in the LSC units has biased the results. Since the USC unit is located close to Quarry 35, where both medium clast and Group III silcrete outcrop, this may also have influenced the composition of the raw material types observed in the assemblage.

The type of cortex at the USC unit indicates that almost one third of the complete flakes were manufactured from gibber cobbles. The colour of the flakes also hints at the source of the stone. Fine clast white silcrete can not be found in the gibber and only occurs as outcrops while flakes with cortex and made from pink or yellow medium clast silcrete were probably manufactured from gibber cobbles. It is more likely that Group III silcrete, independent of colour, originated from a quarry, due to the absence of cortex and the limited availability of this silcrete type found as gibber cobbles.
A surprising number of the flakes in the Lower Stud Creek units have no cortex again suggesting that most of the flakes originated from quarried outcrops and were not manufactured from gibber cobbles. The flakes with no cortex also tend to be made from Group III silcrete which reflects the further reduction of the cores and a preference for this material type. Flakes with a cortical platform probably derive from the gibber pavement and were more likely to be made from medium clast silcrete. Very few cortical platforms occur in all three sampling units indicating that only a limited reduction of gibber cobbles occurred. Most of the platforms in all three units are unifacial, suggesting little core rotation (*cf* Andrefsky 1998:94-96). Few cortical platforms were found on the flakes made from Group III silcrete in all the units especially in the USC unit where only 9% of the flakes had a cortical platform. This result again indicates that cores of Group III silcrete were obtained from a quarried outcrop and transported to the landscape sampling units. Although medium clast flakes with no cortex were also likely to have been obtained from a quarried outcrop.

Further core reduction at the LSC units, especially in Stud 1, is indicated by the high frequency of hinges and bifacial platforms in these two units. The number of flakes with hinge terminations generally increases as a core undergoes further reduction (e.g. Whittaker 1994:109). Likewise, a bifacial platform indicates that core rotation occurred and implies further reduction (e.g. Andrefsky 1998:94-96).

Variation in the frequency of the different flake forms is present in the landscape sampling units. In the Stud 2 assemblage fewer flakes have block or blade forms than in the USC unit indicating that more blade manufacture occurred in the USC unit. This pattern is probably due to Quarry 35 being located nearby, where blades were manufactured, and the reduction of large gibber cobbles producing flakes with a block form. It is interesting to note that across all the units blades are similar in size with no significant difference in the length (*f*=5.8, *df*=2, *p*=0.003). This result suggests a preference for the size of flake made during blade manufacture.

8.6 COMPARISON OF TOOL TYPES MADE FROM QUARRIED AND NON-QUARRIED STONE SOURCES

The proportions of the tool types varied between the landscape sampling units. Most of the tool types in the three sampling units are flakes with notched and scraper retouch or macroscopic usewear damage (Table 8.27). However, a substantial number of adze flakes were found in Stud 1 and Stud 2 (Table 8.27). The highest proportion of tools in an assemblage was found in Stud 1 (Table 8.5; n=3134, 13%). In contrast, the USC unit (n=345, 9%) and Stud 2 (n=2160, 8%) have a smaller proportion of tools. There is a significant difference between the units (χ^2=86.266, *df*=12, *p*<0.0001, *phi*=0.087) based on the frequency of scrapers, endscrapers, denticulates, adze flakes, notched tools, flakes with usewear and backed blades (Table 8.27). A large number of the tools at Stud 1 (32%) and Stud 2 (32%) have usewear damage, while only 63 (18%) were found at the USC unit (Table 8.27). Most of the tools at the USC unit have notched (30%, notch and denticulate) omr scraper retouch (38%). In comparison, Stud 1 (23%) and Stud 2 (23%) have few tools with one or more notches (Table 8.27).

Tool Type	Stud 1		Stud 2		USC	
	Count	%	Count	%	Count	%
Burin	1				3	
Thumbnail scraper	26	1			2	1
Usewear	992	32	683	32	63	18
Endscraper	273	9	128	6	12	3
Denticulate	286	9	190	9	36	10
Pirri Point	9		3		1	
Saw	16	1	5		2	1
Scraper	850	27	676	31	131	38
Notch	449	14	295	14	69	20
Adze Flake	152	5	121	6	8	2
Backed Blade	80	3	59	3	21	6

Table 8.27: The Tool Types in the Landscape Sampling Units

A selection for Group III silcrete to be used as tools is evident in the landscape sampling units. The ratio of complete tools to complete flakes at the USC unit indicates that a higher proportion (1:6) of tools were made from Group III silcrete (Table 8.19, 8.28). In contrast, fewer tools were made from medium clast silcrete (1:12). These ratios suggest that Group III silcrete was preferred for tool manufacture. The same pattern occurs in the Lower Stud Creek units. At Stud 1 (Table 8.21, 8.29) proportionally less tools were made from medium clast silcrete (1:8) and more tools were made from Group III silcrete (1:4). Very few of the complete tools at Stud 2 (Table 8.25, 8.28) were made from medium clast silcrete (1:12) and more were made from Group III silcrete (1:5).

A comparison of the tool types in the landscape sampling locations indicates a consistent difference in the use of high quality raw material types to manufacture of particular types of tools. Chi square tests comparing the frequency of denticulates, endscrapers, scrapers, flakes with usewear damage, notches, adze flakes and backed blades with type of material gives significant differences at Stud 1 (χ^2=202.89, *df*=12, *p*<0.0001, *phi*=0.262974), Stud 2 (χ^2=106.32, *df*=12, *p*<0.0001, *phi*=0.222504) and the USC unit (denticulate, notch, usewear, scraper; χ^2=17.317, *df*=6, *p*=0.0082, *phi*=0.242257). In both Stud 1 and the USC unit most of adze flakes and thumbnail scrapers were made from Group III silcrete (Table 8.28). In contrast, denticulates, notched, unmodified flakes with usewear damage and scrapers were more likely to be made from medium clast silcrete (Table 8.28). While thumbnail scrapers were not recorded at Stud 2 a similar pattern emerges in the other landscape sampling units for this tool type.

Most of the tools at Stud 1 (n=1396, 48%) and Stud 2 (n=1266, 61%) were made from Group III silcrete (Table 8.28). The high numbers of tools made from Group III silcrete in Stud 1 and Stud 2 indicates that most of the stone used to make the tools was not obtained from the gibber pavement but acquired from a quarried outcrop. Hence, a selection for Group III silcrete being obtained and brought to Stud 1 and Stud 2 was evident in the assemblages. In contrast, at the USC proportionately more tools were made from medium clast silcrete (53%) even though a nearby source of Group III silcrete was readily available at Quarry 35. The increased use of medium clast silcrete at the USC unit suggests that the selection of raw material type only influenced the form of particular tool types – most notably, the manufacture of adze flakes.

The use of Group III material for tools at Stud 1 and Stud 2 may be linked more with decisions to acquire higher quality material. Tools made from high quality material types were more likely to be transported, had a longer use-life, were curated and discarded when spent or when it was possible to retool (Hiscock 1986b, 1988; Bird 1985:27). High quality material has more flexibility for manufacturing and use due to its fracturing consistency (Nelson 1991:70-71; Doelman *et al.* 2001; Cotterell and Kamminga 1987:677). It seems likely that cores of quarried material were reduced at the landscape sampling units producing large numbers of Group III and fine clast flakes and flakes with blade forms. Most of these flakes were discarded immediately and only some were selected for tools. It is possible that the discarded flakes were later "mined" and used as expedient tools to produce usewear along the edges or retouched as notched tools or scrapers (*cf* Camilli and Ebert 1992). In contrast, flakes of medium clast silcrete were more likely to be derived from gibber cobbles and less likely to be transported or curated due to the poor quality of the material, as observed by Gould (1977, 1980) and Hiscock (1986b).

Tool Type	Stud 1 Group III	Stud 1 Medium clast	Stud 1 Fine clast	Stud 2 Group III	Stud 2 Medium clast	Stud 2 Fine clast	USC Group III	USC Medium clast	USC Fine clast
Burin	1						2	1	
Denticulate	93	87	74	69	56	51	7	23	6
Notch	185	139	110	160	66	62	13	50	6
Thumbnail scraper	17	3	4				2		
Scraper	290	209	226	407	139	116	47	63	18
Endscraper	123	69	68	73	28	27	2	7	3
Usewear	505	290	171	414	121	130	27	27	9
Saw	8	3	4	3		1	1	1	
Adze flake	144	6	4	109	2	2	5		
Pirri Point	4	1	3	2	1			1	
Backed blade	26	23	25	29	18	8	4	7	3
Total	1396	830	689	1266	431	397	110	180	45

Table 8.28: The Number of Different Tool Types in Relationship to their Raw Material Types in the Landscape Sampling Units

	Material Type	Group III	Medium clast	Fine clast	Total
USC	Count	74	123	34	232
	%	32	53	15	
	Mean Length (mm)	32	44	43	40
	SD (mm)	12	16	12	15
Stud 1	Count	1026	467	554	2125
	%	48	22	26	
	Mean Length (mm)	30	39	36	34
	SD (mm)	12	13	12	13
Stud 2	Count	830	267	234	1365
	%	61	20	17	
	Mean Length (mm)	30	38	36	32
	SD (mm)	12	12	12	13

Table 8.29: The Number of Complete Tools and their Mean Lengths in Relationship to their Raw Material Types inj the Landscape Sampling Units

The limited reduction of the gibber cobbles, observed in the large size of the tools, cores and flakes with a gibber cortex at the USC unit, indicates that the aim of flaking the gibber cobbles was simply to produce a large flake for immediate tool use (Table 8.30). The presence of a readily available, abundant source of stone from the gibber pavement required no need to conserve material and allowed an expedient technological strategy to be employed. The large amount of cortex on the complete tools shows that for all three landscape sampling units the flakes made from medium clast silcrete have considerably more cortex than those made from Group III or fine clast silcrete (Table 8.31). Significant differences were found comparing the amount of cortex and the material types of the complete tools in Stud 1(χ^2=55.617, *df*=6, *p*<0.0001, *phi*=0.165033). Stud 2 (χ^2=59.044, *df*=4, *p*<0.0001, *phi*=0.210857) but not at the USC unit (χ^2=8.729, *df*=4, *p*=0.0682).

Type of Cortex	Gibber		Weathered		No Cortex		Total
Tool Type	Count	%	Count	%	Count	%	Count
Adze flake	2	3	1	2	2	2	5
Backed blade	1	2	1	2	14	12	16
Notch	14	23	12	23	14	12	40
Denticulate	5	8	5	10	13	11	23
Saw	1	2					1
Endscraper	2	3	1	2	6	5	9
Scraper	23	38	20	38	43	36	86
Thumbnail scraper	1	2			1	1	2
Usewear	11	18	12	23	25	21	48
Pirri Point					1	1	1
Total	60	26	52	22	120	52	232
Length of the complete tools (mm)	Mean	SD	Mean	SD	Mean	SD	
	45	16	40	12	37	15	

Table 8.30: The Types of Cortex on the Different Tool Types in the USC Unit

	Medium clast		Fine clast		Group III	
Stud 1	**Count**	**%**	**Count**	**%**	**Count**	**%**
0%	269	58	356	64	747	73
1-49%	138	30	155	28	233	23
50-99%	54	12	39	7	37	4
100%	6	1	4	1	4	
Total	467		554		1021	
Stud 2						
0%	170	64	172	74	686	83
1-49%	71	27	52	22	129	16
50-99%	20	8	10	4	12	1
100%	5	2			1	
Total	266		234		828	
USC						
0%	54	45	18	53	47	68
1-25%	39	33	8	24	18	26
26-49%	13	11	5	15		
50-75%	8	7	2	6	2	3
76-99%	5	4	1	3	2	3
100%	4	3			3	4
Total	119		34		69	

Table 8.31: The Amount of Cortex on the Complete Tools made from Different Types of Silcrete in the Landscape Sampling Units

The type of tool was dependent on the stone source and related to a variety of factors including the raw material type, the size and form of the cobble and the type of cortex. The overall pattern in the use of stone resources suggests that quarried stone was used for more specialised tool manufacture, such as the adze flakes or blades, while tools made from gibber cobbles were more likely to be used expediently. Cochrane (1998:94) notes that there were relatively high numbers of flakes with usewear on the gibber pavement indicating that expedient manufacture and use of flakes from gibber cobbles that were retouched, used and then discarded in the same context.

8.7 REDUCTION STRATEGIES IDENTIFIED IN THE ANALYSIS OF THE LANDSCAPE SAMPLING UNITS

Two different reduction strategies were employed for the reduction of the non-quarried and quarried stone. A generalised reduction sequence (*cf* Hiscock 1988:33-41) of the cores made from the non-quarried stone can be reconstructed for the landscape samples because the source, manufacture and discard occurred in the same location. In contrast, most of the quarried stone arrived at the landscape samples as blade producing cores and were further reduced, sometimes for blade manufacture, until the cores were exhausted and then discarded. A brief outline presents the different strategies used to reduce the cores from different sources.

- **The Reduction of Gibber Cobbles at the Landscape Sampling Units**

Gibber and creek cobbles were used expediently for the production of large flakes. Cores obtained from the gibber pavement underwent little reduction, with only a few flakes removed from each core and little core rotation. Material testing, represented by the presence of cobble cores in the assemblages, was undertaken to assess the quality of the material. This strategy was necessary as the gibber cobbles are completely covered in cortex and the quality of the underlying material is impossible to determine without the removal of cortex from the surface of the cobbles.

Most of the cobble cores at Stud 1 were discarded on the gibber pavement. The gibber pavement was used as a source of stone at this location (Doelman *et al.* 2001). Only a few flakes were removed from each core and these were used and discarded immediately. This process is reflected in the high number of tools with usewear damage on the gibber pavement at Stud 1 (Cochrane 1998).

At the USC unit most of the gibber cores are made of Group II silcrete, with approximately 10% of these cores being made from Group III silcrete. The gibber cores that were made from Group III silcrete had more flakes removed and were smaller than the Group II cores indicating a preference for high quality material. The frequency of the raw material types with gibber cortex probably reflects the composition of the different material types in the gibber pavement.

The cores made from gibber cobbles in the Upper Stud Creek unit are larger than those at Stud 1 and Stud 2. The large size of multi-platform, uni-platform bi-platform and cobble cores is due to the use of the presence of large gibber cobbles, available along the creek, at the USC unit. Consequently, the presence of large cobbles has influenced the size of the cores at the USC unit.
Most of the cores in the landscape sampling units are either uni-platform or bi-platform. These core types in the USC unit have a gibber cortex, retain a large size and only have a few large flakes were removed. The flake scars on the multi-platform, uni-platform and bi-platform cores are the same size indicating that the goal was simply to produce a large flake from these core types and once this was achieved the cores were discarded.

- **The Reduction of Quarried Stone at the Landscape Sampling Units**

Based on the large number of blade cores a considerable emphasis on blade manufacture occurred in the USC unit. This outcome is probably due to the unit's small distance from an outcrop used for blade manufacturing, Quarry 35. A similar pattern in the use of the stone resources occurred at the USC unit and the LSC units. However, unlike the USC unit very few microblade cores occur at Stud 1 or Stud 2 and these were even more heavily reduced, possibly reflecting the distance from the source. The proportions of the different core types found in the LSC units suggest that relatively little blade production occurred in these locations.

Most of the blade cores were made from Group III silcrete indicating a selection for high quality material for this type of reduction strategy. High quality material allows greater control during reduction and enables tools to be fashioned and resharpened (Andrefsky 1994a, 1994b; Kuhn 1992).

Several different types of blade cores, indicating the use of different blade production strategies (*cf* Witter 1992:53-54) was observed in the USC assemblage including prismatic and tranchet cores. These core types were observed in the Quarry 35 assemblage and it is likely that this quarry was the source of the blade cores.

8.8 SUMMARY

The results of this chapter highlight how the quality of the stone, the size and form of the raw material source and the sources available were important factors influencing the manufacture of stone tools. All these factors in varying combinations contribute to assemblage composition at each location. Even within a catchment, where abundant stone was available, differences in the use of material derived from the quarried and non-quarried sources was evident, reflected in the tool type, the use-life, degree of transport and time invested in procurement. The use of gibber cobbles, even at the USC unit where high quality material was available nearby at Quarry 35, indicates an expedient use of a source, where tools were made, used and immediately discarded (*cf* Binford 1977:34).

Gibber cobbles are generally small in size, averaging at approximately 60 mm in length. In contrast, the mean length of the cobbles outcropping at Quarry 35 is 230 mm. Blade production is generally regarded as wasteful, labour intensive and time consuming, in the amount of core preparation and rejection of unsuitable flakes. However, the finished blade can be considered versatile in design, thereby meeting a variety of needs, and enables a greater potential cutting edge for the mass of stone (Nelson 1991:70). Hence, larger cores of high quality material, found at the quarries, would be needed for blade manufacture. Gibber cobbles would not only be unsuitable for blade production based on the raw material but also on the size of the cobble. Consequently, quarried stone was used for specific purposes, such as blade production, or the manufacture of tools that require resharpening, such as adze flakes. The disproportionately high number of flakes with a blade form and microblade cores highlights more blade production in the USC when compared to Stud 1 and Stud 2 and emphasises how the distance from a blade quarry (Quarry 35) impacts on the composition of the assemblages.

A selection for high quality raw material types was evident in the landscape sampling units. High quality stone was used in the manufacture of the adze flakes that required a greater degree of retouch and resharpening. In contrast, the low quality material was generally used for unmodified flakes with usewear damage, notch and scraper tools. The use of particular raw material types for particular tools may be a response to the size and form of the sources. Indeed, Dibble and Rolland (1992:12-13) argued for the Middle Palaeolithic that nearly all assemblage variability could be accounted for by the utilisation of resources, stone quality and availability, proximity to sources, size and shape of nodules and the texture of the available material.

Unlike the USC unit, most of the tools at Stud 1 and Stud 2 were made from Group III silcrete. The extensive use of Group III silcrete for expedient tools in the LSC units may be linked with the acquisition of cores from quarries, which were worked further in these two locations. During later occupations the flakes discarded during reduction may have been 'mined' and used as tools (*cf* Camilli and Ebert 1992). The presence of high numbers of tools made from Group III silcrete may be a good indicator of the amount of time spent at a location, since the opposite pattern occurs in the USC unit.

The assemblage composition at Stud 1 and Stud 2 may be more influenced by transport choices – where high quality stone is more likely to be returned to the 'occupation location'. As Group III material has more consistent fracturing, allowing greater flexibility (*cf* Nelson 1991; Kuhn 1994; Shott 1989) when manufacturing and finer retouch for tool use, it is more efficient. While the presence of large creek cobbles allows large flakes to be made, the manufacturing flexibility is restricted by the material quality as most of the creek cobbles were made from poor quality, medium clast silcrete (Doelman *et al.* 2001). In this way the stone available nearby has also influenced the compositions of the assemblages at the landscape sampling units.

The patterns of raw material use, created over the long-term, indicate two simultaneous organisational strategies that are dependent of the type of stone source. One strategy is the expedient use of gibber cobbles and the other is the curated use of stone obtained from quarries, seen through the transport of microblade cores and the tool types manufactured. This is primarily contingent on the raw material quality: tools of high quality stone allow a longer use-life, are more efficient, could be transported and continually maintained (ie tulas). Expedient tools (ie flakes with usewear) required little effort to manufacture and had a short-use life. Thus the strategies used were dependent on the source location (distance from the source), stone availability and quality, form of raw material and the required use of the stone.

Chapter 9

Time to Quarry

9.1 INTRODUCTION

This chapter examines the similarities and differences between Quarry 27 and Quarry 35 in order to re-evaluate and redefine a 'quarry' and contrast this definition with the use of the gibber sources in the landscape sampling units. The two quarries are compared to determine the factors that influenced the spatial distribution and characteristics of the artefacts, and the formation and composition of the quarry assemblages. Comparisons are made between the reduction strategies employed, the spatial distribution of the artefacts and the time invested in reducing stone at each quarry. These comparisons are used to determine the 'role' the different sources played in the organisation of lithic technology.

The quarries differ based on a number of characteristics, such as the size, artefact density, the raw material types available and their abundance, topography, location, reduction strategies and the degree of "on-site" reduction. However, both quarries share a number of common features that have influenced the placement of different reduction stages. The spatial analyses of the quarry assemblages were used to identify these features. Lastly, the sampling strategies used at quarries are discussed in order to highlight the importance of an appropriate methodological approach to accurately record a quarry assemblage.

9.2 THE FORMATION AND COMPOSITION OF A QUARRY ASSEMBLAGE

Spaulding (1960:37-38) viewed archaeology as "the study of the interrelationships between three dimensions of artefacts: form, spatial locus and temporal locus." Each dimension requires its own measuring device (Spaulding 1960:438). Spaulding (1960:456) concluded that archaeology was meaningless if these three dimensions are not taken into account in the interpretation of the archaeological record. Furthermore, the archaeological record can be viewed at any scale of time and space and at each scale many factors influence the final form and distribution of the artefacts (Stoltman 1978). Hence, each scale also requires a different set of methods to measure variability in assemblage composition and in the distribution of the archaeological record (Wandsnider 1992, 1996, 1998). This observation calls for a need to pursue archaeological inquiry "...which does not ignore the structural properties of its data..." rather than more traditional approaches which produce only broad and often trivial generalisations (Murray 1999:9).

For this study a technological organisational approach was used to determine the types of technological strategies employed by the late Holocene inhabitants of a small area of the Mt Wood Ranges located in the arid zone of Australia. Shott (1989:221) defined a technological organisation approach as "how material technologies [reflected in the form of artefacts] were integrated into the cultural system" and a cultural system as a "set of relationships and activities' that were organised "in time and space" (Shott 1989:236). Each artefact recorded has a unique position in space and a unique set of characteristics, incorporating aspects of non-site distributional archaeology (e.g. Ebert 1992), that enabled the distribution of the artefacts at the quarries to be 'viewed' and analysed at any scale of space with the aid of *ArcView* GIS. The spatial analysis of Quarry 27 and Quarry 35 enabled the distribution of different stages of reduction to be examined and related to the various landscape features present at the quarries. These two approaches helped to identify some of the factors that influenced the composition and distribution of the quarry assemblages.

How time may have impacted on the formation and composition of the quarry assemblages is explored in this chapter. A series of overlapping, short-term behaviours, which are reflected in the characteristics and location of the discarded artefacts, creates a palimpsest of the archaeological record. The very formation of the archaeological record into a palimpsest makes any reconstruction of past behaviour extremely difficult (Stern 1993:215, 1994). Typically, the archaeological record is viewed collectively as a series of short-term behaviours that produces long-term patterns in the discard of artefacts, reflecting the organisational structure of a group and giving an average representation of behaviour (Binford 1981). Alternatively, the archaeological record can be viewed as being time-averaged, resulting from the amalgamation of all behaviours and post-deposition processes, and should not be viewed as providing an 'average' of human behaviour (Stern 1994:101). Hence, it is better to examine the structure of the archaeological record by determining what variability occurs in the formation, distribution and composition of the archaeological record and identifying the reasons why this variability occurs.

According to Fletcher (1992:36) all scales of time are interrelated, interdependent and structure each other, the problem lies in trying to correlate short and long-term behavioural processes to the archaeological record. By bracketing the duration of occupation in the study area to the last 4000 years (discussed in Chapter 3) an attempt is made in this study to determine how short and long term time processes, discussed by Bailey (1983) and Fletcher (1992), have shaped the variability of the archaeological record. In undertaking this goal, the time artefacts spent in the cultural system, the time invested in core reduction and the time spent at Quarry 27 and Quarry 35 are compared in order to determine how these three variables impacted on the distribution, formation and composition of the quarry assemblages. Lastly, how the occupation of the region over the last 4000 years has influenced the formation and composition of the archaeological record is also explored.

In the following sections the reduction strategies used at Quarry 27 and Quarry 35 to work the stone (Section 9.3), the landscape features and their influence on the spatial distribution of assemblages (Section 9.4), and the short and long-term time processes that formed the assemblages (Section 9.5) are compared in order to tease apart the many factors that influenced the composition of their assemblages. Acknowledging the factors behind the formation of a quarry assemblage leads to a better understanding of the role they played in the organisation of lithic technology.

9.3 COMPARISON OF REDUCTION STRATEGIES USED AT THE QUARRIES

The ways stone was extracted, tested, selected and reduced at Quarry 27 and Quarry 35 are compared in order to highlight the similarities and differences between the reduction strategies used at the two quarries. The results of this comparison are used to gain an idea of the factors that influenced the form of the artefacts and composition of the quarry assemblages. By understanding these factors the ways each quarry was used as a resource can be better ascertained.

9.3.1 Comparing the Composition of the Quarry Assemblages

Several differences can be observed in the composition of the technological types in the quarry assemblages. The most obvious difference between the two quarries is the ratio of cores to complete flakes (Table 9.1). These ratios indicate that few flakes were removed from the cores at Quarry 27, Area B and the transects of Quarry 35 (Table 9.1). In contrast, more flakes were removed from the cores in Area A and Area C of Quarry 35, areas where cores were prepared for blade manufacture and blade production occurred.

Complete splits occur when an incorrect or excessive force was used to remove a flake from a core (Hiscock 1988:365-366). The number of complete splits in an assemblage often decreases as reduction continues, particularly in the manufacture of blades, due to increased levels of flaking control (Clarkson and David 1995:33). The highest frequencies of complete splits are at Quarry 27 and in Area B of Quarry 35. These ratios suggest that initial core reduction occurred in these locations. In contrast, Area A and Area C have considerably fewer complete splits, possibly associated with the controlled reduction of cores that is indicative of blade manufacture. The lowest frequency of complete splits are in the transects of Quarry 35 again indicating that little or no core reduction occurred in any of the four transects.
A high number of broken flakes were found in both of the quarry assemblages but especially in Quarry 27. The greater frequency of broken flakes is typical of a quarry assemblage where more treadage and manufacturing errors are likely to occur (Torrence 1986:181-189; Hiscock and Mitchell 1993:28-32). Seemingly atypical of a quarry assemblage is the high frequency of tools in Area A of Quarry 35.

9.3.2 Extraction Strategies

At both Quarry 35 and Quarry 27 similar extraction strategies were used to extract stone. The extraction strategy was dependent on the type of outcrop and the size of the cobbles. Since both quarries have large cobbles exposed as massive silcrete outcrops, extraction was often simply the removal and breakage of the cobbles on anvils. Large cobbles, with points of impact, were used as anvils at Quarry 27 and Quarry 35. Several ethnographic observations of stone extraction and working at a quarry have been made in Australia. Tindale (1965:140) described how the cobbles lying on the surface were deemed unsuitable for reduction requiring cobbles to be extracted from the ground. The largest of these cobbles were raised above the head and smashed onto rocks below. The use of anvils, a block-on-block technique, to split cobbles was also observed in the Western Desert of Australia (Gould 1980:123) and in the eastern highlands of New Guinea (Flenniken and White 1985:134). This technique often occurred on or near the outcrop. The high frequency of blocks (Table 9.1) at Quarry 27 suggests that cobbles may have been split by being thrown against another stationary cobble. The resulting blocks were used as cores.

The size and shape of the cobble influences the way material was reduced (Flenniken and White 1985:131-132; Dibble and Rolland 1992; Andrefsky 1994b). Smaller cobbles could more easily be reduced using freehand percussion (Flenniken and White 1985:134) and it is likely that this technique was more commonly used at both quarries rather than the block-on-block technique. The large size and the amount of cortex on the cores at Quarry 27 indicates that most of the cobbles were reduced using a free-hand percussion technique. The presence of large, broken hammerstones in Area B, on the outcrop, tends to confirm the use of freehand percussion to split the larger cobbles at Quarry 35. In addition, splitting the larger cobbles on anvils was not often required as weathering and spalling of the silcrete cobbles naturally removed the cortex and provided the acute angles needed to remove flakes from a core.

Ratio	Q27	Area A	Area B	Area C	Transects of Quarry 35
Complete tool/complete flake	1:17	1:3	1:18	1:16	1:15
Block/complete flake	1:5	1:21	1:17	1:21	1:36
Complete flake/complete split	1:4	1:9	1:5	1:7	1:60
Complete flake/ broken flake	1:2	1:5	1:4	1:4	1:7
Core/complete flake	1:7	1:17	1:12	1:29	1:7

Table 9.1: The Ratios of the Different Technological Types Indicative of the Amount of Core Reduction and Tool Manufacture

Extraction of large cobbles from the ground was observed in ethnographic studies of quarrying behaviour (Binford and O'Connell 1984:412; Tindale 1965). Similarly, at the Ngilpitji Quarry in Arnhem Land Jones and White (1988:59-60) witnessed how pits were dug and rocks were smashed with a heavy cobble to remove cortex and expose the quality of the material. Binford and O'Connell (1984) also described the use of fire to split large cobbles at a quarry. No evidence of these practices was found at Quarry 27 or Quarry 35.

9.3.3 Testing and Material Selection

Within a single outcrop the quality of the available material may vary depending on the parent lithology. At both quarries investigated cobbles were tested and the quality of the stone was assessed. This process resulted in the presence of large discarded, cobble cores of poor quality material usually found on or near the outcrops. Material testing was a technique often used as a way of economising the time invested in reducing a core of unsuitable material (Flenniken and White 1985:133; Fullagar and Torrence 1992:126; Jones and White 1988; Binford and O'Connell 1984:410). In addition, Tindale (1965:140) observed how cobbles of high quality material were flaked at a quarry until the core was exhausted. The selection of high quality stone was observed at Quarry 27. At this location most of the artefacts were made from Group III silcrete. Here an isolated "patch" of extremely high quality Group III silcrete was found within a massive exposure of Group II silcrete. The cobbles of Group II silcrete remained relatively untouched while the cobbles of Group III silcrete were targeted and worked. At Quarry 27 most of the cores were made from Group III silcrete (98%), in comparison at Quarry 35 only 27% of the cores were from Group III silcrete (Table 9.2). However, a similar selection for high quality material was observed at Quarry 35: cores of Group III silcrete were reduced further than the cores made from Group II silcrete even though the sizes of the original cobbles were the same (Table 9.2).

9.3.4 The Reduction Strategies

All the cores were recorded at Quarry 35 and Quarry 27 and these are the focus of comparisons made between the reduction strategies employed at the quarries. Where possible the results of the flake analysis are used to support the inferences made from the core analysis. It is assumed that the differences in the characteristics of the cores at the quarries was not influenced by the size of the available cobbles as no significant difference (t=0.9307, df=205, p=0.351) occurs between the Group III cobbles at Quarry 27 and Quarry 35. Although the medium clast or fine clast cobbles were not recorded at Quarry 27 it is presumed that there is no difference in their sizes between the two quarries as none occurs at Quarry 35 (discussed in Chapter 7). Hence, the characteristics of the cores reflect the amount of reduction that they underwent, the strategies used to reduce the cores and the goal of core reduction at both Quarry 27 and Quarry 35.

Raw Material Type	Quarry 27				Quarry 35			
	Number		Length (mm)		Number		Length (mm)	
	Count	%	Mean	SD	Count	%	Mean	SD
Group I								
Coarse Clast					2		53	11
Group II								
Medium Clast	2	1	150	85	2254	65	83	29
Fine Clast	1	1	99		242	7	82	30
Group III								
Amorphous	7	4	77	13	66	2	67	18
Banded	21	11	95	24	3		55	1
Matrix-	159	83	116	51	861	25	74	34
Porcellinate	2	1	97	4				
Breccia					4		75	27
Sub-total	189	98			934	27		
Total	**192**				**3432**			

Table 9.2: The Raw Material Types used to make the Cores at Quarry 27 and Quarry 35

A comparison of the reduction strategies used at the two quarries indicates that cores at Quarry 35 were reduced further than those at Quarry 27. The cores at Quarry 35 generally have less cortex, greater numbers of step terminations, more flake scars and are smaller than the cores at Quarry 27 (Table 9.3). These characteristics are considered indicative of the amount of core reduction (Andrefsky 1998:102-103; Arnold 1987a, 1987b).

The frequency, size and diversity of the different types of cores reflect the amount of reduction occurring at the two quarries. A significant difference was found in the frequency of blade cores, cobble, uni-platform, bi-platform and multi-platform cores between the two quarries (χ^2=156.82, df=4, p<0.0001, phi=0.20802). At Quarry 27 a higher number of cobble and bi-platform cores occur in the assemblage than at Quarry 35 (Table 9.3). In contrast, more blade cores were identified in the Quarry 35 assemblage. Furthermore, the equivalent core types at Quarry 27, with the exception of the prismatic cores, are substantially larger than those at Quarry 35 (Table 9.3).

The high number of uni-platform cores at Quarry 35 indicates that most of the cores were reduced from one platform. However, several stages of reduction are reflected in the uni-platform cores. Large uni-platform cores with more than 50% of their surfaces covered in cortex underwent little reduction before being discarded. In contrast, smaller uni-platform cores with no cortex, often having a slab form, were made from large flakes and used to manufacture blades (described in Chapter 7). Additionally most of the uni-platform cores have elongated flakes scars, a characteristic indicative of blade manufacture (Table 7.3)

The size of the cores also reflects the amount of reduction (Andrefsky 1998:12). At Quarry 27 the cores are significantly larger (Table 9.3) than the cores at Quarry 35 (t=16.369, df=3564, p<0.0001) even though boulders of the same size outcrop at both quarries. This result suggests that the cores at Quarry 27 were discarded earlier in core reduction than the cores at Quarry 35. Furthermore the length of the flake scars on cores at Quarry 27 (Table 9.3) are also significantly longer than those measured at Quarry 35 (t=6.095, df=3564, p<0.0001). In addition, At Quarry 27 few cores have no cortex (Table 9.3) and more cores have greater than 50% of their surfaces covered in cortex. In contrast, most of the cores (8?%) in Quarry 35 have less than 50% cortex. A significant difference was found between the amount of cortex on the cores at the two quarries (χ^2=97.857, df=2, p<0.0001, phi=0.164433). The cores at Quarry 27, independent of the amount of cortex on their surfaces, are considerably larger than at Quarry 35 (Table 9.3)

The number of flake scars present on the surfaces of the cores with step terminations often increases as the size of the core decreases (Cotterell and Kamminga 1987:452; Hiscock 1988:117). At Quarry 27 only 12% of the cores have two or more flakes scars with step terminations (Table 9.3). In contrast, at Quarry 35 29% of the cores have two or more flake scars with step terminations. A significant difference in the frequency of step terminations on the flake scars of cores was found between the quarries (χ^2=34.514, df=3, p<0.0001, phi=0.09759) indicating that the cores at Quarry 35 underwent greater reduction. The size of the cores at Quarry 27 decreases as the number of step terminations on the flakes scars increases, however, at Quarry 35 this pattern does not occur and cores with no step terminations on their flakes scars are smaller then cores with step terminations (Table 9.3). It is possible that at Quarry 35 the larger cores with step terminations on their flake scars were discarded early in reduction because it was too difficult to remove further flakes due to the quality of the material.

The number of flake scars on the surface of a core is a measure of the amount of reduction a core underwent before it was discarded (Arnold 1987b). At Quarry 27 most (28%) of the cores have only one flake scar (Table 9.3). In contrast, only 14% of the cores at Quarry 35 have one flake scar. Over 40% of the cores at Quarry 35 have four or more flakes scars on the cores. The high number of flake scars on the cores at Quarry 35 indicates that more flakes were removed from the cores and that these cores were reduced further than the cores at Quarry 27. A significant difference was found in the number of flake scars on the cores at Quarry 27 and Quarry 35 (χ^2=30.098, df=6, p<0.0001, phi=0.091132). The mean number of scars per core also varies between the quarries (Table 9.3). A higher mean number of flake scars occurred on the cores at Quarry 35 than at Quarry 27. The lengths of the cores correlate to the number of flakes scars on the cores at Quarry 35 and Quarry 27, as the number of flake scars increases the size of the cores decreases, until five or more scars are present on the cores (Table 9.3).

The number of cortical platforms on a core can be used to measure the amount of core reduction occurring in an assemblage. Cores with cortical platforms are more frequent in the Quarry 27 assemblage than at Quarry 35 (Table 9.3). A chi-square test gives a significant difference between the two quarries based on the number of cortical platforms on the cores (χ^2=43.985, df=3, p<0.0001, phi=0.111154). In addition, cores with a cortical platform are considerably larger at both quarries than those without a cortical platform (Table 9.3). However, a high number of the cores in both of the quarries do not have a cortical platform (Table 9.3). The lack of cortical platforms may also be due to the weathering and subsequent removal of the cortex from the cobbles and the splitting of the large cobbles on anvils.

At Quarry 27 a larger proportion of the complete flakes have cortical platforms than those at Quarry 35 (Table 9.4). The presence of a cortical platform on complete flakes indicates that it was manufactured early in core reduction (Andrefsky 1998:93-94). At Quarry 27 12% (n=155) of the flakes have cortical platforms, in comparison Area B, an area of initial core reduction at Quarry 35, only 8% (n=128) of the complete flakes have cortical platforms. A chi-square test shows that a significant difference occurs between the platform types in Area B and Quarry 27 (χ^2=272.58, df=4, p<0.0001, phi=0.306161). The relatively low number of cortical platforms supports the argument that naturally weathered cobble surfaces and/or split cobbles were used as platform surfaces.

	Quarry 27				Quarry 35			
Core Type	**Count**	**%**	**Mean Length**	**SD**	**Count**	**%**	**Mean Length**	**SD**
Bi-platform	54	28	100	31	518	15	80	25
Microblade	1	1	45		199	6	60	14
Multi-platform	24	13	105	28	145	4	89	39
Prismatic	1	1	104		73	2	72	13
Cobble	34	18	157	77	136	4	142	83
Uni-platform	78	41	109	43	2154	63	78	22
Bidirectional					89	3	73	19
Tranchet					118	3	76	17
Percentage of Cortex								
0%	9	5	78	23	598	17	70	18
1-49%	96	50	100	30	2254	66	77	22
50-99%	84	44	129	56	578	17	99	39
Number of Cortical Platforms								
0	154	80	108	43	3123	93	79	30
1	33	17	130	67	224	7	91	39
2	4	2	133	55	14		124	54
=>3	1	1	186	69	3		83	14
Number of Step Terminations								
0	112	58	119	54	1365	40	77	27
1	58	30	105	43	1093	32	81	37
2	13	7	97	36	541	16	80	24
3	9	5	108	20	244	7	81	28
4					121	4	86	36
5					39	1	84	28
=>6					29	1	85	21
Number of Flake Scars								
1	53	28	124	66	494	14	90	39
2	37	19	118	58	753	22	83	39
3	42	22	103	29	785	23	77	23
4	22	11	99	38	616	18	75	21
5	14	7	118	51	366	11	75	20
6	9	5	102	28	219	6	77	30
=>7	15	8	121	26	199	6	77	22
Mean Number of Flake Scars	**Mean**	**SD**			**Mean**	**SD**		
	3	2			3	2		
Size Dimensions (mm)								
Length (mm)	114	50			80	27		
Width (mm)	95	41			61	23		
Thickness (mm)	68	33			42	18		
Scar Length (mm)	52	27			45	15		

Table 9.3: The Characteristics of the Cores in the Quarries

	Area A		Area B		Area C		Quarry 27	
Platform Type	**Count**	**%**	**Count**	**%**	**Count**	**%**	**Count**	**%**
Bifacial	87	3	20	1	51	2	76	6
Cortical	175	7	128	8	121	5	155	12
Missing	21	1	78	5	448	19	235	18
Step	16	1	2		39	2	25	2
Unifacial	3384	88	1409	86	1692	72	780	61

Table 9.4: The Platform Types on the Complete Flakes in the Sampling Areas of Quarry 35 and at Quarry 27

Evidence of more blade manufacturing was found at Quarry 35. The diversity of blade producing cores at Quarry 35 is reflected in the number of microblade, tranchet, bi-directional and prismatic cores. These blade cores represent the three different reduction strategies used to manufacture blades at Quarry 35. The analysis of the cores from Quarry 35 shows that both large flakes and blocks were used to manufacture blade cores. In contrast, only two microblade cores and 24 flakes with a blade form were found at Quarry 27. The limited number of blade producing cores at Quarry 27 suggests that blade manufacturing was not the primary goal of core reduction at the quarry. It is likely that the goal of core reduction at Quarry 27 was simply to manufacture large flakes.

A flake with a blade form is characterised as having a length to width ratio of ≥2:1 (Crabtree 1972). The flakes used to prepare the core before the removal of the blades are usually also elongated (Flenniken and White 1985:133). Comparing the length and width of the complete flakes from Quarry 27 and the sampling areas in Quarry 35 provides an idea of the elongation of the flakes at each location. These results are used to show the form of the flakes produced in Quarry 27 and Quarry 35 (Table 9.5). The flakes in both Area A and Area B of Quarry 35 have a length to width ratio of 1:2. This result is consistent with blade manufacturing. In contrast, although the flakes at Quarry 27 are similar in length to those in Area A they are considerably wider (t=20.33, df=3733, p<0.0001). A t-test comparing the length to width ratios between Area A (t=19.465, df=3733, p<0.0001) and Area B (t=1.637, df=3453, p<0.0001) with Quarry 27 also has a significant results.

A way of comparing the size dimensions of the complete flakes in the quarries is using a reduction chart developed by Witter (1992:101-107) to view the correlation between the length, width and thickness. Figure 9.1 show that the size and shape of the complete flakes in Area B of Quarry 35 and Quarry 27 are similar. In contrast, the complete flakes in Area A of Quarry 35 (Figure 9.1) are longer and thinner representing a shape and size more typical of blade manufacturing.

	Area A	Area B	Quarry 27
Count	2379	1076	1271
Mean Length:Width Ratio	1:2	1:2	1:1
SD	0.7	0.6	0.5
Mean Length (mm)	44	51	45
SD (mm)	17	21	23
Mean Width (mm)	29	36	39
SD (mm)	14	19	20
Mean Platform Width (mm)	18	21	22
SD (mm)	13	14	17
Mean Platform Thickness (mm)	7	9	12
SD (mm)	5	6	9
Mean Platform Area	179	274	341
SD (mm)	233	364	419

Table 9.5: The Size Dimensions of the Complete Flakes in Area A and Area B of Quarry 35 and at Quarry 27

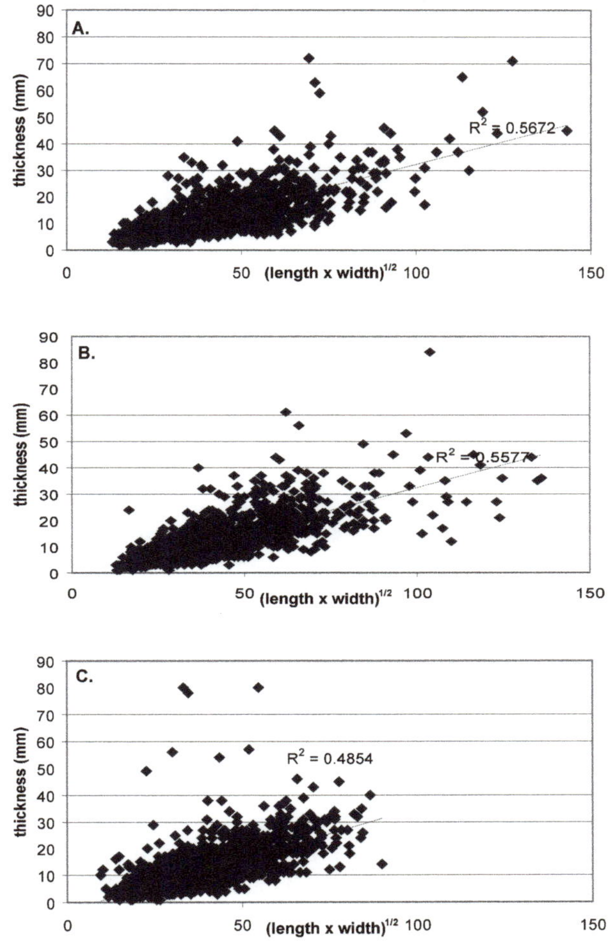

Figure 9.1: A. Reduction Charts of the Complete Flakes at Quarry 27 B. Area B at Quarry 35, C. Area A at Quarry 35

Blades also have a longitudinal ridge formed from the removal of two previous flakes (Whittaker 1994:219-221). Consequently, blades should have evidence of two parallel flake scars on their dorsal surface. Blades are usually struck, but not always, from the same direction and the same platform, and hence come from Quadrant 1. The frequency and direction of the flake scars can be used to establish whether a blade reduction strategy dominated the quarry assemblages. Furthermore, the number of scars on a flake has an inverse relationship to the percentage of cortex on the dorsal surface (Andrefsky 1998:104-106). The complete flakes with a cortical dorsal surface are the first to be removed from the core and the frequency of these can be used as an indicator of the amount of core reduction (Magne and Pokotylo 1981; Odell 1989:185; Mauldin and Amick 1989:70). By comparing the frequency of the complete flakes with a cortical dorsal surface with the number of flakes with one or two scars from Quadrant 1 a significant difference was found between the two quarries (χ^2=27.98, df=2, p<0.0001, phi=0.111961). More complete flakes in Quarry 27 have their dorsal surfaces completely covered in cortex than in Area A and Area B of Quarry 35 (Table 9.6). In contrast, more flakes in Area A have two flake scars originating from Quadrant 1 than at Quarry 27. These results suggest that more core reduction and blade manufacturing was undertaken in Quarry 35 than Quarry 27.

The form and size of a flake can be influenced by the amount of force used to remove the flake and the place the hammerstone impacted on the platform of the core (Hiscock 1986a:49). Small or focalised platforms, created when the core is struck near the edge with a controlled amount of force, are generally more common in blade technologies (Hiscock 1988:371; Fullagar and Torrence 1991:133-134). The area of the platform, gained by multiplying the platform length by the thickness, was used to measure and compare the platforms on the complete flakes at the two quarries. The size of the platforms in Area B and Quarry 27 are larger and more variable, shown by the high standard deviations, than in Area A of Quarry 35 (Table 9.5). A highly significant difference was found between the platform areas in Area A and Quarry 27 (t=14.413, df=3496, p<0.0001). This result is consistent with Area A being an area of blade manufacturing while Area B and Quarry 27 represent locations were initial core reduction and large flake manufacturing occurred.

Whereas core reduction was relatively simple at Quarry 27, the variety of core types, the amount of core reduction, the manufacture of blades and tools has created a far more complex assemblage composition at Quarry 35. Only a small amount of spatial variation in the distribution of the artefact characteristics occurred at Quarry 27. In contrast, the reduction strategies employed, the quarry size and density of the artefacts at Quarry 35 emphasised the locations where different stages of reduction occurred.

9.3.5 Tool Manufacture, Use and Discard at the Quarries

At both Quarry 27 and Quarry 35 discarded tools made from non-local materials were found in the assemblages. These tools may reflect retooling, tools being discarded and replaced at the quarry, or random, accidental losses (Gramly 1980; Shott 1989:17-19).

Most of the tools at Quarry 35 and Quarry 27 represent the expedient use of exhausted cores, large broken or complete flakes and weathered spalls that were retouched. The presence of retouched cores (core tools) was also identified in a quarry assemblage along the Murchison River, Western Australia (Byrne 1980:116). The concentration of tools in Area A of Quarry 35 highlights the expedient use of tools at a quarry. The large number of tools in Area A was surprising and hints at the presence of considerable heavy-duty woodworking being undertaken at Quarry 35. In this location, one in every three complete flakes was a tool (Table 9.1). Although most of the tools in Area A are large with notch or scraper retouch, tools broken or abandoned during manufacturing were identified by the appearance of partially backed points and blades in the assemblage.

9.4 LANDSCAPE FEATURES THAT INFLUENCED THE SPATIAL DISTRIBUTION OF THE ARTEFACTS IN THE QUARRIES

In the past quarry studies have often failed to place the quarry within a landscape or geological context, account for the distribution and abundance of the available sources or the quality of the material, and define the strategy of procurement (Bamforth 1992:131). The placement of quarries can be better understood by first gaining an understanding of the regional geology and the properties of the available stone (Doelman et al. 2001).

In Chapters 3, 6 and 7 it was shown that various landform features influenced the distribution of the artefacts at Quarry 27 and Quarry 35. Common features associated with these quarries include the outcrop (or stone source), landform surfaces and a changing topography. How these features have influenced the placement of different stages of reduction at the quarries are summarised below.

It has been argued that spatial differentiation of core reduction activities is equated to the amount of specialist production undertaken at a quarry (Torrence 1984, 1986; Fullagar and Torrence 1992:135). If a high degree of segregation in different reduction activities occurred at a quarry then this may have been due to specialist knappers who were highly organised and controlled the working and distribution of stone from the quarry. Hence, a better understanding of the ways people organised their "lithic production system" (Ericson 1984) can be gained by studying the spatial variation of artefacts in a quarry in order to determine how the quarries were used as resources.

Location	100% Cortex		1 scar from Quadrant 1		2 scars from Quadrant 1		Total
	Count	%	Count	%	Count	%	
Area A	71	3	679	26	884	34	2575
Area B	51	3	299	18	471	29	1637
Quarry 27	58	5	289	24	298	24	1221

Table 9.6: The Number of Complete Flakes in Area A and Area B of Quarry 35 and Quarry 27 with a Cortical Dorsal Surface and One or Two Flake Scars Originating from Quadrant 1

9.4.1 The Outcrop

The distribution and types of the available stone resources was understood by placing the study area within its geological context. Silcrete dominated the landscape in the study area. An understanding of how silcrete outcropped allowed a targeted surveying strategy to be employed. Hence, the type and position of the outcrops influenced the location of the quarries. Silcrete outcrops were only visible, due to erosion, along the stream gullies or gorges and scarp edges of the Mt Wood Hills. Three forms of silcrete, columnar, concretionary and massive, outcrop in the Mt Wood Hills (Doelman *et al.* 2000; Tait 1998). Silcrete generally incorporates a lower columnar horizon, up to 2 m thick, overlain by a concretionary layer (Watts 1974, 1978; Wopfner 1978; Tait 1998); in relatively few places it has a more massive appearance. Of the numerous outcrops of silcrete in the study area, only those with a massive appearance were quarried; columnar and concretionary outcrops were untouched.

At a finer scale of observation the outcrops at Quarry 27 and Quarry 35 were a major "pull" feature that influenced where different stages of reduction occurred. At both quarries a distance-decay trend was observed in the size and characteristics of the artefacts. Hence, cores underwent further reduction as the distance increased from the outcrops. The frequency of the different termination types, flake forms, amount of cortex and raw material types changed as the distance increased from the outcrop (Tables 6.17, 7.9). Most of the large cores and flakes, indicative of early reduction, were discarded on or near the outcrops. In contrast, the heavily reduced cores and smaller flakes were worked and discarded away from the outcrops. The artefacts in Area B of Quarry 35, which was solely on the outcrop, reflected initial core reduction. In contrast, Area C encompassed part of the outcrop and the analysis of the artefacts in this sampling area reflected a mixture of different reduction strategies. The characteristics of the cores and flakes in Area A, located away from the outcrop indicated further reduction, blade manufacture and tool use.

9.4.2 The Cobbles

Cobbles were often "tested" to assess the quality of the material. Evidence of material testing and selection was closely associated with the presence of outcropping cobbles. As Area C partly encompassed the cobbles of Outcrop 1 at Quarry 35 an overprinting of different stages of reduction was observed within the sampling area.

It is likely that already exposed cobbles at the outcrop were targeted for reduction due to their smaller size and manageability. There was no need to extract the exposed cobbles from the surrounding outcrop and colluvium. However, at Quarry 27 flakes, indicative of material testing, were removed from the large, buried cobbles. Due to the nature of the outcrop, less fully exposed cobbles were available at Quarry 27 than at Quarry 35. The ease of material extraction at Quarry 35 may be another reason why this outcrop was more fully exploited, reflected in the density of the artefacts, than the outcrop at Quarry 27.

9.4.3 Topography

The topography was not correlated with the distribution of the artefacts at Quarry 27 due to its gentle slope gradient. At Quarry 27 most of the flaking occurred on the western slope, below the outcrop. Little reduction occurred on the flat, upper area above the outcrop and only a small number of artefacts were discarded along the creek to the north. Only on the boundary of the southern transect, also along the creek, the density of artefacts increased.

At Quarry 35 the area of greatest slope gradient occurred on the outcrop and below on the scarp face towards Stud Creek. In this area, particularly on the outcrop, evidence of initial extraction and reduction occurred. After large flakes or prepared blocks were produced they were removed from the outcrop and most were transported to the flat, upper area north of the outcrop. In this location most of the blade manufacturing was undertaken. The height of the upper quarry area allowed the entire catchment to be viewed but also afforded a better surface on which to knap. These two factors in combination may have influenced the distribution of different stages of reduction at the quarry.

The influence of topography on the distribution of different stage in core reduction has been noted in other quarry studies. Three 'activity areas' were distinguished for the Robesonia quarry in south-eastern Pennsylvania; an extraction and preliminary block reduction occurred near the outcrops and later stages of core reduction was undertaken on the ridge above the outcrop (Snethkamp *et al.* 1982, described in Arnold 1987a). Torrence (1986) noted that the final working of quarry products was sometimes conducted on hilltops as these granted a suitable lookout.

9.4.4 Landform Surface

The gibber pavement influenced the distribution of the artefacts at both Quarry 27 and Quarry 35. The gibber pavement formed a natural boundary to the artefact distribution at Quarry 35 and Quarry 27 as only a few artefacts were found on this landform surface. Most of the flaking debris at Quarry 35 occurred on the flat, sandy depositional unit north of the outcrops. A small number of artefacts were recorded in the transects which cross the gibber pavement at Quarry 27. It seems likely that the gibber formed an unsuitable surface on which to knap or cross. Most of the artefacts found in the transects were found either in the transects along the creek (seen at Quarry 27) or towards the creek (seen at Quarry 35). The higher density of artefacts in these transects suggests that the creeks formed the transport routes by which people repeatedly travelled to and from the quarries.
The presence of streams and vegetation, providing water, shade and wood resources, near the quarries, following similar arguments made by Reeves (1997), may also have influenced decisions to quarry a particular outcrop.

9.5 TIME TO QUARRY

In this section an attempt is made to tease apart the influence of time on the formation and composition of the quarry assemblages. Time at the quarries was defined as having three scales; Scale 1, the length of time each individual artefact had within the cultural system; Scale 2, the time invested in core reduction and preparation and

Scale 3, the total time spent at the quarry (Table 9.7). By comparing these three scales further similarities and differences between the two quarry assemblages were identified that reflected the reduction strategies employed and the duration of occupation. Scale 4 acknowledges how the occupation of the study area may also have influenced the formation and composition of the archaeological record.

Measuring the Scales of Time

- Scale 1 - Time at the scale of the artefact
- Scale 2 - Time invested in core reduction
- and preparation
- Scale 3 - Time spent at the quarry
- Scale 4 - The occupation of the study area

Table 9.7: Scales of Time

9.5.1 Scale 1 - The Life Span of an Artefact

Gould (1977), expanding on Schiffer's model of the life Gould (1977), expanding on Schiffer's model of the life history of an artefact, compared the life history of quarried and non-quarried stone in the Australian context. The complexity of Gould's system suggests that quarried stone spent more time in the system in comparison to non-quarried stone. By using Foley's (1981a:167) scheme for stone moving through a cultural system a much more precise comparative picture of stone use and time within the system can be identified. Foley (1981a) observed seven stages at which material was discarded from the system: extraction debris, raw material lost, raw material abandoned, manufacturing debris, preparation debris, repair debris and artefact discard. At each stage more time was accumulated in the cultural system. The smallest duration of an artefact's life span within the cultural system occurred during procurement while the longest artefact time may be a tool that was transported, modified and repaired and then discarded. By placing an arbitrary time scale on the systemic formation process of an artefact a relative classification of an artefact's time in the cultural system can be established.

The life span of an artefact is dependent on the form, time of discard, time produced within the manufacturing process and place of discard. At a quarry typically large amounts of flaking debris occur through the processes of core preparation and reduction. Furthermore, the formation of a quarry assemblage was created from a complex amalgamation of many overlapping knapping events from repeated visits.

Most artefacts discarded at a quarry have a short life span in the cultural system. Artefacts produced during material testing were discarded immediately and would typically include large, heavy, cortical cobble cores found on or near the outcrops. Testing and initial core decortification would commonly produce large numbers of flakes with cortex on their dorsal surfaces, terminations or platforms. These flakes were found in close association with cobble cores and large uni-platform and bi-platform cores at both Quarry 27 and Quarry 35.

Cores at Quarry 35 were transported away from the outcrop and prepared as blade cores. Some of the blade cores were removed from the quarry while others were used to manufacture blades at the quarry. These cores have a longer life span in the cultural system but most of the flakes removed from them were immediately discarded during blade core preparation and blade manufacture, and only a few were selected as tools. Some tools were used and immediately discarded and others were fashioned and removed from the quarry. The blade cores, fashioned tools, and flake blanks removed from the quarry were transported throughout the landscape accumulating still more time in the cultural system. In addition, a spent and discarded core can also re-enter the cultural system as a core tool.

Discard processes of tools, as defined by Shott (1989:171), occur as six major times; breakage during production, abandonment during or after production, tool breakage in use, abandonment in use and depletion. Depleted or wasted tools are discarded at a quarry and have been considered examples of retooling (Gramly 1980). The longer a tool stays in the system the more chance it will be spent and then discarded. The best example of time invested in a tool type would be a adze flake (tula and burren) which were hafted, continually resharpened and transported until only a slug remained and the tool was discarded (Hiscock and Veth 1991).

9.5.2 Scale 2 - Time Invested in Core Reduction and Preparation

Often a measure of volume or rates of production output have been used to reconstruct the amount of material removed from a quarry and determine the importance of the quarry in the wider settlement-system (e.g. Soles 1983; Gramly 1980, 1984; Root 1992; Arnold 1987a 1987b, Torrence 1982, 1984, 1986; Singer and Ericson 1977). It may be more helpful to view the quarry assemblage as a product of the time invested in core reduction and preparation. The rate of production can be influenced by either the number of people quarrying or how often the quarry was visited. Furthermore, the difficulties of measuring the potential amount of available material, the volume reduced and the volume removed is dependent on the sample size, amount of "on-site" reduction, the product removed and the reduction strategies employed at the quarry.

The results of the core analysis from the quarries in the previous chapters indicates a relative timeframe which spans from the short use-life of a cobble core to the procurement, preparation, transport, manufacturing and final discard of spent, blade cores. The use-life of a core reflects the amount of reduction a core underwent, the amount of labour invested in reduction and reflects the time spent in the cultural system before it was discarded. In this section the core types are used to express time at a location, as different core types also have a greater or lesser time within the cultural system (Figure 9.2).

Time Frame Based on Core Types

Time ↑
Blade cores
Multi-platform
Bi-platform
Uni-platform
Cobble core

Figure 9.2: A Relative Timeframe of Core Use-life

By giving each core a relative life-span in the systemic context and comparing the ratios of the different core types at a quarry the results can be related to the time spent at each location. The low number of blade cores at Quarry 27 indicates less time at the location and time invested in reduction. This pattern was supported by other characteristics of the cores that reflect the amount of reduction, such as the amount of cortex on the surface of the cores, the number of flakes removed and size of the cores (Table 9.3). In contrast, the high number of blade cores reflects a greater time invested in preparation and reduction at Quarry 35.

By comparing the results from the three sampling areas within Quarry 35 to Quarry 27 a relative time scale can be identified for the time at each of the quarries. For arguments sake, if a relative time unit (rtu) can be applied to the core types that reflects the amount of core preparation and reduction (manufacturing time and time in the cultural system) then a cobble core would have the least amount of time and a blade core more time in the cultural system (Figure 9.2). A cobble core could equal one unit of time while a blade core may equal five units of time. By looking at the frequency of the different core types in each of the quarries it is possible to gain an idea of the time involved in reducing cores at each of the quarries and the time spent at the quarry. Although this comparison is simplistic the results do give the expected outcomes (Table 9.8). Since more blade cores are present in the Quarry 35 assemblage, more time was invested in their preparation and reduction, hence more time was spent at Quarry 35. It must be remembered, that the cores discarded at the quarries may only represent a portion of the cores produced since cores, particularly prepared blade cores, were often removed from a quarry. This scenario is especially relevant to Quarry 35 where blade cores from this quarry were found in the USC unit.

The amount of cortex can also be related to the time invested in core reduction. Typically flakes with no cortex are produced later in core reduction (Andrefsky 1998:102-104). The higher the frequency of flakes with no cortex the more core reduction has occurred. The results show that relatively less time was invested in reducing the cores at Quarry 27 and Area B of Quarry 35 while more time was invested in core reduction at Area A and Area C of Quarry 35.

9.5.3 Scale 3 - Time Spent at the Quarry

Variation between assemblages is not only the product of interactions between activity and post-depositional formation processes (as argued by Schiffer (1975) but also the occupation span at a particular place (Shott 1997:198). The underlying premise of Shott's argument is that two occupations, which are identical in all formational respects, can produce a very different assemblage composition based on the occupation span. This observation was called the "Clarke Effect" by Schiffer (1975:265).

Shott (1997) used the diversity of tool types in an assemblage as a measure of occupation time based on the premise that different tool types have different life-spans in the cultural system. Shott (1997:206) argued that it was possible to characterise the composition of an assemblage by comparing the total assemblage size to the diversity of the tool types. This approach is based on the premise that the assemblage composition is dependent on the duration of occupation at a place. If the use-life of a particular tool type is constant, than the relationship between tool number and discard rates can be simply expressed as Figure 9.3. This figure shows that as discard rates increases the frequency of different tool types also increases exponentially until stability is reached and this relationship was related to the degree of mobility. If mobility is high, then tool diversity is low, while the opposite trend occurs for a low mobility (Figure 9.3). This argument implies that a place was occupied for a long duration will have high tool diversity (Shott 1986:25).

		Quarry 27		Quarry 35	
Core Type	RTU	Count	Count*RTU	Count	Count*RTU
Cobble	1	34	34	136	136
Uni-platform	2	76	152	2154	4308
Bi-platform	3	54	162	518	1554
Multi-platform	4	24	96	145	580
Blade Core	5	2	10	479	2395
Total RTU			454		8973

Table 9.8: The Relative Time Units Invested in the Reduction of Cores at Quarry 27 and Quarry 35

		Quarry 27		Quarry 35	
Cortex	RTU	Count	Count*RTU	Count	Count*RTU
100%	1	58	58	72	72
50-99%	2	256	512	1048	2096
1-49%	3	560	1680	279	837
0%	4	347	1388	1170	4680
Total RTU		3627	3569	8477	7685

Table 9.9: The Relative Time Units Invested in Core Reduction Indicated by the Amount of Cortex on the Complete Flakes in the Sampling Units of Quarry 35 and Quarry 27

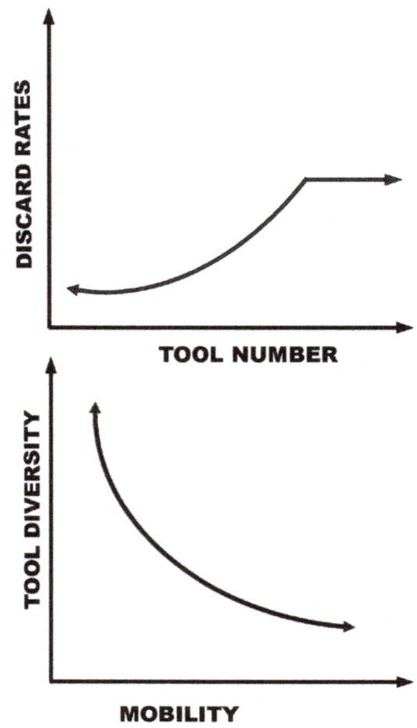

Figure 9.3: The Relationship between Discard and Tool Number (top) and the Relationship Between Tool Rates, Diversity and Mobility (bottom)

Distribution analysis was used by Shott (1997) as a method to characterise the time spent at a location based on the frequencies of different tool types within the assemblage. Shott (1997:206) defines distribution analysis as measures of tool richness, evenness and heterogeneity (defined in Table 7.10). The same method, using Holdaway and Frankel's *Diverse* program (1993), was used to compare whether the diversity of the different types of tools and cores can be used to make inferences about the time spent at the two quarries.

The results show that Quarry 27 has a lower diversity of core types than Quarry 35, indicating that the frequency of the different core types was more evenly spread at Quarry 27 (Table 9.10). Shott (1997) argues that an increased 'richness' equals an increased time of occupation at any particular place. The 'richness' values also indicate that more time was spent at Quarry 35. A similar pattern emerges for the tool types. While the heterogeneity figures are similar for the sampling locations in Quarry 35 and tools in Quarry 27, the 'richness' figures give very different results. It seems likely that more time was spent in Area A than at the other sampling locations in Quarry 35 or at Quarry 27.

If the richness of the different tool types in an assemblage is dependent on the size of the assemblage, and the size of the assemblage is dependent on the time of occupation, then richness values can be used to compare the time involved in the formation of the assemblages. Bootstrapping is a method used to measure richness in an assemblage by repetitively resampling of an assemblage at regular intervals until the total population has been sampled. This method takes into account assemblage size and 'evenness' (refer Table 7.10). The mean 'richness' values calculated for each sample size can be plotted on a graph (refer to Table 7.10). Depending on the resulting curve the time span of occupation can be inferred as "assemblages differ in size and composition purely from a relationship between tool use lives and settlement occupation spans" (Shott in press). A bootstrapping program written by Holdaway (1998) based on the formulas used by Lipo *et al.* (1997) compares the different types of cores, rather than the tool types, at Quarry 27 and Quarry 35.

Lipo *et al.* (1997:316) defines an assemblage with a high 'richness' of different artefact types as having a Type I curve (Figure 9.4). The curve is initially steep but quickly reaches a stable plateau before the final sample size is calculated indicating that a considerable amount of time was accumulated in the formation of the assemblage. A similar profile occurs for Type II curves but the curve's gradient is less defined and more gradual (Figure 9.4). In comparison, a Type III curve gradually inclines and does not plateau or reach stability before the final sample size is measured indicating that not enough time was accumulated for the assemblage to become stable (Figure 9.4).

A bootstrapping test was undertaken for the core types at Quarry 27 and those in the sampling areas of Quarry 35. The results show that for all the sampling areas in Quarry 35 Type I curves were calculated (Figure 9.5). These curves suggest that considerable time was spent at the quarry, either as a series of repeated visits or as an extended episode of stone working, to give high richness values. In contrast, a Type III curve was calculated for the frequency of the different core types at Quarry 27 indicating that it was still in the process of 'becoming' stable, having low richness values, and had not accumulated much time in the formation of its assemblage. However, these results may simply be dependent on the different types of reduction strategies employed at the two quarries. Indeed, Quarry 27 may have been visited many times but the simplistic nature of the core reduction (refer 9.3.4) means that the richness of the core types is low.

	Richness	Evenness	Heterogeneity
Core Types			
Quarry 27	0.441	0.749	0.711
Quarry 35	0.141	0.616	0.567
Tool Types			
Quarry 27	0.714	0.772	0.756
Area A	0.280	0.581	0.699
Area B	0.587	0.774	0.745
Area C	0.475	0.684	0.681

Table 9.10: The Diversity Results for the Different Tool and Core Types at Quarry 35 and Quarry 27

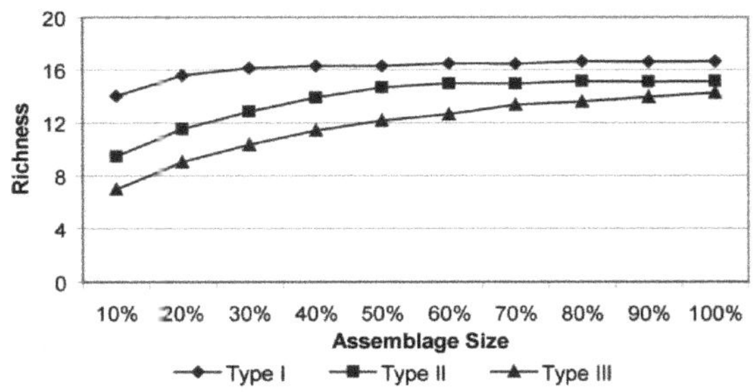

Figure 9.4: Types of Curves that Represent the Formation of an Assemblage

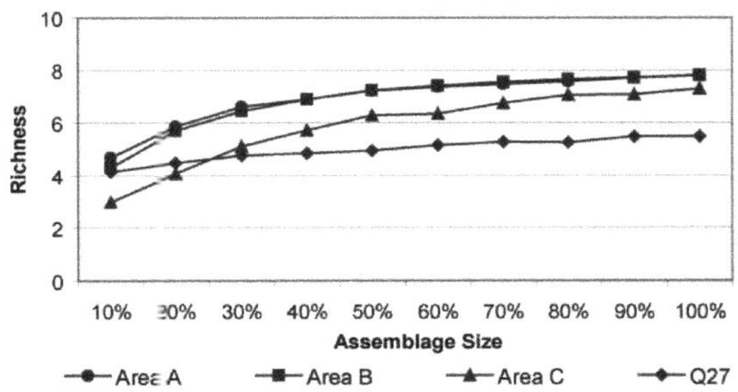

Figure 9.5 Results of a Bootstrapping Analysis of the Cores in Quarry 35 and Quarry 27

The size of the two quarries and the density of the artefacts also reflect the manufacturing time invested at the quarry. Quarry 35 is considerably larger than Quarry 27 with approximately 700 000 artefacts present with a density of ~190 artefacts per m^2 based on the three sampling locations. Only 3384 artefacts are present at Quarry 27 with an artefact density of ~7 per m^2.

At Quarry 27 the number of tools was low but discard rates of the manufacturing debris are high. The lack of strong spatial patterning at Quarry 27 and the presence of individually defined knapping events, based on the number of refits, indicates that little overprinting from multiple visits occurred. The low density and artefact number also indicates that little concentrated use of the outcrop occurred. These results may indicate that Quarry 27 was used for a short time. A far more complex picture emerges from Quarry 35. The high number of tools in Area A, the diversity of core types and the amount of time invested in core preparation and reduction indicates that repeated visits probably occurred. The manufacturing time and tool use suggests that during each visit a longer time was spent at Quarry 35 in contrast to Quarry 27.

9.5.4 Scale 4 - Wider Implications to the Time of Occupation in the Study Area

T The time of occupation at the Stud Creek catchment falls into the late Holocene period (Reeves 1997, Holdaway *et al.* 2000b, in press). Simulation studies based on radiocarbon dates obtained from heat retainer ovens statistically implied that occupation at Stud Creek was episodic in nature with two distinct periods of occupation (Reeves 1997:97). The model he used to explain this pattern was grounded in the presence/absence of floral resources. Plants are a critical resource providing food, fuel and material for tools (Allen 1972; Gould 1980; O'Connell *et al.* 1983). By using an ethnographic analogy as a heuristic device Reeves (1997:105-107) argued that once floral resources were stripped a location would no longer be suitable for occupation. Being an arid environment revegetation of an area is particularly difficult as a combination of different environmental and climatic conditions are necessary in order for seeds to germinate (Cunningham and Walker 1973). Furthermore, if the tree density is thin or nonexistent then only a few seedlings regenerate (Burrows 1973:57). The difficulty in re-establishing floral resources resulted in areas being abandoned for long periods of time, until favourable conditions occurred and reoccupation could again take place (Reeves 1997:108-109).

The building of heat retainer ovens is associated with cooking meat and vegetables (Roth 1897:104-105; Spencer 1914:27; Brock 1975:25). Spencer (1914:27) observed ethnographically that ovens were only built when a place was occupied for a considerable duration. The movement of most faunal species in the arid zone is nomadic prompted by rains and favourable feeding conditions. The population numbers in Sturt National Park increase or decrease depending on the amount of water and feed available causing animal migrations and increased breeding (Denny 1975). Favourable environmental conditions do not occur very often in northwestern New South Wales. Only a rare set of climatic circumstances brings rain to the area and provide enough water to reactivate the ephemeral creek systems (Reardon 1995:76-77). Combining these variables, floral and faunal resources, water availability and the presence of people limits the likelihood that a combination of these conditions occur together at one time and at one place in the landscape. Such a scenario may have occurred in the Stud Creek catchment. The combination of 'ideal' environmental conditions may have prompted the reoccupation of the catchment.

The occupation in the Stud Creek catchment was highly visible due to the density of artefacts and the presence of hearths but the actual formation of the assemblage was probably rapid. Previous research has used an increase in the number of 'sites,' artefact discard rates and grinding stone implements post 2000 BP to support the notion of intensification in the arid zone of Australia (Smith 1986, 1988, 1996; Ross *et al.* 1992). It seems likely that the Stud Creek catchment was occupied throughout a considerable period of time (~1500 years) but in a series of "hits" where resources were used and reduced until occupation was unfavourable and the catchment abandoned. This does not indicate that the whole region was abandoned but small, localised areas were disused for long periods while other areas were exploited. At this scale it is impossible to view whether a period of intensification, as described by other archaeologists (e.g. Lourandos 1983, 1984, 1985a, 1985b; Ross 1981, 1985; Ross *et al.* 1992; Lampert and Hughes 1974), occurred or not in the region without further radiocarbon dating of heat retainer hearths over a wider area. However, it seems likely that the technological strategies employed for the acquisition, reduction and use of stone obtained from quarried and non-quarried stone sources led to a rapid and dense accumulation of artefacts in places that were considered favourable for occupation. These locations may only have been favourable for a short time causing the continual relocation of residences and increasing the distribution of the archaeological record. In addition, occupation may simply have become much more visible within the landscape (*cf* Hiscock 1986a) during the late Holocene, after the advent of the 'Australian Small Tool Tradition', and employing this technology may also have aided in the exponential rise of quarrying during this time.

At a larger time scale of occupation within the eastern margin of the arid zone in Australia it tentatively appears that the area was occupied in the Pleistocene, abandoned and then reoccupied during the late Holocene. Radiocarbon and thermoluminescence dating of archaeological deposits in the area indicates that occupation of the area fell within the last two thousand years (Lance 1992; Reeves 1997; Holdaway *et al.* 2000a). A hearth at Lake Yantara, 60 km south of the study area, was dated to 22 000 BP (Dury and Langford-Smith 1965). These results may simply reflect the lack of research undertaken within the area or hint at similar patterns of abandonment and reoccupation, in tune with climatic conditions, already observed in the arid zone of Australia (e.g. Veth 1989, 1993, 1995; Smith 1986, 1988, 1989, 1993, 1996). Again, without further research within the area any conclusions remain purely speculative.

9.6 THE ROLE OF THE STONE RESOURCES IN THE STUDY AREA

A comparison of the types of stone resources, quarried outcrops and gibber cobbles, identified in the study area was undertaken to establish the 'role' of these sources in the organisation of lithic technology. Quarries have been widely recognised in the literature as being important in defining prehistoric organisational patterns (Ataman *et al.* 1992; Hiscock 1984, 1988, 1993; Bamforth 1990, 1992; Root 1992; Torrence 1986; Ericson 1984). Understanding how a quarry was used can provide useful insights into how people in the past moved across a landscape, occupied the landscape, traded goods or organised their labour force (Ataman *et al.* 1992:2). Although the study area does not encompass a region, by covering the entire distribution and discard of artefacts from the quarries found in the Mt Wood Ranges, we can gain some idea of how the quarried stone was used by comparing the artefacts at the quarries and landscape sampling units. In addition, studies in Australia have shown that the greatest amount of change in the number, size and form of artefacts occurs within one to five kilometres of the stone source and inversely relates to the increase in the number of retouched tools as distance increases from the source (e.g. Hiscock 1986b; Byrne 1980).

Information gained from the analysis of the landscape sampling units is used to further differentiate the 'role' of quarried and non-quarried stone sources in the organisation of lithic technology. To distinguish the 'role' of the quarry we must recognise the reduction strategies employed at the quarry, identify the form material was removed from the quarry and determine how the stone obtained at a quarry was used in the landscape sampling units. Needless to say, the 'role' of the stone resource is used as a term to describe patterns of acquisition, reduction and use that occurred in an envelope of time – the last 4000 years of occupation within the study area.

Arnold (1987a:109) argued that quarries could be classified based on the patterns of behaviour represented within their assemblages. Quarry 25 and Quarry 27 were similar in terms of the type of outcrop and the types of available material, yet they were vastly different in the size and artefact density, degree of spatial variation, the composition of the assemblages, the amount of "on-site" core reduction and the products manufactured and removed from the quarries. Why do these differences occur? How does the use of these two stone sources compare to the use of gibber cobbles in the landscape sampling locations? To answer these two questions an overview of each stone resource is presented which describes how stone from each source was procured, manufactured, distributed and discarded in the study area.

9.6.1 The Role of Gibber Cobbles as a Stone Resource in the Organisation of Lithic Technology

In Chapter 8 the use of the gibber pavement as a stone In Chapter 8 the use of the gibber pavement as a stone resource was examined at the Upper and Lower Stud Creek landscape sampling units. The results indicate that approximately one third of the artefacts found in these locations were manufactured from gibber cobbles.

Most of the cores in the USC unit were made from gibber cobbles (60%) and a large number of these were made from medium clast silcrete (89%). In contrast, only a small number of the cores with a gibber cortex were made from Group III silcrete (~10%). These results indicate that medium clast silcrete is the dominant silcrete type within the gibber pavement. The cores made from the gibber cobbles represent a random sample of the proportion of the different silcrete types found in the pavement since there is no way of viewing the quality of the material without removing the outer cortex. The characteristics of the cores with a gibber cortex indicate that they underwent only a limited amount of reduction. These cores are large, have only a few flake scars and have large amounts of cortex on their surfaces. Hence, the gibber cores maintained their original cobble form. As the cobbles were completely covered in cortex material "testing" was often undertaken, seen in the presence of cobble cores, to assess the quality of the underlying material.

The use of the gibber pavement as a stone source is clearly reflected in the core types. Most of the cores in the USC unit are uni-platform, bi-platform, multi-platform and cobble cores and have a gibber cortex. These cores are generally larger than the quarried cores indicating less reduction and an early discard. In addition, the expedient use of gibber cobbles to manufacture large flakes is shown by the long length of the flake scars on the cores made from gibber cobbles.

Of the complete tools made from gibber cobbles in the USC unit, 58% were made from medium clast silcrete, a considerably smaller proportion than the number of cores with a gibber cortex (89%). This result indicates a selection of Group III gibber cores for tool manufacture, resulting in the further reduction of these cores.

Analysis of the Stud 1 assemblage by Cochrane (1998:78-80) suggests that gibber cobbles were used as a source of stone to produce expedient tools. On the gibber pavement the tool types are generally unmodified flakes with usewear or have notched retouch. Similarly, the cores and complete flakes are larger in size, have highly cortical surfaces and large numbers of cobble cores are present on the gibber pavement in comparison to the artefacts on the adjacent lagged surface. The frequency of these characteristics diminished in the assemblage as the distance increased from the pavement (Cochrane 1998). Hence, the gibber pavement at Stud 1 influenced the distribution of the artefacts within the assemblage.

Although no gibber pavement was surveyed at Stud 2 a similar pattern emerged for the use of gibber cobbles. Consequently, the degree of gibber use can be inferred – large gibber cobbles were selected, a large flake was removed, used and then discarded on or near the gibber pavement. The presence of a readily available, highly abundant source of stone (gibber cobbles) meant that there was no need to conserve material and allowed an expedient technological strategy to be employed.

It seems likely that gibber cobbles were used as a secondary source of stone to the material acquired from the quarried outcrops. Consequently, the procurement of gibber cobbles (and creek cobbles) was "embedded" (Binford 1973:273-275, 1979) in other activities, such as residential activities, suggested by the numerous hearths and the presence of grinding equipment. The location of the landscape sampling units also provides suitable topography, vegetation and water sources for a more permanent occupation. These findings support Gould's (1977) earlier observation that artefacts of non-quarried stone are uncommon at occupation areas unless there is a nearby surface [gibber] source. It seems likely that the gibber pavement was used as a local, convenient source of stone for ready tool use.

Gibber cobbles were readily available at the landscape sampling locations and required a low investment of time to manufacture a tool that could cater for an immediate need. Ethnographic accounts describe the importance of having a sharp cutting edge (Binford and O'Connell 1984:415; Horne and Aiston 1924:91). Horne and Aiston (1924:91) described any flake with a sharp edge as being a "causal stone" which were used until the edge became blunt and then discarded.

Various behavioural implications can be inferred from the use of the gibber pavement as a stone source. The high number of tools with usewear damage on the gibber pavement indicates that these tools have a short use-life. At the landscape sampling units tools made from the gibber cobbles were made, used and discarded in the same context. However, ethnographic evidence suggests that unmodified flakes were used as tools and sometimes maintained a long time in the cultural system (Hayden 1977:179), although this scenario is probably the exception rather than the 'rule'. Since the aim of reducing a gibber cobble was simply to manufacture a large flake, little effort was required as material was plentiful and could be used expediently or "informally" (*cf* Andrefsky 1994a:30). Wilke and Schroth (1989:171), using the term defined by Gould (1977:164), described the use of similar stone as sources of "ephemeral stone acquisition" meaning tools were made quickly and used to satisfy the "needs of the moment".

The continuous and/or repeated use of the gibber pavement as a stone source created subtle patterns in the distribution of artefacts within Stud 1. Cochrane (1998) applied the contingent/necessity arguments of Shermer (1995) to the results of his analysis at Stud 1. Shermer (1995:69-73) defined a necessity as a "constraining set of circumstances compelling a certain cause of action," whereas a contingency is a "conjunction of events without design" with both factors interacting to form the archaeological record. The use of gibber cobbles can be considered a "contingency". By its very nature it does not constrain behaviour but simply produces behaviour, as it was a source of stone used opportunistically when required by people who repeatedly occupied the location. Artefacts made from gibber cobbles have a spatially redundant, short and simple life-span within a cultural system. The gibber pavement represents an easily accessible stone source for immediate use. Material was tested, suitable cobbles were selected and large flakes, with usewear or notched retouch, were used in an expedient manner involving minimal labour costs to manufacture. Repeated use of the gibber pavement as a stone source has emphasised patterns of stone procurement and use in the landscape sampling units. However, the gibber use has far wider implications as it not only influences the composition of the assemblage at each location but also structures the distribution of the artefacts within each landscape sampling unit.

The discard of artefacts made from gibber cobbles accumulated over a maximum time span of ~4000 years (Holdaway *et al.* 2000a:17). As a result, the assemblages at the landscape sampling units are "time-averaged" (Stern 1994:101, 1995:49-50). Stern (1994:101-102) has argued that due to the complexity of formation of archaeological assemblages over many different time spans and processes of artefact discard, human behaviour can in no way be considered constant or unchanging.

However, as the gibber pavement was a stable resource it structured the discard patterns of material obtained from it and created stability in the long-term use of a small area of the landscape.

9.6.2 The Role of Stone Acquired from Quarry 27 in the Organisation of Lithic Technology

Due to the small size of Quarry 27, the low density of Due to the small size of Quarry 27, the low density of artefacts and the limited amount of core reduction little technological variation could be observed in the reduction strategies employed at the quarry. It seems likely that minimal labour input and time was invested in the extraction and reduction of the cores at Quarry 27. The labour and time involved in using the material from Quarry 27 was invested elsewhere, during transport, later stage manufacturing, tool use and resharpening. Hence, fundamental differences in the way stone obtained from the gibber pavement and Quarry 27 occur in the transportation of stone away from the source and the time the products from each of the stone sources had within the cultural system.

The Group III silcrete at Quarry 27 is very high quality and was targeted for extraction over the surrounding lower quality, medium clast silcrete. High quality material, with a homogenous and isotropic structure, was used for tulas, endscrapers or any other tool types requiring fine retouch (Cotterell and Kamminga 1987:677). Based on the results of the landscape sampling units and the removal of large, wide flakes from the quarry it is probable that the material from Quarry 27 was used in such a fashion. Tools, such as tulas, were often transported, curated and discarded throughout the landscape (Hiscock and Veth 1991).

"Lithic raw material prospect" or "assay site" were terms used by Wilke and Scroth (1989:171) to describe a source of stone that was occasionally used in conjunction with the collection of another resource. These sources were generally typified as having a low abundance of low quality material. While the material at Quarry 27 was of extremely high quality the quantity of the material was low and possibly did not warrant extensive exploitation.

The limited degree of spatial segregation in the reduction at Quarry 27 suggests that specialists did not undertake core reduction. Torrence (1984, 1986) argued that the lack of spatial variation in the quarries at the island of Melos was due to the exploitation of the obsidian resources by fishermen who required the stone for their own use. The obsidian was not efficiently reduced, no standardisation of the product was found, and hence, no labour organisation or specialisation occurred at the quarry. If a considerable amount of spatial variation in the assemblage occurred at Quarry 27 then this would be a mark of a more organised, efficient stone procurement system. However, the lack of strong patterns of spatial variation at Quarry 27 may simply represent the small amount of time spent at the quarry or the limited reduction of cores.

The standardisation of the product manufactured at a quarry is another marker used to assess the organisation of production (Torrence 1984, 1986:171-186; Arnold 1987a; Root 1992). Surprisingly, the form of the material removed from Quarry 27 was standardised. Large, wide flakes of a similar size were manufactured and then removed from the quarry. No difference was found in the size of the flake scars on the uni-platform, bi-platform and multi-platform cores. These results indicate that while little effort was invested in preparing and reducing the cores at Quarry 27 a particular product was manufactured indicating that the reduction of the cores was orientated towards a planned goal. Minimal time and labour costs were invested in the reduction of cores to produce the desired product making the organisation of production highly efficient. Core reduction at Quarry 27 involved low costs but had a high return due to the high quality of the material and the large size of the flakes which allowed resharpening to extend their use-lives.

If 'risk' is defined as the probability of failing to obtaining subsistence goals than the technological 'costs' employed to avoid failure can be identified in the ways stone was procured, reduced and used (Bamforth and Bleed 1997:126). Hence, different technological solutions to reduce the chance of failing can result. The solutions employed at Quarry 27 are presented in Table 9.11. Using Bamforth and Bleeds' (1997:127, Figure 7.1) heuristic model, at Quarry 27 procurement costs are relatively low since material is readily available in the Mt Wood Ranges, equally the costs involved in production are also low as manufacturing the large flakes required little skill which limited the failure rates. The application costs of failure (the 'risk' resulting from a tool that does not work when required) are also low as tools are large and made from high quality material which could be continually resharpened to extend its use-life. Furthermore, it is likely that these tools were used for processing rather than being used directly in the acquisition of subsistence resources. Hence, the failure costs of not having these tools may be lower than tools employed in the direct acquisition of subsistence resources.

Quarry 27 may have had little impact on the structure of the technological organisation of the people who inhabited the Mt Wood Ranges during the late Holocene. Although Quarry 27 was a particularly good source of stone, the 'role' of the source was probably an unplanned procurement resulting in an opportunistic exploitation of an available resource. The outcrop of Group III silcrete was not large or located near any "support facilities", a prerequisite to be defined as a quarry by Wilke and Scroth (1989:168-169). While a standardised flake form was produced and removed from the quarry the time spent at the quarry was either a series of short-term occurrences or even a single collection. The composition of the assemblages makes it unlikely that considerable time was spent working the cores during each visit to the quarry or that the quarry was continually visited. However, it seems likely that the quarry was visited more than once due to the presence of blade manufacturing at the outcrop, a strategy seemingly atypical to the rest of the quarry assemblage.

Procurement
• Long use-life of tools
• Tool resharpening to economise raw material use
• Use of durable stone
Production
• Advanced production of tools
• Production at the quarry
• Use of easily workable stone
Application
• Careful design for specific purposes
• Easy repair - standardised form

Table 9.11: The Technological Options used at Quarry 27 to Reduce the Failure Probabilities ('Risks') involved in Procurement, Production and the Tool Application (adapted from Bamforth and Bleed 1997:128, Table 7.3)

9.6.3 The Role of Stone Acquired from Quarry 35 in the Organisation of Lithic Technology

The use of Quarry 35 as a stone resource was far more complex than at Quarry 27. Intensive use of the quarry was observed by the degree of spatial variation found in the distribution of the artefacts which Torrence (1984, 1986) argued was a marker of labour specialisation. Strong spatial patterns of extraction, material testing and initial manufacturing, blade manufacture and tool manufacture were observed in the quarry assemblage. The obvious amalgamation of times both as a result of repeated visits, the length of time spent at the quarry during each visit and the considerable time invested in the manufacturing of blades has added to the complexity of the assemblage but time accumulation has also aided in the spatial recognition of these patterns.

In comparison to Quarry 27, the entire reduction continuum was observed at Quarry 35. The expedient use of large flakes as tools, core tools and spall tools, the discard of expended tools and the manufacture of tools was observed in the assemblage. The discard of these different tool types indicates that a whole range of different behaviours was being undertaken at the quarry. The high number of large, expedient tools with usewear damage, notch or scraper retouch in Area A of Quarry 35 indicates that heavy-duty processing of wood working, possibly to manufacture spear shafts, may have occurred in this area. Gould (1980:128) has observed that spear hafts needed to be replaced every three weeks as the wood becomes dry and brittle. The need to replace spears may have prompted the manufacture of large, expedient tools at Quarry 35.

Quarry 35 was a focal point in the organisation of the stone production shown by the presence of residential support facilities, such as evidence of food processing, seen at the landscape sampling units. The presence of "support facilities" was used by Wilke and Scroth (1989:170-171) to define a typical quarry where the material was seen as valuable enough to warrant a "direct" procurement strategy as originally described by Binford (1979). However, Binford (1979:259) argued that the procurement of stone was always "embedded" in other scheduled subsistence activities unless "*things have gone wrong*". The structure, degree of technological organisation and the complexity of the assemblage formation and composition observed at Quarry 35 suggests that the procurement of stone from the quarry was goal driven. Nevertheless, it is difficult to determine whether the stone source at Quarry 35 prompted the occupation of the Stud Creek catchment or was simply a convenient and readily available outcrop of suitable stone while the catchment was occupied.

Without dating Quarry 35 it is impossible to directly relate the time it was used to the landscape sampling units. However, given the closeness (<2 km) of Quarry 35 and other smaller quarries in the immediate vicinity of the landscape sampling units and the compositions of their assemblages it is probable that material was procured from Quarry 35 and used at the landscape sampling locations. However, statistical analysis of the radiocarbon dates from Stud 1 and Stud 2 suggests two distinct phases of occupation separated by at least 200-300 years (Reeves 1997, Reeves *et al.* 1998; Holdaway *et al.* 2000a:12). Each phase of occupation was probably very short (a decade or less) but several occupations possibly occurred within each phase until the plant resources were exhausted and the area abandoned (Reeves *et al.* 1998). It is possible that these locations were reoccupied ~300 years later without prior knowledge of the area. The implications of these results for the use of Quarry 35 as a stone resource suggests that the mass of cores and flaking debris discarded at the quarry were potentially produced during one or both phases of occupation by a series of repeated visits in a very short time (less than ten years). Alternatively, the quarry was continually use by people who resided elsewhere or did not construct hearths.

Scheduling stone procurement allows a play-off between the costs and benefits of allowing the time to acquire stone and avoiding the 'risk' of not having stone in the appropriate form (Torrence 1989b:64; Elston 1990). The stone at Quarry 35 may have been considered valuable due to the amount of available material and the quality of the material. Although stone is highly abundant as outcrops and as gibber pavements in the Mt Wood Hills away from the hills only gibber cobbles occur on the plains. Even within the Mt Wood Hills not all the outcrops were suitable for quarrying due to the form of the outcrop or the quality of the material (Doelman *et al.* 2001). While silcrete has created the gibber pavement near the Mt Wood Hills towards the contact metamorphic and igneous plutonic zone, surrounding the Tibooburra township, milky quartz cobbles dominate the gibber pavement. It is not until the Grey Range, approximately 25 km west of Tibooburra, that silcrete is again available. Although this it not a large distance, it is possible that

Quarry 35 was not only used to repeatedly "gear-up" for hunting forays while based at Stud Creek but also to retool before leaving the catchment and crossing the plains – although this scenario tends to delve into the realms of a creating an ethnographic reality. Nevertheless, the need to have the appropriate tool at the appropriate time may have prompted the time and labour invested in the core reduction observed at Quarry 35.

It seems likely that Quarry 35, on the whole, represents a place for retooling and as a result the blades and points, which were probably used as projectile points, produced at the quarry were widely dispersed across the landscape. Prepared cores and large flakes were also transported away from the quarry to the landscape sampling units for further reduction or tool use. The high proportion of blade cores, fashioned points, unretouched points, large flakes with a weathered cortex and blade blanks obtained from a quarried outcrop in the USC landscape sample supports the observations made at Quarry 35.

A distance-decay in core reduction could be observed from quarry to sampling locations in the Stud Creek catchment. Even within a small distance (from Quarry 35 to the Upper Stud Creek landscape sample is ~200 m, while from Quarry 35 to the Lower Stud Creek Landscape samples is ~2 km) a considerable decrease in core, scar length (Figure 9.6) and the length of the complete flakes was found (Figure 9.7). The largest cores occur at Quarry 27 and Quarry 35. The USC unit also has a large core size due to the presence of large gibber cobbles found along the creek.

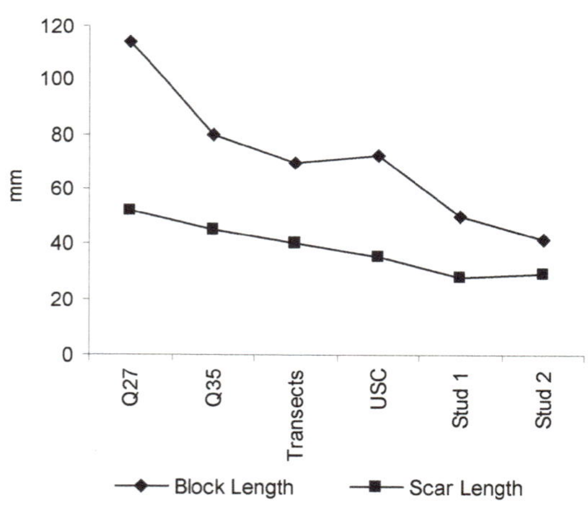

Figure 9.6: The Block Length and Scar length of the Cores Recorded in the Study Area

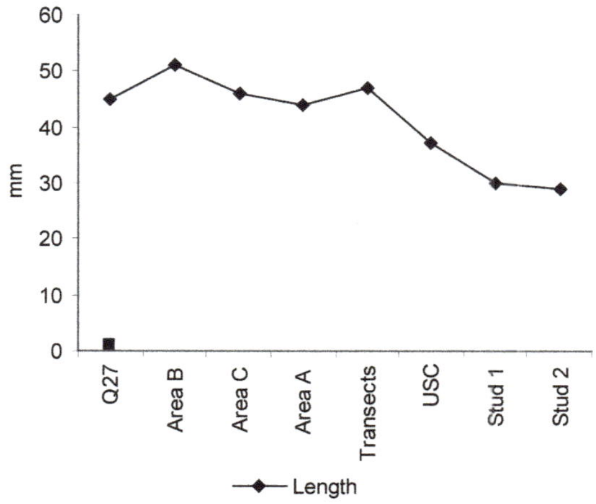

Figure 9.7: The Length of the Complete Flakes Recorded in the Study Area

Binford and O'Connell (1984:428) described how blade blanks for "mens' knives" were manufactured at the quarries while prepared cores were often removed and used to manufacture expedient tools for "everyday" use at the residential camps. However, if a source of stone was readily available near the residential camps, such as the gibber pavement, than this source would be used for expedient tool manufacture (Gould 1977).

Ethnographic accounts about quarries have described how stone from certain quarries have mystical properties, high prestige or mythological associations (Jones and White 1988:51; Gould 1977:164). The value of these quarries warranted special trips to collect stone (Gould 1977:164). At quarries described by Thomson (1983:70) Jones and White (1988:70-73), Aiston (1924:96-97) and Binford and O'Connell (1984:422) points for spears or mens' knives for weapons or circumcision ceremonies were manufactured. While artefacts fitting these ethnographic descriptions were found at Quarry 35 it is impossible to place any sort of ritual interpretation onto them.

Several different strategies were used to extend the use-life of the stone obtained at Quarry 35. Suitable material of a high quality was selected and the difficult task of preparing blade cores was undertaken at the quarry. Moreover, most of the blade manufacturing was generally undertaken at the quarry rather than at the landscape sampling units. Blade production is regarded as wasteful, labour intensive and time consuming in the amount of core preparation and rejection of unsuitable flakes (Hayden *et al.* 1996:37; Fullagar and Torrence 1991:134-135). However, once the blade core is prepared the core can be reduced efficiently to maximise the amount of material available (Sheets and Muto 1972; Nelson 1991:68). A flake with a blade form enables a long, sharp cutting edge per unit weight and facilitates the design of many different tool types (Hayden *et al.* 1996:37). Hence, more time and labour, increasing the 'venture risk' (Elston 1990:154), was invested in the preparation of cores to produce blades at Quarry 35 than to manufacture large flakes at Quarry 27. Furthermore, high material quality allowed the controlled removal of flakes (Flenniken and White 1985:132), tools to be resharpened (Dibble 1984, 1987) and reduced the 'contingency risk' of not having an appropriate tool at the appropriate time (Torrence 1989b:64; Elston 1990:154). Hence, the input of considerable time and labour in preparing the cores at Quarry 35 ultimately outweighed the costs involved and allowed for greater efficiency in the acquisition of resources such as game. This reasoning is especially relevant to the blades and points produced at Quarry 35 as they are likely to have been used as spearheads, as described by Mulvaney (1960) and McCarthy (1976).

In contrast to Quarry 27 a different set of technological options, including design principals, were used at Quarry 35 to reduce the probability of failure (or 'risk') to secure subsistence resources. These options are summarised in Table 9.12. Again using Bamforth and Bleeds' (1997:127, Figure 7.1) heuristic model, the procurement costs at Quarry 35 are relatively low since material is readily available in the Mt Wood Ranges. In contrast, the costs involved in production are high as blade manufacturing has high failure rates and requires considerable preparation time. However, the production costs are reduced if blade core preparation and the manufacture of blade blanks were undertaken at a quarry where ample, high quality material is available. Furthermore, time employed in the manufacture of blades equates to less time spent undertaking other activities, increasing the 'production risk of failure', and may even demand the need for specialists (Bamforth and Bleed 1997:127). The application costs of failure are low as blade blanks can be fashioned into the required form to replace broken parts.

Procurement
- Residence near to the quarry
- Multi-functional tools to economise raw material use
- Use of durable stone

Production
- Advanced production of tools
- Production at the quarry
- Use of easily workable stone
- Production by specialists (?)

Application
- Careful design for specific purposes
- Easy repair – standardised blade form
- Transport of extra tools

Table 9.12: The Technological Options used at Quarry 35 to Reduce the Failure Probabilities ('Risks') involved in Procurement, Production and the Tool Application (adapted from Bamforth and Bleed 1997:128, Table 7.3)

9.6.4 Identifying the Influence of Form, Space and Time by Comparing the Formation and Composition of the Quarry Assemblages

A number of similarities and differences occur between the two quarries. The similarities include the geological characteristics of the quarries, common features, such as the outcrop and landform surfaces, that influenced the distribution of artefacts, the selection for high quality Group III silcrete and the presence of non-local tools and large "expedient" flake tools, core tools and spall tools in the quarry assemblages. The differences between the two quarries reflect the time and labour invested in core reduction.

The goal of core reduction at Quarry 27 was simply to produce large flakes for removal. The manufacture of large flakes at Quarry 27 required little skill or time invested in core preparation and reduction. The flakes, due to the quality of the material and their size, could be resharpened, maintaining a long use-life. The limited amount of reduction, its small size and low density of artefacts at Quarry 27 suggests that the outcrop was not frequently used and little time was spent at the quarry. While extremely high quality material was available and targeted at Quarry 27 the amount and possibly the difficulty of extracting the cobbles limited the use of the quarry. It is likely that its location, away from larger creeks, and the presence of ample material in the Twelve Mile Creek Gorge also reduced the use of the outcrop. The assemblage at Quarry 27 may have been the result of an opportunistic collection of material, embedded in the acquisition of another resource, as its location is relatively isolated.

Material suitable for blade manufacture was readily available in the Twelve Mile Creek Gorge. Many of the quarries in the gorge have evidence of blade manufacture (refer Appendix 1). In contrast, at Quarry 27 little blade manufacture occurred and this could be the result of more easily assessable outcrops being present in the gorge. Furthermore, the size of the boulders at Quarry 27 allowed large flakes, possibly for the manufacture of tulas, to be produced.

It must be remembered that the reduction strategy employed at Quarry 27 did not require a large amount of core preparation or reduction to manufacture the large flakes removed from the quarry. As a result, less flaking debris were produced at Quarry 27 than at Quarry 35 where blade manufacture typically produces large amounts of flaking debris. Hence, caution is needed about making correlations between numbers of artefacts present and the importance of a stone resource. The limited amount of cores and flaking debris at Quarry 27 may not reflect the true value of the quarry as it may have been of vital importance to have large flakes made from high quality material. However, this scenario is unlikely as high quality material is available in more accessible locations.

In comparison to Quarry 27 a more complex pattern of core reduction was present at Quarry 35. Cores were selected and prepared, transported away from the outcrop and used to manufacture blades and points. The blades and points were fashioned into tools, some of which were broken and discarded during manufacturing, and others were used to retool. The discard of non-local blades and tools also implies that retooling occurred at the quarry.

Large discarded broken and complete flakes, spalls and spent cores were used expediently as tools. The presence of large scraper and notch tools associated with the manufacture of blades indicated wood working, possibly of spear shafts, occurred at the quarry. These tools were made, used and discarded in the same context. The focus of blade manufacturing and the degree of spatial segregation in manufacturing stages suggests that Quarry 35 had a more significant role in the organisation of lithic production than Quarry 27. It is possible that specialists obtained stone from the quarry due to the degree of spatial segregation of the different core reduction strategies, although these patterns may simply be a product of the time spent at the quarry, and the skill involved in blade manufacturing.

Quarry 35 is located near areas where 'support facilities' occurred and this may have influenced the use of the quarry. It seems likely that more time was spent at Quarry 35 during each visit and that the quarry was visited repeatedly over time. The degree of "on-site" core reduction and the intensity of quarry use were shown by the artefact density, the size of the quarry and the clear spatial segregation of the different core reduction strategies within Quarry 35. Considerable evidence of tool use and blade manufacturing was also observed at Quarry 35. The products of Quarry 35 were removed in several forms, as prepared cores, blade blanks, points, retouched tools and large flakes (seen not only in the quarry but also in the survey transects and the USC landscape sampling location). While most of the products removed from the quarry underwent considerable reduction, the large flakes removed from the quarry were produced early in core reduction.

Common features have influenced the distribution of artefacts at both quarries. The primary influence was the outcrop. On the outcrop most of the extraction, material testing and initial reduction occurred. A secondary influence was the topography. Its influence was especially evident at Quarry 35 where blade manufacturing tended to occur on the flat, high area behind the outcrop. At both quarries the gibber pavement formed an almost natural boundary to the distribution of the artefacts as few artefacts were discarded in the survey transects that crossed the gibber pavement. Post-depositional influences at Quarry 27 and Quarry 35 were the growth of vegetation and areas of kangaroo disturbance which moved or buried the artefacts. On a wider scale, massive, high quality silcrete outcrops were targeted for quarrying in the study area (Doelman *et al.* 2001).

A preference for Group III silcrete was found in the selection of material at the quarries. Adze flakes recorded in the study area were almost exclusively made from Group III silcrete (Doelman *et al.* 2001). In this way the source of the stone has dictated the types of tools manufactured. The use of Group III silcrete suggests a functional requirement for a material type that could be retouched easily and hold a working edge (Doelman *et al.* 2001).

In summary, the processes of artefact discard that influenced the formation, distribution and composition of the quarry assemblages were far more complex at Quarry 35 than at Quarry 27. There are several reasons why this has occurred: the raw material types available, the reduction strategies employed, the goal of procurement, the location of the quarry, the landscape features of the source location and the time invested in quarrying. All these factors have interacted to influence the quarry's 'role' in the organisation of lithic technology.

9.7 REDEFINITION OF A QUARRY

The same problems and arguments that have arisen with the definition of a 'site' can be applied to the definition of a quarry. The term 'site' has been criticised by many archaeologists (e.g. Dunnell 1992; Ebert 1992; Thomas 1975; Robins 1997; Holdaway *et al.* 1998). A quarry, like a "site", also becomes subjective, based on the eye of the observer, difficult to quantify and interpret. In removing the term quarry, the quarry has no boundary but becomes continuous, seen as an ever-decreasing dispersal of material, in whatever size or form, away from the source of the stone. In this way the quarry becomes part of the landscape allowing the researcher to view the quarry, people's strategies and interactions with the quarry, the ways they acquired and used the available stone and how material from that quarry was discarded across the landscape. The quarry becomes an important place in the landscape that due to its obvious attractions has influenced the distribution of the archaeological remains over time and across space. However, a quarry is a distinctive place in landscape where particular activities are carried out and needs to be recognised as such due to its 'role' in the organisation of lithic technology.

Stone procurement ranges from extremely complex with many behavioural and organisational implications, such as the Mt William Quarry (McBryde 1978, 1984a, 1984b, 1986) to the somewhat random, opportunistic local gathering of stone described by Gould (1977, 1980) and Hayden (1977). Axes from the Mt William Quarry were traded over hundreds of kilometres, crossed social boundaries and established extensive networks of interaction (McBryde 1978, 1986). Hence, the term quarry is often dependent on how liberal a definition of the term was used by the researcher. A quarry researcher may define a quarry as simply a source of stone or as having an underlying set of behavioural implications. If a quarry is defined as a source of stone, as described by Nelson (1987:121) and Hiscock (1988:327), then the gibber pavement is a quarry. However, based on previous theoretical ideas, such as Ericson (1984), Torrence (1984, 1986, 1989b), Elston (1990), Binford (1979), Hiscock (1988), Gould (1977), the term quarry has a far wider application to the structure and organisation of the lithic technology system. The quarry is the beginnings of the 'lithic production system', how stone was procured and what it was used for has far wider implications as it can be used to identify the organisational strategies employed by the people who used it as a resource (Ericson 1984).

Based on the detailed examination of Quarry 27 and Quarry 35 a quarry can be defined as having several criteria:

- A source of stone
- Evidence of extraction, material testing and selection
- Initial reduction
- The removal of cores for further reduction and/or flakes for tool use

The above definition of a quarry includes not only the place of material extraction but also the associated artefacts produced during core reduction. While Hiscock and Mitchell (1993:22) make a clear distinction between quarry and reduction site they acknowledge that these two "site types" often overlap in a single location. It may be more helpful to firstly isolate and define the geological context and characteristics of the available stone resources and then describe the human modification of the stone.

It is also important to note that Quarry 35 and Quarry 27 may represent extremes in quarrying behaviour or two examples of many types of different quarries. Here it is only suffice to say that quarries do vary in the degree of complexity and are likely to play different roles in the organising principles that govern stone procurement. The extensive trade in leilira (large) blades highlights the fact that quarries in the region vary in terms of their sphere of influence and the goals of stone procurement. Ritual trade in leilira blades has been well documented for northern and central Australia (Allen 1996a, 1996b). The most southerly leilira blade quarry is the Mt Stuart Quarry which is located ~10 km south of the Stud Creek catchment (Figure 4.1). The nearest place these blades appear in "sites" is in northwestern Queensland, 1500 kilometres away from Mt Stuart (Smith and Cundy 1985:36).

Use of Quarry 35

-Complex organisation
-Considerable manufacturing time invested
-High quality stone
-Long life span
-Time in system
-Blade tools
-Focused on outcrops
-Curated and transported
-Specialised Direct procurement?
-Planned
-High quality stone
-High manufacturing risk
-Intensive labour investment

Product Dependent

Location Dependent

Quality Dependent

Use of the Gibber Cobbles

-Simple organisation
-Short time invested in manufacture
-Low quality stone
-Short lifespan/ limited time in system
-Utilised flakes
-Mainly associated with LSU*
-Expedient tool use
-Opportunistic procurement
-Embedded procurement
-Low quality raw material
-Low manufacturing risk
-Low labour investment

Figure 9.8: Contrasting the Use of Quarry 35 and the Gibber Pavement as Sources of Stone

9.7.1 Gibber Procurement Versus Quarrying

By comparing the use of Quarry 35 and the gibber cobbles (Figure 9.8) several differences occur in the way the stone from each type of stone source was procured, worked, used and the place/s the material were discarded in the landscape. More procurement and manufacturing time was invested in working the stone from Quarry 35 than the gibber cobbles. Following Torrence's arguments (1983, 1989b) less 'costs' would have been involved in procuring or using stone from the gibber as the use of this source was not essential in procuring food resources. In contrast, the 'risk' of not having a particular tool at a particular time, such as when hunting, outweighed the time and labour invested in manufacturing blades and prepared cores at a quarry. Hence, the 'costs' involved in manufacturing blades were outweighed by the benefit of having them. 'Risk' was also reduced by using higher quality material derived from the quarry allowing greater versatility, reliability, flexibility, transportability and maintainability to obtain a greater efficiency (*cf* Nelson 1991, Bleed 1986, Torrence 1989b, Gamble 1986, Kuhn 1994).

The time invested in the reduction of quarried and non-quarried stone is also implicit in the use-life of the tool types manufactured from the different stone sources. Tools, such as backed blades, made from quarried stone were manufactured at the quarry, removed, used and then discarded elsewhere in the landscape. In contrast, the tools made from gibber cobbles were made, used and discarded in the same context. Whereas procuring stone from Quarry 35 was planned, labour intensive and structured, even undertaken by specialists, the use of the gibber reflects an expedient technological strategy (*cf* Binford 1973, 1979). The intensive use of the gibber cobbles at Stud 1 and the USC unit was simply due to the ready availability of the resource. Hence, the use of these two types of stone sources represents two different problem solving strategies that were used simultaneously to achieve different goals that nonetheless helped to accomplish the tasks needed to live within an often difficult environment.

Is the gibber pavement a quarry? Based on the above definition the gibber pavement did not act as a quarry because material was not removed from the source. Typically, the use of gibber cobbles included manufacturing, immediate use and the discard of both modified and unmodified flakes in the same context. In contrast, Quarry 35 has a far wider sphere of influence as material was removed from the quarry, transported and reworked before being discarded across a wide area.

9.8 PROBLEMS OF SAMPLING QUARRIES

Quarry analysis has been fraught with methodological
Quarry analysis has been fraught with methodological
problems (Hiscock and Mitchell 1993:72). Quarries are often very large, forming a dense volume of cores and flaking debris (Ericson 1984). Often fieldwork methodologies have been neglectful of the problems of gaining a representative quarry sample for analysis or were unable to gain a representative sample due to the immense size and density of artefacts at the quarry.

The analysis of Quarry 27 and Quarry 35 shows that it is important to understand the distribution of artefacts across the entire quarry. The spatial analysis of the quarry assemblages indicates that variation in the distribution of the artefacts appears within a quarry. This spatial variation needs to be taken into consideration or a sampling bias will occur that presents a misleading view of the quarry and prejudices the interpretation of the quarry assemblage. With the aid of a GIS a quarry assemblage can be further defined by detecting the placement of different stages of core reduction. The stages of reduction have become spatially redundant as common features, such as the outcrop, have influenced the distribution of the artefacts. For this study a GIS was a valuable tool which enabled patterns in the distribution of the artefacts at a quarry to be identified.

If the quarry is part of the landscape, incorporating a non-site distributional approach, then there is no such thing as a quarry boundary. The quarry simply becomes a component of the continuous distribution of artefacts, where particular locations due to their position or resources have created a higher density of artefact discard or "nodes" (Isaac and Harris 1980:21). Practically, it is impossible to trace the distribution of stone from one quarry across the entire landscape. For this study a boundary was simply used as a methodological tool to encapsulate the entire quarry for the analysis. By following the density of artefacts and using natural features as boundaries an effort was made to encompass all the stages of core reduction undertaken at each quarry. Furthermore, the survey transects were used to gain a more comprehensive understanding of the artefact distribution and characteristics of areas immediately surrounding the quarry.

Problems occur using random or systematic sampling methods as the distribution of the entire quarry assemblage and the scale of spatial variation are unknown. Variation in the distribution of artefact at a quarry is due to many different factors, such as the degree of core reduction, the core reduction strategies employed, the location of the quarry and the landform features present. How all these factors interact ultimately impinges on when and where spatial variation occurs. Hence, it is important to understand the distribution of the quarry assemblage and use an appropriate sampling method. By using the artefact as the smallest unit of analysis, with each artefact having a unique position in space and a unique set of characteristics, the scale of artefact variation can be sought using a GIS. By comparing the artefacts' characteristics in space the location of the different stages of reduction were identified at Quarry 27 and Quarry 35.

An approach used successfully at Quarry 35 was to define the entire quarry based on the distribution and characteristics of the cores. This approach was achieved by using a systematic transect survey of the entire quarry for cores, combined with a spatial analysis in a GIS to search for overlapping patterns of core characteristics. The core morphology highlighted where different stages in the core reduction occurred. The results of the targeted sampling strategy, focusing on the cores, directed the location of the sampling areas for analysis of the flaking enabling a comprehensive assessment of core reduction at the quarry. The core distribution was also helpful in defining the quarry boundaries by following the density of the cores and allowed a comparison, using a GIS, of the core distribution with the landform features, topography and outcrop.

9.9 SUMMARY

By using an appropriate analytical and methodological framework data obtained from the analysis of individual artefacts can be used to tease apart the factors, such as the reduction strategies employed, the landscape features and time invested in procurement, which influence the composition and distribution of a quarry assemblage. There is a real need to define the entire quarry assemblage requiring both a technological and spatial analysis, and use of suitable sampling strategies. Recording the distribution of the cores and their characteristics has allowed not only a total understanding of the entire quarry assemblage but enabled the location of the various stages of core reduction to be identified. Lastly, it must be remembered that a quarry does not exist on its own but is part of a wider system of lithic organisation. The key to understanding the role of the quarry in the organisation of technology is to firstly understand the quarry. In the following chapter the main findings of this s are presented.

Chapter 10

Conclusion

The goal of this study was to identify the 'role' the different stone resources played in the organisation of lithic technology seen in the Mt Wood Ranges of northwestern New South Wales. Two types of stone resources are present in the study area, quarried outcrops and an almost continuous pavement of gibber cobbles defined as a non-quarried stone source. The detailed study of two quarries, Quarry 27 and Quarry 35, enabled the many factors that contributed to the formation, composition and distribution of the assemblages to be identified in order to view the complex interplay between people and the environment. The results of the quarry analysis were contrasted with the strategies used to acquire and reduce the stone obtained from the non-quarried stone resources at the landscape sampling units. Identifying the technological strategies employed at two quarries and three lithic scatters reflects the problem-solving strategies people undertook in the past to organise their use of the available stone resources and survive in arid zone Australia.

The theoretical framework was drawn from aspects of non-site distributional archaeology, an approach to lithic analysis known as technological organisation and the formation of an archaeological assemblage described by Schiffer (1972, 1976) as a behavioural chain analysis. The results of the analysis shows that many factors contribute to the formation of a quarry assemblage including, the landscape features and the goal of stone procurement, which dictated the types of reduction strategies employed and the placement of different stages of reduction. Each of these factors interact and influence the formation of an assemblage ranging from the manufacture and discard of an individual artefact to the spatial patterning of many artefacts created through time and distributed across space.

- ***An Approach to the Analysis of a Quarry Assemblage***

This study has contributed to the development of quarry studies in Australia. The lack of detailed study of quarry assemblage has created a breach in our understanding of an important part of the archaeological record. Information derived from the analysis of quarries can further our knowledge about the ways stone was acquired, reduced, transported and used in the wider landscape. One of the reasons quarry studies may not have flourished in Australia is that they are notoriously difficult to record due to the volume of artefacts they contain. This study has presented a methodological approach that takes into account the problems of sampling a quarry by undertaking both a technological and spatial analysis to characterise the entire quarry assemblage.

An understanding of the geological context is vital in identifying the types of raw materials available, their location and their occurrence (i.e. an outcrop) and enabled the survey for the quarries to focus on particular landform features. In the study area, quarries concentrate along the bluffs and within the creek gullies where the outcrops of silcrete are exposed.

The framework used in this study allowed comparisons to be made between the individual components of the assemblages (cores, flaking debris and tools) at the quarries and landscape samples. Hence, the smallest scale of analysis was the individual artefact, encapsulating aspects of non-site, distributional archaeology. Each artefact has a position in space and a unique set of characteristics. The characteristics of each artefact are indicative of its place of formation in the core reduction process. Decisions were made as to which characteristics of the artefacts best reflected the types of reduction strategies employed (Chapter 4). These characteristics were recorded for each artefact using an 'on-site' recording program at the quarries. A spatial analysis of the quarry assemblages was undertaken to investigate whether the reduction stages identified in the technological analysis occurred in different locations. The types of reduction strategies employed at the two quarries, their geological features, the results of the spatial analysis of the quarries were compared to determine the similarities and differences between the assemblages and establish their 'role' in the organisation of lithic technology.

Researchers who examine quarry assemblage are faced with many methodological problems due to their large size and high density of artefacts. The analysis of all the cores and flaking debris at a quarry is often impossible and leads to the fundamental problem of gaining a representative sample. At Quarry 35 and Quarry 27 spatial variation in the characteristics of the artefacts reflects where different stages of core reduction were undertaken at the quarry. Hence, depending on the sampling strategy potentially very different results can be gained from the analysis and interpretation of a quarry assemblage. Furthermore, landscape features, such as topography and the location of the outcrop, have influenced where different stages of core reduction occurred. Spatial variation within a quarry assemblage needs to be considered or a sampling bias may occur in the results. The analysis of the quarries in this study highlights the importance of understanding the distribution of artefacts in the entire quarry before making decisions about where and how to sample. The spatial analysis of the entire quarry assemblage at Quarry 27 shows that the characteristics of the cores and flakes discarded in close spatial association reflect the same stage of core reduction. Accordingly, at the much larger Quarry 35 cores were used to identify the extent of the quarry, define the reduction strategies employed, search for spatial variation in the reduction stages and determine the size and placement of the sampling areas for the recording of the flaking debris.

- ***The Factors that Influence the Formation, Distribution and Composition of a Quarry Assemblage***

The regional geology, seen in distribution of the available stone resources and the types of stone resources, has influenced how people in the past organised their lithic technology. In the Mt Wood Ranges an abundance of stone is available leading to the presence of large quarries. The form of the outcrop, its location and the quality of the material has contributed to decisions about which outcrop should be quarried. It seems likely that the outcrops were systematically tested since only the massive outcrops of high quality silcrete were quarried, indicating a knowledge of the how silcrete can outcrop and the properties of silcrete.

At each of the quarries, the landscape features, such as the outcrop, have influenced the placement of different core reduction activities. It is essential to identify the features that influenced the discard of artefacts at a quarry as these impinge on the way a quarry assemblage should be recorded. A primary contributor to the characteristics and distribution of artefacts within a quarry assemblage is the outcrop. A distance-decay profile was observed at both quarries: as the distance increased from the outcrop characteristics of artefacts reflected a greater amount of core reduction. Moreover, it seems likely that the movement of people was influenced by the topography, landform surface, such as the gibber pavement, and the location of waterways as these affected the discard of the cores and flaking debris at the quarries.

The distance from occupation areas to the quarries, although these distances are relatively small, also influenced the use of the available stone resources. Quarry 35 is close to large lithic scatters in the Stud Creek Catchment and was probably used in conjunction with its occupation for retooling. In contrast, Quarry 27 is located away from any lithic scatters and its use may have been more opportunistic.

Time has also played a role in the formation and composition of a quarry assemblage. The underlying premise is that if more time is invested in core preparation and reduction at a quarry than the time spent at a quarry during each visit increases. Furthermore, the number of times a quarry was revisited also adds to its size and density depending on the reduction strategy employed. By comparing Quarry 27 and Quarry 35 the influence that time had on their assemblage composition can be seen. In comparison to Quarry 27, more time was spent preparing and reducing cores at Quarry 35. Hence, more time was spent at the Quarry 35 during each visit. Quarry 35 was probably used for a greater duration of time in conjunction with the occupation of the Stud Creek Catchment. The amount of time spent at Quarry 35 has created stronger spatial patterns of in the distribution of different stages of core reduction than at Quarry 27.

The reduction strategies used to procure stone from Quarry 27 and Quarry 35 and the products removed from each quarry are very different. More time and labour was invested in core preparation and reduction to produce blades at Quarry 35 than to manufacture the large flakes at Quarry 27. Furthermore, blade manufacture results in the discard of large amounts of waste material, quickly increasing the size and density of the quarry assemblage. In contrast, the limited reduction of cores at Quarry 27 meant that little waste would be discarded. Hence, the density and size of the quarry and the composition of its assemblages also reflects the technological strategies employed.

- ***The 'Role' of the Stone Resources in the Organisation of Lithic Technology***

By understanding how material was reduced and removed from a quarry a better understanding of the 'role' each quarry played within the organisation of technology was gained. While some quarries, such as the large lelira blade quarries at Mt Stuart, may have been involved in long distance trade and exchange networks it seems more likely that the acquisition of material from Quarry 35 and Quarry 27 was prompted by utilitarian needs. As a result, data obtained from these quarries has far reaching applications to the organisation of lithic technology in the area. However, it must be remembered that these two quarries are only two examples of the many quarries found in the Mt Wood Ranges.

Since both quarries have very similar geological features, such as the form of the outcrop and the quality of the material, the differences in the procurement and reduction strategies observed at both quarries relates to its 'role' of the quarry in the organisation of technology. The 'role' of the quarry has a direct interrelationship with the time invested at the quarry. So while the geological features influenced the position of the quarry and the positioning of different reduction activities within the quarry it ultimately influenced by the strategies people undertook to utilise the outcrop. These strategies governed the time invested in core reduction and the form material was acquired and removed from the quarry.

The acquisition of stone from the quarries was goal driven. At both quarries specific products, designed for particular purposes, were manufactured and removed. The form material was removed from the quarries reflects peoples' problem-solving approaches undertaken to successfully utilise their stone resources and live in a difficult environment. Particular design variables were deemed important for the manufacture of tools at each quarry. Hence, stone obtained from both quarries was curated but curated in different ways.

It is likely that a more opportunistic collection of stone was made from Quarry 27 due to its relatively isolated location, small size and low artefact density. At Quarry 27 the large of the flakes and their high quality material allowed resharpening and extended their uselife. In addition, the standardised size of the flakes manufactured at Quarry 27 meant that wasted or broken tools, such as adze flakes, could be replaced in the same haft. The Group III silcrete at Quarry 27 is ideal for the manufacture of tula adzes. Nearly all the adze flakes recorded in the landscape sampling units were made from the same Group III silcrete available at Quarry 27. Furthermore, high quality material anticipates future tool use and increases the reliability of a tool thereby reducing the 'contingency risks' (Odell 1996:65-67; Elston 1990:154). Hence, little time and labour was invested in the reduction of stone at Quarry 27 but more time was invested in the maintenance of tools made from its material, such as the adze flakes, to gain a reliable tool design.

Quarry 35 was probably repeatedly used as a stone source, in conjunction with the occupation of the Stud Creek catchment, for retooling. At Quarry 35, considerable time and labour was invested in the preparing blade cores. Prepared blade cores of high quality material were removed, transported and used to manufacture blades or another flake forms when required providing a reliable and versatile form of material. The costs required to prepare a blade core is outweighed by its ability to conserve stone by carefully controlling the removal of blades (Nelson 1991:68). Blade blanks were also removed from Quarry 35. The advantage of transporting blade blanks is being able to be fashion these into a variety of tools, such as points or geometric microliths, depending on the task-orientated goal. Hence, blade blanks are an example of a versatile design variable. If a replacement part was required a blade blank could be manufactured into whatever form was needed and this is another example of how blade technology is reliable, a trait that was required in the acquisition of game. Fashioned blades and points were probably used as spearheads and having replacement parts for these when broken was vital to the successful capture of food resources. The production costs invested in the manufacture of blade cores, blade blanks and fashioned blades were outweighed by the benefits of having stone in these forms.

By contrasting quarried and non-quarried stone sources several differences can be observed in their 'roles' within the organisation of lithic technology. While planning, standardisation and an investment of time and energy are integral parts of quarrying behaviour at Quarry 35 these do not occur in the procurement of stone from non-quarried stone resources. Approximately one third of the artefacts in the landscape sampling units were manufactured from non-quarried stone. Gibber cobbles were readily available in the Stud Creek Catchment as a source of stone for the manufacture of expedient tools, such as large tools with usewear damage. These tools represent a situational response to an immediate demand. This problem-solving strategy involved randomly acquiring a cobble from the gibber pavement and removing a large flake for a tool. Consequently, tools were made, used and discarded in the same context.

It was more time and energy efficient to use high quality material found in a suitable form at a quarried outcrop, already known, than to test gibber cobbles for suitable high quality material. Furthermore, the frequency of high quality cobbles in the gibber pavement is low at approximately 10% reducing the likelihood of obtaining high quality material from this source. Using a quarried outcrop to obtain stone in the required form involves skill, and an investment of time and labour, but the products becomes more efficient and reduces the 'risk' of failing to achieving subsistence goals.

What is a quarry? The detailed analysis of the available stone resources in the study area has allowed a close examination of the term quarry. A quarry is defined as being a source of stone with evidence of extraction, material testing and selection, initial reduction and the removal of core and/or flakes for later use. Quarry 27 and Quarry 35 have very similar landscape features yet they have very different assemblage compositions. These differences reflect their location, the time invested in core reduction and the selection of particular aspects of tool design. However, both quarry assemblages indicate the anticipation of future needs. The procurement of stone is goal-orientated and involves long-term planning and risk-reduction strategies. In contrast, the procurement of gibber cobbles is an unplanned response to a situation that demands a tool with a sharp working edge and as a result has smaller realm of influence in the organisation of technology.

- ***Wider Implications***

Two different scales of occupation can be viewed in the study area – the occupation of the Stud Creek Catchment and the occupation of the arid zone in northwestern New South Wales. Radiocarbon dating of the hearths in the Stud Creek catchment indicates that it was occupied, abandoned and reoccupied during the late Holocene (Reeves 1997; Fanning *et al.* 2000a). Similarly, at a larger spatial scale the eastern margin of the arid zone in Australia tentatively appears to have been occupied in the Pleistocene (*cf* Dury and Langford-smith 1965), abandoned and then reoccupied the late Holocene (discussed in Chapter 9). Potentially, the combination of 'ideal' environmental conditions may have prompted both the reoccupation of the region during the late Holocene and the reoccupation of the Stud Creek Catchment. However, without further radiocarbon dating over a wider area and an explicit environmental record any conclusions about the occupation of the area remain academic.

Previous researchers have argued that the increased in the number of 'sites,' artefact discard rates and grinding stone implements indicate a late Holocene intensification in the arid zone of Australia (Smith 1986, 1988, 1996;

Ross *et al.* 1992; Lourandos and Ross 1994). Measuring intensification by the number of sites and the density of artefacts was questioned by Hiscock (1986a:47; 1994:277) who argued that the increase in discard rate of artefacts reflects changes in the nature and structure of settlement, technology and resource use resulting in a more visible record. From this research it seems likely that the abundance of stone and technological strategies employed for the acquisition, reduction and use of stone obtained from quarried and non-quarried stone sources led to a rapid and dense accumulation of artefacts in places that were considered favourable for occupation. However, based on the occupation model proposed by Reeves (1997), places may only have been favourable for a short time and these locations were repeatedly occupied until faunal and floral resources were exhausted and the area was abandoned. This process, through time, created a widely dispersed archaeological record where lithic scatters concentrated in many areas.

It has been argued that the abundance and quality of available stone conditions the types of technological strategies employed (e.g. Andrefsky 1994a, 1994b). In the study area there is an almost continuous blanket of gibber pavements and numerous outcrops suitable for exploitation. Andrefsky (1994a) predicted that if there was an abundance of high quality material in an area it was used both expediently and 'formally'. This scenario does not hold true for the study area. At the USC unit, an area located only 200 m from Quarry 35, creek and gibber cobbles were reduced expediently to manufacture 'informal' tools. In contrast, the high quality material obtained from quarried outcrops was primarily used for blade manufacture. This outcome indicates that the size and shape of the cobbles and the raw material quality available at an outcrop certainly *allowed* blade manufacture but these factors do not negate the manufacture of 'informal' tools. 'Informal' tools at the USC unit were simply manufactured from non-quarried stone sources. These results support the observation already made by Gould (1977:164-165) that different stone sources were used to manufacture different kinds of tools. Furthermore, he adds that quarried stone had ritual associations for men. Whatever the case, the results of the lithic analysis from the USC unit reflect how people utilised the stone resources that were immediately available to them.

Contrary to the use of the stone resources at the USC most of the 'informal' tools were made from high quality silcrete in the Lower Stud Creek units. The distance from the LSC units to the nearest quarry is approximately 2 km. This pattern may be evidence of recycling, reflecting the multiple occupations at the LSC units. Flakes discarded during the reduction of blade cores obtained at a quarry and made from high quality silcrete may have been later 'mined' and used as 'informal' tools. This pattern would only accumulate if the locations were repeatedly occupied or occupied for a longer time during each visit.

- ***Future Directions***

It is hoped that more detailed quarry studies will be undertaken in Australia as they have been sadly neglected. Only two quarries were analysed in-depth for this study and this in itself presents a sampling bias. By widening the study of quarries, even looking at the 'role' of the large lelira blade quarries, a better understanding of how stone was procured and why is was procured in such a manner can be established.

The Mt Wood Ranges represents an area where a large amount of high quality stone resources are available, placing little restriction on the need to conserve stone. Away from the ranges in the plains surrounding the Tibooburra Ranges no silcrete is present. However, large numbers of silcrete artefacts have been transported to the Tibooburra Ranges and discarded. To comprehensively establish how the stone resources were used in northwestern New South Wales the distribution of silcrete artefacts, their form and quality, discarded away from the Mt Wood Ranges needs to be considered.

Lastly, the gap in occupation phases in the Stud Creek catchment and at a greater temporal scale, the episodic nature of arid zone occupation in northwestern New South Wales remain as two puzzling enigmas. Whether these patterns are real or not can only be examined by undertaking an extensive radiocarbon dating program over a much larger area.

- ***A Final Recapitulation***

Quarries are pivotal points in the landscape as they are the beginnings of the lithic reduction system providing information about how stone was acquired, reduced and in what form material was removed from a quarry. Furthermore, the types of reduction strategies employed at each quarry can be used to make inferences how stone was used as a resource and why it was used in such a fashion. In addition, by comparing the costs and benefits of using stone acquired from quarried and non-quarried sources the different problem-solving strategies used by peoples in the past to live successfully within their environment can be identified.

The key to successfully unravelling the complexity of a quarry assemblage is the design of an appropriate analytical and methodological framework. Many factors, interacting at different scales of time and space, influence the formation, distribution and composition of the archaeological record. By reducing the archaeological record to the smallest scale, the artefact, the factors that influence its location and characteristics can be used to view the archaeological record at any scale of time and space. In this way the factors that contribute to variability in the archaeological record were identified and used to identify the 'role' of the stone sources in the organisation of lithic technology.

Appendix 1: The Characteristics of the Quarries in the Study Area

ID	Location AMG	Material	colour	Outcrop Form	Lithology	Exposure Location	Artefact Types Present	Size (m^2)	Density (per m^2)
Q1	54 622506E 6754079N 12 Mile Creek	medium clast fine clast	grey grey white	massive, boulder	sandstone	waterway	large blades, microblades, large flakes	>1000	75-100
Q2	54 621608E 6754962N 12 Mile Creek	medium clast	yellow red grey	boulder pillar	sandstone	mesa bluff	expanded flakes cores	<100	<25
Q3	54 621472E 6754449N 12 Mile Creek	medium clast fine clast	grey	massive, boulder and pillar	sandstone	gully	flakes	100-500	25-75
Q4	54 595061E 6740851N Tibooburra	hornfels	grey black	bedded exposures		hill	axe, flake production	500-1000	25-75
Q5	54 619460E 6754011N 12 Mile Creek	medium clast	grey	massive, boulder and nodular	sandstone	bluff	flake	100-500	25-75
Q6	A 54 621349E 6755647N B 54 621382E 6756077N 12 Mile Creek	amorphous banded matrix- dominated	grey red	boulder	siltstone	bluff	expanded flakes reduced cores	500-1000	75-100
Q7	54 617043E 6745237N Acoola Stream	medium clast amorphous	grey yellow red	boulder	sandstone siltstone	gully	flake, core	<100	<25
Q8	54 618570E 6747999N Acoola Stream	medium clast	grey yellow	boulder	sandstone	mesa	flakes, core	<100	<25
Q9	A54 621403E 6741324N B 54 622096E 6741324N Stud Creek	amorphous	grey	massive boulder	siltstone	bluff	flakes, cores, blade production	<100	<25
Q10	54 621030E 6742040N Stud Creek	amorphous banded	grey	boulder, nodular	siltstone	bluff	blade flake	100-500	75-100
Q11	54 620822E 6741741N Stud Creek	amorphous banded	grey	boulder	siltstone	hillslope	blade	100-500	>100
Q12	54 6020568E 6741387N Stud Creek	coarse clast	grey	massive, boulder	pebble conglomerate	waterway	large flakes	<100	<25
Q13	54 619869E 6749430N Acoola Stream	medium clast	grey yellow	massive, boulder	sandstone	waterway/gully	expanding flakes blades	500-1000	75-100
Q14	54 619108E 6749606N Acoola Stream	medium clast	grey yellow	boulder, nodular	sandstone	gully	flakes	<100	<25
Q15	54 619180E 6749374N Acoola Stream	medium clast	grey	massive, boulder and nodular	sandstone	gully	large flake large blade microblade	500-1000	>100
Q16	54 621554E 6751042N Acoola Stream	medium clast	grey	boulder	sandstone	gully	flakes	<100	<25

Q17	54 617707E 6751864N Accoa Stream	medium clast	grey yellow	boulder, nodular	sandstone	gully/waterway	flakes blade cores	<100	<25
Q18	54 618072E 6748646N Accoa Stream	medium clast	grey yellow	boulder, nodular	sandstone	gully/waterway	flakes	<100	<25
Q19	54 617680E 6748998N Accoa Stream	medium clast	grey	boulder	sandstone	bluff	large flakes	<100	<25
Q20	54 618174E 6749643N Accoa Stream	medium clast	yellow	cobbles	sandstone	gully/waterway	large flakes	<100	<25
Q21	54 622117E 67643N 12 Mile Creek	medium clast amorphous	grey yellow grey	a massive, boulder, cobble	sandstone siltstone	mesa	tula? flakes microblade blade	>1000	75-100
Q22	54 621547E 6754962N 12 Mile Creek	amorphous medium clast	grey	boulder	sandstone	bluff	flake, large flake	<100	>100
Q23	54 622523E 6754780N 12 Mile Creek	medium clast	grey	large boulder	siltstone	buff	large flakes, large blades	<100	<25
Q24	54 622616E 6654824N 12 Mile Creek	medium clast	grey	boulder	sandstone	gully	flake micriblade	<100	25-75?
Q25	54 622850E 6754689N 12 Mile Creek	medium clast	grey	boulder	sandstone	bluff	flake microblade	<100	25-75?
Q26	54 623138E 6754674N 12 Mile Creek	medium clast	grey	boulder	sandstone	bluff	flake microblade	<100	25-75?
Q27	54 623323E 6754674N 12 Mile Creek	amorphous	grey	boulder	siltstone	gully	flake	100-500	25-75?
Q28	54 621810E 67463N 12 Mile Creek	medium clast	grey	large boulders	sandstone	receded bluff, at base	large flakes, large blades	<100	<25
Q29	54 620919E 6754317N 12 Mile Creek	fine clast	grey white	large boulders	sandstone	mesa	large flakes	<100	<25
Q30	54 616272E 6728429N Mt Stuart	fine clast	white	large boulders	sandstone	bluff	large blades (leilira), large flakes microblades	>1000	>100
Q31	54 622210E 6757594N	fine clast	white	cobble	sandstone	mesa	blade	100-500	25-75
Q32	623853E 6758961N	fine clast medium clast	white grey	cobble	sandstone	mesa	blade	500-1000	25-75
Q33	54 620137E 6765927N Narcoola	Fine clast	white	boulder	sandstone	bluff	large blades (leilira),	100-500	25-75
Q34	54 616308E 6744013N Stud Creek	matrix- domainted medium clast	yellow grey	massive exposure	siltstone	bluff	blade	100-500	25-75
Q35	54 616458E 6743777N Stud Creek	medium clast fine clast matrix- dominated	Grey	cobble	sandstone	bluff	large flakes blade	500-1000	75-100
Q36	54 622672E 6754144N 12 Mile Creek	medium clast	grey	cobble	sandstone	gully	blade	100-500	25-75
Q37	54 622476E 6753970N 12 Mile Creek	medium clast	grey	cobble	sandstone	gully	blade	500-1000	75-100

Q38	54 622458E 6754052N 12 Mile Creek	medium clast	grey	cobble	sandstone	gully	blade	100-500	25-75
Q39	54 622520E 6753962N 12 Mile Creek	medium clast	grey	cobble	sandstone	gully	blade	100-500	25-75
Q40	54 622308E 6754048EN 12 Mile Creek	medium clast	grey	cobble	sandstone	gully	blade	500-1000	25-75
Q41	54 622171E 6754091N 12 Mile Creek	medium clast	grey	cobble	sandstone	gully	blade	<100	25-75
Q42	54 622228E 6753937N 12 Mile Creek	medium clast	grey	cobble	sandstone	gully	blade	<100	25-75
Q43	54 622191E 6753918N 12 Mile Creek	medium clast	grey	cobble	sandstone	gully	blade	<100	25-75
Q44	54 622187E 6753911N 12 Mile Creek	medium clast	grey	cobble	sandstone	gully	blade	<100	25-75
Q45	54 622048E 6753992N 12 mile Creek	medium clast	grey	cobble	sandstone	gully	blade	100-500	75-100
Q46	54 621726E 6753992N 12 Mile Creek	medium clast	grey	cobble	sandstone	gully	blade	100-500	25-75
Q47	54 621998E 6753865N 12 Mile Creek	medium clast	grey	cobble	sandstone	gully	blade	<100	<25
Q48	54 621813E 6753938N 12 Mile Creek	medium clast	grey	cobble	sandstone	gully	blade	<100	25-75
Q49	54 621754E 6753822N 12 Mile Creek	medium clast	grey	cobble	sandstone	gully	blade	>1000	500-1000
Q50	54 621595E 6754000N 12 Mile Creek	medium clast	grey	cobble	sandstone	gully	blade	<100	<25
Q51	54 622063E 6753806N 12 Mile Creek	medium clast	grey	cobble	sandstone	gully	blade	500-1000	25-75
Q52	54 622068E 6753854N 12 Mile Creek	medium clast	grey	cobble	sandstone	gully	blade	<100	25-75
Q53	54 622219E 6753899N 12 Mile Creek	medium clast	grey	cobble	sandstone	gully	blade	<100	<25
Q54	54 622300E 6753889N 12 Mile Creek	medium clast	grey	cobble	sandstone	gully	blade	<100	<25
Q55	54 614609E 6739680N Thomsons creek	fine clast medium clast	yellow pink grey	cobble	concretionary sandstone	bluff	large flake	<100	25-75
Q56	54 614528E 6739455N Thomsons Creek	matrix-dominated	grey	boulder	siltstone	bluff	small flake	<100	<25
Q57	54 614825E 6739784N Thomsons Creek	matrix-dominated	grey yellow pink	siltstone	cobbles	bluff	large flake	100-500	25-75
Q58	54 615061E 6739863N Thomsons Creek	medium clast	red grey yellow	sandstone	boulder	bluff	large flake blade	500-1000	25-75

Q59	54 6 4997E 6739884N Thomsons Creek	matrix-dominated	grey	siltsone	cobbles	buff	large flakes	<100	25-75
Q60	54 6 5644E 6739985N Thomsons Creek	medium clast	grey yellow	sandstone	boulders	buff	large flake blade	500-1000	>1000
Q61	54 6 5066E 6740069N Thomsons Creek	medium clast	grey	sandstone	no outcrop	buff	large flake large blades	<100	25-75
Q62	54 6 5256E 6740024N Thomsons Creek	matrix-dominated	grey	siltstone	cobbles	buff	small flakes	<100	25-75
Q63	54 6 5389E 6740127N Thomsons Creek	medium clast	grey	sandstone	boulders	buff	large flake	100-500	25-75
Q64	54 6 5796E 6740390N Thomsons Creek	medium clast	yellow	sandstone	cobbles	buff	large flakes	<100	25-75
Q65	54 6 7085E 6741614N Stud Creek	medium clast	grey	sandstone	massive exposure	eastern slope of bluff	large flakes	<100	25-75

Bibliography

Adams. R. M 1965. *Land Behind Baghdad. A History of Settlement on the Diyala Plains.* University of Chicago Press, Chicago.

Ahler, S. A. 1989. Mass analysis of flaking debris: studying the forest rather than the trees. In *Alternative Approaches to Lithic Analysis*, D. O. Henry and G. O. Odell (eds), pp. 85-118. Archaeological Papers of the American Anthropological Association No 1, American Anthropological Association, Washington D. C.

Aiston, G. 1928. Chipped stone tools of the Aboriginal tribes east and north-east of Lake Eyre South Australia. *Papers and Proceedings of the Royal Society of South Australia,* pp. 123-131.

Allen, H. 1972. Where the Crow Flies Backwards: Man and Land in the Darling Basin. Unpublished Ph.D. thesis, Australian National University, Canberra.

Allen, H. 1996a. The distribution of large blades: evidence for recent changes in Aboriginal ceremonial exchange networks. In P. McConvell and N. Evans (eds), *Archaeology and Linguistics: Aboriginal Australian in Global Perspective,* pp. 357-376. Oxford University Press, Melbourne.

Allen, H. 1996b. Ethnography and Prehistoric archaeology in Australia. *Journal of Anthropological Archaeology* 15:137-159.

Allen, K. M., S. W. Green and E. B. W. Zubrow (eds). 1990. *Interpreting Space: GIS and Archaeology.* Taylor and Francis, London.

Amick, D. S. 1987. *Lithic Raw Material Variability in the Central Duck River Basin: Reflection of the Middle and Late Archaic Organisational Strategies.* TVA Publications in Anthropology 50, Norris, Tennessee.

Ammerman, A. J. and W. Jr Andrefsky 1982. Reduction sequences and the exchange of obsidian in Neolithic Calabria. In J. E. Ericson and T. K. Earle (eds), *Contexts for Prehistoric Exchange,* pp.149-172. Academic Press, New York.

Andrefsky, W. Jr 1991. Inferring trends in prehistoric settlement behaviour from lithic production technology in the southern plains. *North American Archaeology* 12:129-144.

Andrefsky, W. Jr 1994a. Raw-material availability and the organization of technology. *American Antiquity* 59:21-34.

Andrefsky, W. Jr 1994b. The geological occurrence of lithic material and stone tool production strategies. *Geoarchaeology: An International Journal* 9:375-391.

Andrefsky, W. Jr 1998. *Lithics: Macroscopic Approaches to Analysis.* Cambridge University Press, New York.

Arnold, J. E. 1985. Economic specialisation in prehistory: methods of documenting the rise of lithic craft specialisation. In S. C. Vehik (ed.), *Lithic Raw Material Procurement: Proceedings from the Second Conference on the Prehistoric Chert Exploration*, pp. 37-58. Centre for Archaeological Investigation, Occasional Papers 4. Southern Illinios University, Carbondale.

Arnold, J. E. 1987a. *Craft Specialisation in the Prehistoric Channel Islands, California.* University of California Publications in Anthropology 18. University of California Press, Berkeley.

Arnold, J. E. 1987b. Technology and economy: microblade core production from the Channel Islands. In J. K. Johnson and C. A. Morrow (eds), *The Organization of Core Technology,* pp. 207-238. Westview Press, Boulder.

Ataman, K., K. Carambelas, R. Elson, E Ingbar, M. Leach and C. Raven 1992. In R. Elston and C. Raven (eds), *Archaeological Investigation at Tosawihi, A Great Basin Quarry, Part 3: A Perspective from Locality 36,* pp. 1-14. Intermountain Research report for the Bureau of Land Management, Nevada.

Bailey, G. N. 1981. Concepts, time-scales and explanations in economic prehistory. In Sheridan and G. N. Bailey (eds), *Economic Archaeology: Towards an Integration of Ecological and Social Approaches,* pp. 97-117. BAR International Series 96, Oxford.

Bailey, G. N. 1983. Concepts of time in Quaternary prehistory. *Annual Review of Anthropology* 12:165-92

Bailey, G. N. 1987. Breaking the time barrier. *Archaeological Review from Cambridge* 6:5-20.

Baker, R. 1987. Bedrock to blank: archaeological investigation of the Lowes Mount Hatchery quarry, Oberon. Unpublished BA (Hons) thesis, University of Sydney, Sydney.

Barker, B. and R. Schon 1994. A preliminary assessment of the spatial distribution of stone artefacts from the South Molle Island Aboriginal quarry, Whitsunday Islands, Central Queensland. *Memoirs of the Queensland Museum* 37(1):5-12.

Bamforth, D. B. 1986. Technological efficiency and tool curation. *American Antiquity* 51:38-50.

Bamforth, D. B. 1990. Settlement, raw material and lithic procurement in the central Mojave Desert. *Journal of Anthropological Archaeology* 9:70-104.

Bamforth, D. B. 1991. Technological organization and hunter-gatherer land use: a Californian example. *American Antiquity* 56:216-234.

Bamforth, D. B. 1992. Quarries in context: a regional perspective on lithic procurement. In J. E. Arnold (ed.), *Stone tool Procurement, Production and Distribution in California Prehistory,* pp.131-150. Institute of Archaeology, University of California, Los Angeles.

Bamforth, D. B. and P. Bleed 1997. Technology, flaked stone technology, and risk. In C. M Barton and G. A Clark (eds), *Rediscovering Darwin: Evolutionary Theory and Archaeological Explanation,* pp. 109-139. Archaeological Papers of the American Anthropological Association No. 7, American Anthropological Association, Arlington.

Barker, B. and R. Schon 1994. A preliminary assessment of the spatial distribution of stone artefacts from the South Molle Island Aboriginal quarry, Whitsunday Islands, Central Queensland. *Memoirs of the Queensland Museum* 37(1):5-12.

Barton, C. M. 1991. Retouched tools, fact of fiction? In G. A. Clark (ed.), *Perspectives on the Past: Theoretical Biases in Mediterranean Hunter-gatherer Research*, pp.143-163. University of Pennsylvania, Philadelphia.

Bar-Yosef, O. 1991. Raw Material Exploitation in the Levantine Epi-Paleolithic. In A. Montet-white and S. Holen (eds) *Raw Material Economies Among Prehistoric Hunter Gathers,* pp. 235-250. Publications in Anthropology No. 19, University of Kansas, Lawrence

Baumler, M. F. 1988. Core reduction, flake production and the Middle Palaeolithic industry of Zobiste (Yugoslavia). In H. L. Dibble and A. Monet-White (eds), *Upper Pleistocene Prehistory of Western Eurasia*, pp. 255-274. The University Museum, University of Pennsylvania, Philadelphia.

Beck, C. and G. T. Jones 1994. On-site artefact analysis as an alternative to collection. *American Antiquity* 59:304-315.

Bettinger, R. L. 1980. Explainitory/predictive models of hunter-gatherer adaptation. In M. B. Schiffer (ed.), *Advances in Archaeological Method and Theory*, volume 3, pp. 189-255. Academic Press, New York.

Bettinger, R. L. 1991. *Hunter-Gatherers: Archaeological and Evolutionary Theory.* Plenum Press, New York.

Binford, L. R. 1973. Inter assemblage variability: the Mousterian and the 'functional' argument. In C. Renfrew (ed.), *The Explanation of Culture Change*, pp. 227-254. Duckworth Press, London.

Binford, L. R. 1975. Sampling, judgement and the archaeological record. In J. A. Mueller (ed.), *Sampling in Archaeology*, pp. 251-257. University of Arizona Press, Tucson.

Binford, L. R. 1977. Forty-seven trips: A case study in the character of archaeological formation processes. In R. V. S. Wright (ed.), *Stone Tools as Cultural Markers: Change, Evolution and Complexity*, pp. 24-39. Australian Institute of Aboriginal Studies, Canberra.

Binford, L. R. 1978a. *Nunamiut Ethnoarchaeology.* Academic Press, New York.

Binford, L. R. 1978b. Dimensional analysis of behaviour and site structure: learning from an Eskimo hunting stand. *American Antiquity* 43:330-361.

Binford, L. R. 1979. Organization and formation processes: looking at curated technologies. *Journal of Anthropological Research* 35:255-273.

Binford, L. R. 1980. Willow smoke and dogs' tails: hunter-gatherer settlement systems and archaeological site formation. *American Antiquity* 45:4-20.

Binford, L. R. 1981. Behavioral Archaeology and the 'Pompeii premise'. *Journal of Anthropological Research* 37:195-208.

Binford, L. R. 1983. The archaeology of place. In L. R. Binford (ed.), *Working at Archaeology*, pp.357-378. Academic Press, New York.

Binford, L. R. and. Binford, S. 1966. A preliminary analysis of functional variability in the Mousterian of Levallois facies. *American Anthropologist* 68 (2):238-295.

Binford, L. R. and O'Connell, J. F. 1984. An Alyawara day: the stone quarry. *Journal of Anthropological Research* 40:406-432.

Binford, L. R. and N. M. Stone 1985. "Righteous rocks" and Richard Gould: some observations on misguided "debate". *American Antiquity* 50:151-153.

Binns, R. and I. McBryde 1972. *A Petrological Analysis of Ground-edge Artefacts from Northern New South Wales*. Australian Institute of Aboriginal Studies, Canberra.

Bird, C. M. F. 1985. Prehistoric lithic resource utilisation: a case study from the southwest of Western Australia. Unpublished Ph.D. thesis, University of Western Australia, Perth.

Blanton, D. 1985. Lithic raw material procurement and use during the Morrow Mountain phase in South Carolina. In S. C. Vehik (ed.), *Lithic Raw Material Procurement: Proceedings from the Second Conference on the Prehistoric Chert Exploration*, pp. 115-134. Center for Archaeological Investigation, Southern Illinois University, Carbondale.

Bleed, P. 1986. The optimal design for hunting weapons: maintainability or reliability. *American Antiquity* 51:737-747.

Bleed, P. 1997. Contents as variability, result as selection: towards a behavioural definition of technology. In C. M Barton and G. A Clark (eds), *Rediscovering Darwin: Evolutionary Theory and Archaeological Explanation,* pp. 95-104. Archaeological Papers of the American Anthropological Association, Number 7. American Anthropological Association, Arlington.

Bobrowsky, R. D. and B. F. Ball 1989. The theory and mechanics of ecological diversity in archaeology. In R.D. Leonard and G.T. Jones (eds), *Quantifying Diversity in Archaeology*, pp. 4-12. Cambridge University Press, Cambridge.

Boot, P. 1987. Trampling damage on stone artefacts – some experimental results. *Australian Archaeology* 24:10-15

Bordes, F. 1961. Mousterian cultures of France. *Science* 134:803-810.

Bordes, F. 1968. *The Old Stone Age*. McGraw-Hill, New York.

Bordes, F. 1969. Reflections on typology and technique in the Palaeolithic. *Arctic Anthropology* 6:1-29.

Bordes, F. and Crabtree, D. 1969. The Corbiac blade technique and other experiments. *Tebiwa* 12:1-21.

Bradley, B. A. 1975. Lithic reduction sequences: a glossary and discussion. In E. Swanson (ed.), *Lithic Technology: Making and Using Stone Tools*, pp.5-14. Mouton, The Hague.

Brock, D. 1975. *To the Desert with Sturt: A Diary of the 1844 Expedition.* Royal Geological Society of Australasia, Adelaide.

Burrows, W. H. 1973. Regeneration and spatial patterns of Acacia Aneura in south west Queensland. *Tropical Grasslands*, 7(1):57-68.

Butler B. M. and E. E. May (eds) 1984. *Prehistoric Chert Exploitation: Studies from the Midcontinent.* Occasional Papers Number 2. Centre for

Archaeological Investigations, Southern Illinois University at Carbondale, Illinois.

Byrne, D. 1980. Dynamics of Dispersion: the place of silcrete in archaeological assemblages from the Lower Murchison, Western Australia. *Archaeology and Physical Anthropology in Oceania,* 15:110-119.

Campbell, T. and R. Edwards 1966. Stone implements. In B. C. Cotton (ed.), *Aboriginal Man in South and Central Australia*, pp. 159-220. South Australian Branch of the British Science Guild Handbooks Committee, Adelaide.

Camilli, E. L. and J. I. Ebert 1992. Artefact reuse and recycling in continuous surface distributions and implications for interpreting land use patterns. In J. Rossignol and L. Wandsnider (eds), *Space, Time and Archaeological Landscapes,* pp.113-136. Plenum Press, New York.

Carr, C. 1991. Left in the dust: contextual information in model-focused archaeology. In E. M. Kroll and T. D. Price (eds), *The Interpretation of Archaeological Spatial Patterning,* pp.221-256. Plenum Press, New York.

Carr, P. J. 1994. The organization of technology: impact and potential. In P. J. Carr (ed.), *The Organization of North American Chipped Stone Tool Technologies*, pp. 1-8. Archaeological Series 7, International Monographs in Prehistory, Ann Arbor, Michigan.

Chang, K. C. 1968. Toward a science of prehistoric society. In K. C. Chang (ed.), *Settlement Archaeology,* pp.1-9. National Press Books, Palo Alto.

Chatters, J. C. 1989. Hunter-gatherer adaptations and assemblage structure. *Journal of Anthropological Research* 6:336-375.

Clark, J., W. C. Dix, C. E. Dortch and K. Palmer 1978. Aboriginal sites on the Millstream Station, Pilbara, Western Australia. *Records of the Western Australian Museum* 6(2): 221-257.

Clarke, D. L. 1977 Spatial Information in Archaeology. In D.L. Clarke (ed.), *Spatial Archaeology*. Academic Press, New York.

Clarkson, C. and B. David 1995. The antiquity of blades and points revisited: investigating the emergence of systematic blade production south-west of Arnhem Land, northern Australia. *The Artefact* 18:22-44.

Cleghorn, P. L. 1986. Organisational structure of the Mauna quarry complex, Hawaii. *Journal of Archaeological Science*. 13(4):375-387.

Cochrane, G. W. 1998. Toward an unfamiliar past, new approaches to spatial analysis at Stud Creek, northwest NSW. Unpublished B.A. (Hons) thesis, La Trobe University, Melbourne.

Collins, M. 1975. Lithic technology as a means of processual inference. In E. Swanson (ed.), *Lithic Technology: Making and Using Stone Tools*, pp. 15-34. Mouton, The Hague.

Cotterell, B. and J. Kamminga 1979. The mechanics of flaking. In B. Hayden (ed.), *Lithic Use-Wear Analysis*, pp. 97-112. Academic Press, San Diego.

Cotterell, B. and J. Kamminga 1987. The formation of flakes. *American Antiquity* 52:675-708.

Crabtree, D. E. 1966. A stoneworkers approach to analysing and replicating the Lindenmeir Folson. *Tebiwa* 10(1):3-39.

Crabtree, D. E. 1968. Mesoamerican polyhedral cores and prismatic blades. *American Antiquity* 33:446-79.

Crabtree, D. E. 1972. *An Introduction to the Technology of Stone Tools.* Occasional Papers of the Idaho State University Museum, Number 28. Pocatello, Idaho.

Cramsie, J. N. amd J. M. Hawke 1985. Introduction. In J. M. Hawke and J. N. Cramsie (eds), *Contributions to the Geology of the Great Australian Basin in New South Wales*, pp.1-12. Bulletin No. 31, Geological Survey of NSW, Department of Mineral Resources.

Cunningham, G. M. and P. J. Walker, 1973. Growth and survival of mulga (*Acacia Anerua F. Muell. Ex Benth*) in western New South Wales. *Tropical Grasslands* 7(1):69-77.

Curran, M. and J. R. Grimes 1989. Ecological implications for Paleoindian lithic procurement economy in New England. In C. J. Ellis and J. C. Lothrop (eds), *Eastern Paleoindian Lithic Resource Use*, pp. 41-74. Westview Press, Boulder.

Dallas, M, P. Mitchell, H. Barton and J. Beckett 1995. Archaeological and anthropological study of the proposed connection of grid power to Sturt and Mootwingee National Parks environmental impact statement. Report of Mandis Roberts Consultants for NSW National Parks and Wildlife Service.

Davis, R. S. 1983. Theoretical issues in contemporary Soviet Palaeolithic archaeology.*Annual Review of Anthropology* 12:403-428.

Denny, M. 1994. Investigating the past: an approach to determining the changes in the fauna of the western division of NSW since the first explorers. In D. Cunney, S. Hand, P. Reed and D. Butcher (eds), *Future of the Fauna in Western New South Wales*, pp. 53-63. Transactions of the Royal Zoological Society of New South Wales, Mosman.

Dibble, H. L. 1984. Interpreting typological variation of Middle Palaeolithic scrapers: function, style or sequence of reduction? *Journal of Field Archaeology* 11:431-436.

Dibble, H. L. 1987. The interpretation of Middle Palaeolithic scraper morphology. *American Antiquity* 52:109-117.

Dibble, H. L. 1991. Local raw material exploitation and its effects on Lower and Middle Paleolithic assemblage variability. In A. Montet-white and S. Holen (eds), *Raw Material Economies Among Prehistoric Hunter Gathers,* pp. 33-47. Publications in Anthropology No. 19, University of Kansas, Lawrence.

Dibble, L. H. 1995. Middle Paleolithic scraper reduction: background, clarification and review of evidence to date. *Journal of Archaeological Method and Theory* 2:299-368.

Dibble, L. H. and A. Debenath 1991. Paradigmatic differences in a collaborative research project. In G. A. Clark (ed.), *Perspectives on the Past: Theoretical Biases in Mediterranean Hunter-gatherer Research,* pp. 217-226. University of Pennsylvania Press, Philadelphia.

Dibble, H. L. and A. Pelcin 1995. The effects of hammer mass and velocity on flake mass. *Journal of Archaeological Science* 22:429-439.

Dibble, H. L. and N. Rolland 1992. On assemblage variability in the Middle Palaeolithic of Western Europe: history, perspectives and a new synthesis. In H. Dibble and P. Mellars (eds), *The Middle Palaeolithic Adaptation, Behaviour and Variability*, pp. 1-28. The University Museum, Philadelphia.

Dickman, C. R. 1994. Native mammals of Western New South Wales: past neglect, future rehabilitation. In D. Cunney, S. Hand, P. Reed and D. Butcher (eds), *Future of the Fauna in Western New South Wales*, pp. 81-91. Transactions of the Royal Zoological Society of New South Wales, Mosman.

Dobson, J. E. 1974a. Vegetation and climatic history from Lake Keilambete, West Victoria. *Australian Journal of Botany* 22:709-717

Dobson, J. E. 1974b. Vegetational history and water fluctuations at Lake Leake, South East Southern Australia. *Australian Journal of Botany* 22:719-741.

Domanski, M. and J. A. Webb 2000. Flaking properties, petrology and use of polish flint. *Antiquity* 74: 822-832.

Domanski, M., J. A. Webb and J. Boland 1994. Mechanical properties of stone artefact materials and the effect of heat treatment. *Archaeometry* 36:177-208.

Dortch C. E. 1984. *Devils Lair, A Study in Prehistory*. Western Australian Museum, Perth.

Dortch, C. E. and G. Gardiner 1976. Archaeological investigations in the Northcliffe District, Western Australia. *Records of the Western Australian Museum* 4(3):257-293.

Dunnell, R. C. 1971. *Systematics in Prehistory*. Free Press, New York.Dunnell, R. C. 1992. The notion site. In J. Rossignol and L. Wandsnider (eds), *Space, Time and Archaeological Landscapes,* pp. 21-41. Plenum Press, New York.

Dunnell, R. C. and W. S. Dancey. 1983. The siteless survey. A regional scale data collection strategy. *Advances in Archaeological Method and Theory* 6:267-287.

Dury, G. and T. Langford-Smith 1965. A Pleistocene Aboriginal camp fire from Lake Yantara, northestern new South Wales. *Search* 1(2):73.

Ebert, J. I. 1992. *Distributional Archaeology*. University of New Mexico Press, Albuquerque.

Edmonds, M. R. 1987, Rocks and Risk: Problems with lithic procurement strategies. In A. G. brown and M. R. Edmonds (eds), *Lithic Analysis and Later British Prehistory: some problems and Approaches*, pp. 155-179. BAR International Series 162, Oxford.

Ellis, C. S. 1989. The explaination of northeastern paleoindian lithic procurement patterns. In C. S. Ellis and J. C. Lothrop (eds), *Eastern Paleoindian Lithic Resource Use,* pp. 139-164. Westview press, Boulder.

Ellis, C. S. and J. C. Lothrop (eds) 1989. *Eastern Paleoindian Lithic Resource Use*. Westview press, Boulder.

Elston, R. G. 1990. A cost-benefit model of lithic assemblage variability. In R. G. Elston and E. E. Budy (eds), *The Archaeology of James Creek Shelter*, pp. 153-163. University of Utah Anthropological Papers No. 13, University of Utah Press, Salt Lake City.

Elston, R. and C. Raven 1992. *Archaeological Investigation at Tosawihi, A Great Basin Quarry, Part 3: A Perspective from Locality 36*. Intermountain Research report for the Bureau of Land Management, Nevada.

Ericson, J. 1981. Exchange and production systems in California. BAR International Series 110, Oxford.

Ericson, J. 1982. Production for obsidian exchange in California. In J. E. Ericson and T. K. Earle (eds), *Contexts of Prehistoric Exchange,* pp. 129-148. Academic Press, New York.

Ericson, J. E. 1984. Toward the analysis of lithic production systems. In J. E. Ericson and B. A. Purdy (eds), *Prehistoric Quarries and Lithic Production*, pp.1-9. Cambridge University Press, Cambridge.

Ericson, J. E. and B. A. Purdy (eds) 1984. *Prehistoric Quarries and Lithic Production.* Cambridge University Press, Cambridge.

Fanning, P. C. 1994. Long-term erosion rates in an arid rangelands environment in western New South Wales, Australia. *Journal of Arid Environments* 28:73-187.

Fanning, P.C. 1999. Recent landscape history in arid western NSW, Australia: a model for regional change. *Geomorphology* 35:445-453.

Fanning, P. C., S. Holdaway, D. Witter, J. Martin, G. Nicholls, J. Reeves and J. Shiner 2000a. Temporal patterns of aboriginal occupation in northwestern New South Wales, Australia, during the late Holocene: a reflection of environmental or socio-political change? Paper presented at the Quaternary Studies Meeting, Australian National University.

Fanning, P. C., D. Witter and S. Holdaway 2000b. Use of a backhoe for stratigraphic test pits. In S. Holdaway, D. Witter and P. C. Fanning (eds), report of the Western New South Wales Archaeological Project, pp.15-18. Unpublished report, University of Auckland, Auckland.

Fanning, P. C. and S. Holdaway 2001a. Stone artefact scatters in western NSW, Australia: geomorphic controls on artefact size. *Geoarchaeology: an International Journal* 16(6), 667-686.

Fanning, P. C. and S. Holdaway 2001b. Temporal limits to the archaeological record in arid western NSW, Australia: lessons from OSL and radiocarbon dating of hearths and sediments. In P. Sheppard and M. Jones, Australasian Connections and New Directions: Proceedings of the 7th Australasian Archaeometry Conference, Research in Anthropology and Linguistics 5, 85-105. University of Auckland, Auckland.

Farmer, I. W. 1968. *Engineering Properties of Rocks.* Spoon and Spoon, London. Findlow,

Findlow F. J. and M. Bolognese 1982. Regional modelling of obsidian procurement in the American Southwest. In T. K. Earle and J. E. Ericson (eds),

Exchange Systems in Prehistory, pp. 53-81. Academic Press, New York.

Fladmark, K. R. 1984. Mountain of glass: archaeology of the Mt Edziza obsidian source, British Columbia Canada. *World Archaeology,* 16(2):139-156.

Flenniken, J. 1985. Stone tool reduction technologies as cultural markers. In M. G. Plew, J. C. Woods and M. G. Pavesic (eds), *Stone Tool Analysis: Essays in Honour of D. E Crabtree*, pp. 265-276. University of New Mexico Press, Albuquerque.

Flenniken, J. and J. P. White 1985. Australian flaked stone tools: a technological perspective. *Records of the Australian Museum* 36:131-151.

Fletcher, M. and G. Lock 1994. *Digging Numbers: Elementary Statistics for archaeologists.* Oxford University Committee for Archaeology, Oxford.

Fletcher, R. 1992. Time perspectivism, Annales and the potential of archaeology. In A. Knapp (ed.), *Archaeology, Annales and Ethnohistory,* pp. 35-49. Cambridge University Press, Cambridge.

Foley, R. 1981a. Off-site Archaeology: an alternative approach for the short-sited. In I.Hodder, G, Isacc and N. Hammond (eds), *Pattern of the Past: Studies in Honour of David Clarke*, pp. 157-183. Cambridge University Press, Cambridge.

Foley, R. 1981b. *Off-site Archaeology and Human Adaptation in Eastern Africa. An Analysis of Regional Artefact Density in the Amboseli, Southern Kenya.* BAR International Series 97, Oxford.

Foley, R. 1981c. A model of regional archaeological structure. *Proceedings of the Prehistoric Society* 47:1-17.

Fox, C. 1922. *The Archaeology of the Cambridge Region.* Cambridge University Press, Cambridge.

Frankel, D. 1988. Characterising change in prehistoric sequences: A view from Australia. *Archaeology in Oceania* 23:41-48.

Frankel, D. 1995. The Australian transition: real and perceived boundaries. In J. Allen and J. O'Connell (eds), Transitions, Special Number 265, *Antiquity* 69:649-655.

Frankel, D. and S. Holdaway 1993. Diverse: a computer program to calculate various measures of diversity. Department of Archaeology. La Trobe University, Melbourne.

Freeman, L. G. 1991. What mean these stones? Remarks on raw material use in the Spanish Paleolithic. In A. Montet-white and S. Holen (eds*), Raw Material Economies Among Prehistoric Hunter Gathers,* pp. 73-125. Publications in Anthropology No. 19, University of Kansas, Lawrence.

Frison, G. 1978. *Prehistoric Hunters of the High Plains.* Academic Press, New York.

Frison, G. 1968. A functional analysis of certain chipped stone tools. *American Antiquity* 22:140-155.

Fuller, S. 1995. *Rocks and Minerals.* Dorling Kingsley Ltd, London.

Fullagar, R. L. K., and Torrence, R. 1991. Obsidian exploitation at Umleang, Lou Island. In J. Allen and C. Gosden (eds), *Report of the Lapita Homeland Project*, pp. 113-143. Occasional Papers in Prehistory 20. Department of Prehistory, Research School of Pacific Studies, Australian National University, Canberra.

Gamble, C 1986. *The Palaeolithic Settlement of Europe.* Cambridge University press, Cambridge.

Gerritson, J. 1981. *Tibooburra – Corner Country.* Tibooburra Press.

Gibson, E. 1984. Reconstructing Corbiac: the context of manufacturing at an Upper Paleolithic quarry. In J. E. Ericson and B. A. Purdy (eds), *Prehistoric Quarries and Lithic Production*, pp. 139-146. Cambridge University Press, Cambridge.

Gifford-Gonzalez, P., D. Damrosch, J. Pryer and R. Thunen. 1985. The third dimension in site structure: an experiment in trampling and vertical dispersal. *American Antiquity* 50:803-818.

Goodyear, A. C. 1989. A hypothesis for the use of cryptocrystalline raw materials among Paleoindian groups of North America. In C. J. Ellis and J. C. Lothrop (eds), *Eastern Paleoindian Resource Use,* pp. 1-9. Westview Press, Boulder.

Gould, R. A. 1977. Ethno-archaeology; or where do the models come from? In R. V. S. Wright (ed.), *Stone Tools and Cultural Markers*, pp. 162-168. Australian Institute of Aboriginal Studies, Canberra.

Gould, R. A. 1980. *Living Archaeology.* Cambridge University Press, Cambridge.

Gould, R. A. and S. Saggers 1985. Lithic procurement in central Australia: a closer look at Binford's idea of embeddedness in archaeology. *American Antiquity* 50:117-136.

Gould, R. A. and P. Watson 1982. A dialogue on the meaning and use of analogy in ethnoarchaeological reasoning. *Journal of Anthropological Archaeology* 1(4):355-381.

Gramly, R. M. 1980. Raw material source areas and 'curated' tool assemblages. *American Antiquity* 45:823-833.

Gramly, R. M. 1984. Mount Jasper: a direct-access lithic source area in the White Mountains of New Hampshire. In J. E. Ericson and B. A. Purdy (eds*), Prehistoric Quarries and Lithic Production*, pp. 11-21. Cambridge University Press, Cambridge.

Greenwood, S. 1997. Fluvial Processes and Refitting Stone Artefacts at Stud Creek, Sturt National Park. Unpublished B.A. (Hons) thesis, Department of Archaeology, Latrobe University, Melbourne.

Hall, J. and W. Love 1985. Prickly Bush: a site with backed blades on the Brisbane River: a pilot study towards the measurement of site "disturbance". *Queensland Archaeological Research* 2:71:81.

Hartwell, W. T. 1994. Prehistoric landuse patterning at the Yucca Wash Quarry site. Unpublished paper presented at the 59th Annual Meeting of the Society for American Archaeology, Anaheim, California.

Hartwell, W. T. and D. S. Amick, 1993. Archaeological investigation of early land use at Midway Valley site: an opportunistic quarry on the Nevada Test Site, South Nye Counry, Nevada. *Current Research in the Pleistocene* 10:57-59.

Hawke, J. M. and D. Bourke. Stratigraphy. In J. M. Hawke and J. N. Cramsie (eds), *Contributions to the*

Geology of the Great Australian Basin in New South Wales, pp. 13-21. Bulletin No. 31, Geological Survey of NSW, Department of Mineral Resources.

Hayden, B. 1977. Stone tool functions in the Western Desert. In R. V. S. Wright (ed.), *Stone Tools as Cultural Markers: Change, Evolution and Complexity*, pp. 178-188. Australian Institute of Aboriginal Studies, Canberra.

Hayden, B. 1989. From chopper to celt: the evolution of resharpening techniques. In R. Torrence (ed.), *Time, Energy and Stone Tools*, pp. 7-16. Cambridge University Press, Cambridge.

Hayden, B., N. Franco and J. Spafford 1996. Evaluating lithic strategies and design criteria. In G.H. Odell (ed.), *Stone Tools: Theoretical Insights into Human Prehistory*, pp. 9-45. Plenum Press, New York.

Henry, D. O. 1989. Correlations between reduction strategies and settlement patterns. In D. O. Henry and G. H. Odell (eds), *Alternative Approaches to Lithic Analysis*, pp. 139-212. Archaeological Papers of the American Anthropological Association, Number 1, American Anthropological Association, Washington D. C.

Hietala, H. J. 1984. Intrasite Spatial Analysis in Archaeology. Cambridge University Press, Cambridge.

Hill, S. M., G. Taylor, R. A. Eggleton 1994. *Field Guide and Notes on the Regolith And Landscape of the Broken Hill Region, Western New South Wales*. Australia Regolith Conference, Australian Geological Survey Organisation Record 57, Canberra.

Hiscock, P. 1986a. Technological change in the Hunter River Valley and the interpretation of late Holocene change in Australia. *Archaeology in Oceania* 21:40-50.

Hiscock, P. 1986b. Raw material rationing as an explanation of assemblage differences: a case study of Lawn Hill, northwest Queensland. In G. K. Ward (ed.) *Archaeology at ANZAAS, 1984*, pp. 178-190. Canberra Archaeological Society Incorporated, Canberra.

Hiscock, P. 1988. Prehistoric settlement patterns and artefact manufacture at Lawn Hill, Northwest Queensland. Unpublished Ph.D. thesis, Queensland University, Brisbane.

Hiscock, P. 1993. Bondaian technology in the Hunter Valley, New South Wales. *Archaeology in Oceania* 28:65-76.

Hiscock, P. 1994. Technological responses to risk in Holocene Australia. *Journal of World Prehistory* 8:267-292.

Hiscock, P. 1996. Mobility and technology in the Kakadu coastal wetlands. *Bulletin of the Indo-Pacific Prehistory Association* 15:151-158.

Hiscock, P. and V. Attenbrow. 1998. Early Holocene backed artefacts from Australia. *Archaeology in Oceania* 33:49-62.

Hiscock, P. and S. Mitchell, 1990. Type profiles: stone artefact quarries, stone production and ochre quarries. Unpublished report, Australian Heritage Commission.

Hiscock, P. and S. Mitchell 1993. *Stone Artefacts Quarries and Reduction Sites in Australia: Towards a Type Profile.* Australian Government Publishing Service, Canberra.

Hiscock, P. and P. Veth, 1991. Change in the Australian desert culture; a reanalysis of tulas from Puntutjarpa rockshelter. *World Archaeology* 22:332-345.

Hodder, I. 1987. The meaning of discard: ash and domestic space in Baringo. In S. Kent (ed.), *Method and Theory for Activity Area Research: an ethnoarchaeological Approach*, pp. 424-448. Columbia University Press, New York.

Holdaway, S. 1998. Bootstrapping program. Department of Archaeology. La Trobe University, Melbourne.

Holdaway, S., P. Fanning and D. Witter 2000a. Report of the Western New South Wales Archaeological Project. Unpublished report, University of Auckland, New Zealand.

Holdaway, S., P. Fanning and D. Witter 2000b. Prehistoric Aboriginal occupation of the rangelands: interpreting the surface archaeological record of far Western New South Wales, Australia. *The Rangelands Journal*, 22(1):58-71.

Holdaway, S., P. Fanning, D. Witter, M. Jones, G. Nicholls and J. Shiner 2002. Variability in the chronology of late Holocene Aboriginal occupation on the arid margin of southeastern Australia. *Journal of Archaeological Science*, 29:351-363.

Holdaway S, Fanning P. and J. Shiner 2005. Absence of evidence or evidence of absence? Understanding the chronology of indigenous occupation of western New South Wales, Australia. *Archaeology in Oceania* 40: 33-49.

Holdaway S, Shiner J and Fanning P. 2004. Hunter-gatherers and the archaeology of the long term: An analysis of surface, stone artefact scatters from Sturt National Park, New South Wales, Australia. *Asian Perspectives* 43:34-72.

Holdaway, S., D. Witter, P. Fanning, R. Musgrave, R., G. Cochrane, T. Doelman, S. Greenwood, D. Pigdon and J. Reeves 1998. New approaches to open site spatial archaeology in Sturt National Park, New South Wales, Australia. *Archaeology in Oceania* 33:1-19.

Holdaway, S. and N. Stern 2004. Written in Stone: Decoding the Australian Flaked Stone Record. Museum Victoria, Melbourne.

Horne, G. and G. Aiston 1924. *Savage life in Central Australia.* MacMillian and Company Limited, London.

Huang, J. and S. Wang. 1985. An experimental investigation concerning the comprehensive fracture toughness of some brittle rocks. *International Journal of Rock Mechanics, Mining Science and Geomechanics* 22:99-104.

Hughes, P. J., M. E. Sullivan and R. J. Lampert, 1973. The use of silcrete by aboriginals in southern coastal New South Wales. *Archaeology and Physical Anthropology in Oceania* 8:220-225.

Ingbar, E. E. 1994. Lithic material selection and technological organisation. In P. J.Carr (ed.), *The Organisation of North American Prehistoric Chipped Stone Tool Technologies*, pp. 45-56. International

Monographs in Prehistory, Archaeological Series 7, Ann Arbor, Michigan.

Ingbar, E. E., M. L. Larson and B. A. Bradely 1989. A nontypological approach to debitage analysis. In D. S. Amick and R. P. Mauldin (eds), *Experiments in Lithic Technology*, pp. 117-136. BAR International Series 528, Oxford.

Inizan, M. L., H. Roche and J. Tixier 1992. *Technology of Knapped Stone.* Préhistoire de la Pierre Taile'e, Tome 3. Cercle de Récherches et d'Etudes Préhistoriques, Meudon.

Irons, N. and E. Quinlan 1988. *A Review of Information Relevant to the Mulga Rangelands of Western NSW.* Technical Report Number 11, Soil Conservation Service of NSW, Chatswood.

Isaac, G. and J. Harris 1975. The scatter between the patches. Unpublished paper presented to the Kroeber Anthropological Society, Berkeley.

Isaac, G. and J. Harris 1980. A method for determining the characteristics of artefacts between sites in the Upper Member of the Koobi Fora Formation, East Lake Turkana. In R. E. Leakey and B. A. Ogot (eds), *Proceedings of the 8th Panafrican Congress of Prehistory and Quaternary Studies, Nairobi, 5 to 10 September 1977,* pp.19-22. The International Louis Leakey Memorial Institute for African Prehistory, Nairobi.

Jelinek, A. J. 1991. Observations on reduction patterns and raw materials in some Middle Paleolithic industries in the Perigord. In A. Montet-white and S. Holen (eds), *Raw Material Economies Among Prehistoric Hunter Gathers,* pp. 7-31. Publications in Anthropology No. 19, University of Kansas, Lawrence.

Jeske, R. J. 1989. Economies in raw material use by prehistoric hunter-gatherers. In R. Torrence (ed.), *Time, Energy and Stone Tools*, pp. 34-45. Cambridge University Press, Cambridge.

Jeske, R. J. 1992. Economics in raw material use by prehistoric hunter-gathers. In *The Organisation of North American Prehistoric Chipped Stone Tool Technologies*, pp. 34-45. International Monographs in Prehistory, Archaeological Series 7, Ann Arbor, Michigan

Jeske, R. J. 1992. Energetic efficiency and lithic technology: an upper Mississippian example. *American Antiquity* 57:467-481.

Jochim, M. A. 1983. Optimisation models in context. In J. A. Moore and A. S. Keene (eds), *Archaeological Hammers and Theories*, pp. 157-172. Academic Press, New York.

Jochim, M. A. 1989. Optimisation and stone tool studies: problems and potential. In R. Torrence (ed.), *Time, Energy and Stone Tools*, pp. 106-111. Cambridge University Press, Cambridge.

Johnson, I. 1984. Cell frequency recording and analysis of artefact distributions. In H. J. Hietala (ed.), *Intrasite Spatial Analysis in Archaeology*, pp. 75-96. Cambridge University Press, Cambridge.

Johnson, J. K. 1979. Archaic blade manufacture: production failures, a chronicle of the misbegotten. *Lithic Technology* 8:23-35.

Johnson, J. K. 1984a. Measuring prehistoric quarry site activity in northeastern Mississippi. In B. M. Butler and E. E. May (eds), *Prehistoric Chert Exploitation: Studies from the Midcontinent*, pp 225-235. Occasional Papers Number 2. Centre for Archaeological Investigations, Southern Illinois University at Carbondale, Illinois.

Johnson, J. K. 1984b. Patterns of raw material procurement in Colbert Ferry Park, northwest Alabama. In S. C. Vehik (ed.), *Lithic Raw Material Procurement: Proceedings from the Second Conference on the Prehistoric Chert Exploration,* pp. 153-164. Centre for Archaeological Investigation, Occassional Papers 4. Southern Illinois University, Carbondale.

Johnson, J. K. 1987a. Introduction. In J. K. Johnson and C. A. Morrow (eds), *The Organization of Core Technology*, pp. 1-12. Westview Press, Boulder.

Johnson, J. K. 1987b. Cahokia core technology in Mississippi: the view from the south. In J. K. Johnson and C. A. Morrow (eds), *The Organisation of Core Technology*, pp. 187-206. Westview Press, Boulder.

Johnson, J. K. 1989. The utility of production trajectory modelling as a framework for regional analysis: the view from the south. In D. Henry and G.H. Odell (eds), *Alternative Approaches to Lithic Analysis*, pp. 119-138. Archaeological Papers of the American Anthropological Association 1, American Anthropological Association, Washington D. C.

Johnson, J. K. and C. A. Morrow (eds) 1987. *The Organization of Core Technology.* Westview Press, Boulder.

Johnson, H. and P. Clark 1998. Willandra Lakes archaeological investigations 1968-98. *Archaeology in Oceania* 33:105-119.

Jones, G. T. and C. Beck. 1992. Chronological resolution in Distributional Archaeology. In J. Rossignol and L. Wandsnider (eds), *Space, Time and Archaeological Landscapes,* pp.167-192. Plenum Press, New York.

Jones G. T. and R. D. Leonard 1989. The diversity concept: an introduction. In R. D. Leonard and G. T. Jones (eds), *Quantifying Diversity in Archaeology,* pp. 1-2. Cambridge University Press, Cambridge.

Jones, K. L. 1984. Polynesian quarrying and flaking practices at the Samson Bay and Falls Creek quarries, Tasman Bay, New Zealand. *World Archaeology* 16(2):248-266.

Jones, R. and N. White, 1988. Point Blank stone tool manufacture at the Ngilipitji quarry, Arnhem Land, 1981. In B. Meehan and R. Jones (eds), *Archaeology with Ethnography: An Australian Perspective*, pp. 51-87. Research School of Pacific Studies, Australian National University, Canberra.

Karlin, C. and N. Julien 1994. Prehistoric technology: a cognitive science? In C. Renfrew and Z. B. W. Zubrow (eds), *The Ancient Mind, Element of Cognitive Archaeology*, pp. 152-164. Cambridge University Press, Cambridge.

Kelly, R. L. 1988. The three sides of a biface. *American Antiquity* 53:717-734.

Kelly, R. L. 1994. Some thoughts on future direction in the study of stone tool technological organisation. In P. J. Carr (ed.), *The Organisation of North American Prehistoric Tool Technology*, pp. 132-136. International Monographs in Prehistory, Archaeological Series 7, Ann Arbor, Michigan.

Kozlowski, J. K. 1991. Raw Material procurement in the Upper Paleolithic of Central Europe. In A. Montet-white and S. Holen (eds), *Raw Material Economies Among Prehistoric Hunter Gathers,* pp. 187-196. Publications in Anthropology No. 19, University of Kansas, Lawrence.

Kuhn, S. L. 1989. Hunter-gather foraging organisation and strategies of artefact replacement and discard. In D.S. Amick and R. P. Mauldin (eds), *Experiments in Lithic Technology*, pp. 33-48. BAR International Series 528. Oxford

Kuhn, S. L. 1991. 'Unpacking' reduction: lithic raw material economy in the Mousterian of west-central Italy. *Journal of Anthropological Archaeology* 10:76-106.

Kuhn, S. L. 1992. On planning and curated technologies in the Middle Palaeolithic. *Journal of Anthropological Research* 48:185-214.

Kuhn, S. L. 1994. A formal approach to the design and assemblage of mobile tool kits. *American Antiquity* 59:426-442.

Kuhn, S. L. 1995. *Mousterian Lithic Technology, An Ecological Perspective*. Princeton University Press, Princeton, New Jersey.

Kvamme, K. L. 1989. Geographic Information Systems in regional archaeological research and data management. *Archaeological Method and Theory* 1:139-203.

Lamb, L. 1995. A methodology for the analysis of backed artefact production on the South Molle Island Quarry, Whitsunday Islands. *Tempus* 6:151-159.

Lampert, R. 1981. *The Great Kartan Mystery*. Terra Australis 5. Australian National University, Canberra.

Lampert, R. and P. Hughes 1974. Sea level changes and Aboriginal adaptations. *Archaeology and Physical Anthropology* 9:226-235

Langford-Smith, T. (ed.) 1978. *Silcrete in Australia*. Department of Geography Armidale, University of New England,

Langford-Smith, T. and G. H. Dury 1965.Distribution, character and altitude of the duricrust in the northwest of New South Wales and the adjacent areas of Queensland. *American Journal of Science* 263:170-190.

Lance, A. 1992. Further studies of an aboriginal archaeological site near Tibooburra, New South Wales. Unpublished report, National Heritage Studies, Adelaide.

Lapidus, D. F. 1990. *Collins Dictionary of Geology*. Harpercollins, Glasgow

Larson, M. 1994. Towards a holistic analysis of chipped stone assemblages. In P. J. Carr (ed.), *The Organisation of North American Prehistoric Chipped Stone Tool Technologies*, pp. 57-69. International Monographs in Prehistory, Archaeological Series 7, Ann Arbor, Michigan.

Leach, H. M. 1984. Jigsaw: reconstructive lithic technology. In J. E. Ericson and B. A. Purdy (eds), *Prehistoric Quarries and Lithic Production*, pp. 107-118. Cambridge University Press, Cambridge.

Leonard, R. D, F. E. Smiley and C. M. Cameron 1989. Changinf strategies of Anasazi lithic procurement on Black Mesa. In Leonard, R. D. and G. T. Jones (eds), Quantifying Diversity in Archaeology, pp.100-108 Cambridge University Press, Cambridge.

Lewarch, D. E. and M. J. O'Brien. 1981. The expanding role of surface assemblages in archaeological research. *Advances in Archaeological Method and Theory* 4:297-342.

Lipo, C. P., M. E. Madsen, R. C. Dunnell and T. Hunt. 1997. Population structure, cultural transmission, and frequency seriation. *Journal of Anthropological Archaeology* 16:301-333.

Lothrop, J. C. 1989. The organisation of Paleoindian lithic technology at the Potts site. In C. J. Ellis and J. C. Lothrop (eds), *Eastern Paleoindian Lithic Resource Use,* pp. 99-138. Westview Press, Boulder.

Lourandos, H. 1983. Intensification: a late Pleistocene-Holocene archaeological sequence from southwestern Victoria. *Archaeology in Oceania* 18:81-97.

Lourandos, H. 1984. Changing perspectives in Australian prehistory: a reply to Beaton. *Archaeology in Oceania* 19:29-33.

Lourandos, H. 1985a. Intensification and Australian prehistory. In T.D. Price and J.A. Brown (eds), *Prehistoric Hunter-Gatherers: the Emergency of Cultural Complexity,* pp. 385-426. Academic Press, Orlando.

Lourandos, H. 1985b. Problems with the interpretation of late Holocene changes in Australian prehistory. *Archaeology in Oceania* 20:37-39.

Lourandos, H. and A. Ross 1994. The 'great intensification debate': it's history and place in Australian archaeology. *Australian Archaeology* 39:54-63.

Luedtke, B. E. 1984. Lithic material demand and quarry production. In J. E. Ericson and B. A. Purdy (eds), *Prehistoric Quarries and Lithic Production*, pp. 65-76. Cambridge University Press, Cambridge.

Lurie, R. 1989. Lithic technology and mobility strategies: the Koster site Middle Archaic. In R. Torrence (ed.), *Time, Energy and Stone Tools*, pp. 46-56. Cambridge University Press, Cambridge.

McBryde. I. 1973. Stone Arrangements and a quartzite quarry site at Brewarrina. *Mankind* 9:118-121.

McBryde, I. 1974. *Aboriginal Prehistory in New England.* Sydney University Press, Sydney.

McBryde, I. 1978. Wil-im-ee Moor-ing: or where do the stone axes come from? *Mankind* 3:354-382.

McBryde, I, 1979. Petrology and prehistory: lithic evidence for the exploitation of stone resources and exchange systems in Australia. In T. H. McClough and

W.A. Cummings (eds), *Stone Axe Studies*, pp. 113-126. CBA Research report No. 23, London.

McBryde, I. 1984a. Exchange in Southeastern Australia: an ethnohistorical perspective. *Aboriginal History* 8:132-153.

McBryde, I. 1984b. Kulin greenstone quarries: the social contexts of production and distribution of the Mt William site. *World Archaeology* 16:267-285.

McBryde, I. 1986. Artefacts, language and social interaction: a case study from southeastern Australia. In G.N. Bailey and P. Callow (eds), *Stone Age Prehistory*, pp. 77-93. Cambridge University press, Cambridge.

McBryde, I. and G. Harrison 1981. Valued good or valuable stone: a consideration of some aspects of the distribution of greenstone artefacts in south-eastern Australia. In F. leach and J. Davidson (eds), *Archaeological Studies of Pacific Stone Resources*, pp. 183-208. BAR International Series No. 104, Oxford.

McBryde, I. And A. Watchman, 1976. The Distribution of greenstone adzes in southeast Australia: a preliminary report. *Mankind* 10:163-174.

McCarthy, F. D. 1976. *Australian Stone Impliements*. Australian Museum, Sydney.

McCourt, T. 1975. *Aboriginal Artefacts*. Rigby, Adelaide.

McCoy, P. C. and R. Gould 1977. Alpine archaeology in Hawaii. *Archaeology* 30:234-243.

McNiven, I. J. 1992. Shell middens and mobility: The use of off-site faunal remains, Queensland, Australia. *Journal of Field Archaeology* 19:495-508.

McNiven, I .J. 1994. Technological organistion and settlement in southwest Tasmania after the glacial maximum. *Antiquity* 68:75-82.

McPherron, S. and Holdaway, S. 1996. *Entrer Trois Data Entry Software*. Department of Anthropology, University of Pennsylvania.

Magne, M. 1989. Lithic reduction stages and assemblage formation processes. In D. S. Amick and R. P. Mauldin (eds), *Experiments in Lithic Technology*, pp. 15-31. BAR International Series 528, Oxford.

Magne, M. and D. Pokotylo 1981. A pilot study in bifacial lithic reduction sequences. *Lithic Technology* 10:34-47.

Marks, A. E. 1988. The curation of stone tools during the Upper Pleistocene: A view from the Central Negev, Israel. In H. L. Dibble and A. Montet-White (eds), *Upper Pleistocene Prehistory of Western Eurasia*, pp. 275-286. University Museum Press, Philadelphia.

Marks, A. E., J. Shoker and J. Zilhao 1991 Raw material usage in the Paleolithic: the effects of local availability on selection and economy. In A. Montet-white and S. Holen (eds), *Raw Material Economies Among Prehistoric Hunter Gathers,* pp. 127-139. Publications in Anthropology No. 19, University of Kansas, Lawrence.

Mauldin, R. P. and D. S. Amick 1989. Investigating patterning in debitage from experimental bifacial core reduction. In D. S. Amick and R. P. Mauldin (eds), *Experiments in Lithic Technology*, pp. 67-88. BAR International Series 528, Oxford.

Meltzer, D.J. 1989. Was stone exchanged among Eastern North American Paleoindians? In C. J. Ellis and J. C. Lothrop (eds), *Eastern Paleoindian Lithic Resource Use*, pp. 99-138. Westview Press, Boulder.

Montet-White, A. and S. Holen (eds) 1991. *Raw Material Economic Among Prehistoric Hunter Gathers*. Publications in Anthropology No. 19, University of Kansas, Lawrence.

Morrow, C .A. and R. W. Jefferies 1989. Trade or embedded procurement? A test case from Southern Illinois. In R. Torrence (ed.), *Time, Energy and Stone Tools*, pp. 27-33. Cambridge University Press, Cambridge.

Mueller-Wille, C. S. and D. B. Dickson 1991. An examination of some models of Late Pleistocene society in southwestern Europe. In G. A. Clarke (ed.), *Perspectives on the Past: Theoretical Biases in Mediterranean Hunter-Gatherer Research*, pp. 25-55. University of Pennsylvania Press, Philadelphia.

Mulaney, D. J. 1960. Archaeological excavations at Fromm's Landing on the lower *Murray River. Proceedings of the Royal Society of Victoria* 72:53-85.

Mulvaney, D. J. and E. B. Joyce 1965. Archaeological and geomorphological investigations on Mt. Moffatt Station, Queensland. *Proceedings of the Prehistoric Society* 31:147-212.

Murray, T. 1999. A return to the 'Pompeii premise'. In T. Murray (ed.), *Time and Archaeology*, pp. 8-27. Routledge, London.

Murray, T. and M. J. Walker 1988. Like what? A practical question of analogical inference and anthropological meaningfulness. *Journal of Anthropological Archaeology* 7:248-287.

Myers, A. 1989. Reliable and maintainable technological strategies in the Mesolithic of mainland Britain. In R. Torrence (ed.), *Time, Energy and Stone Tools*. pp.78-91. Cambridge University Press, Cambridge.

Nash, S. E. 1996. Is curation a useful heuristic? In G. H. Odell (ed.), *Stone tools: Theoretical Insights into Human Prehistory,* pp. 81-99. Plenum Press, New York.

Nelson, M. C. 1987. Site content and structure: metate quarries and workshops in the Maya Highlands. In B. Hayden (ed.), *Lithic Studies Among Contemporary Highland Maya*, pp. 120-147. University of Arizona Press, Tucson.

Nelson, M. C. 1991. The study of technological organization. *Method and Theory in Archaeology* 3:57-100.

New South Wales National Parks and Wildlife Service, 1996. *Sturt National Park: Plan of Management*. NSW National Parks and Wildlife Service, Hurstville.

O'Connell, J. F. 1977. Aspects of variation in Central Australian lithic assemblages. In R.V.S. Wright, (ed.), *Stone Tools as Cultural markers: Change, Evolution and Complexity*, pp. 269-81. Australian Institute of Aboriginal Studies, Canberra.

O'Connell, J. F., P. K. Latz and P. Barnett 1983. Traditional and modern plant use among the Alyawara of Central Australia. *Economic Botany* 37 80-109.

Odell, G. H. 1981. The morphological express at function junction: searching from meaning in lithic tool types. *Journal of Anthropological Research* 37:319-342.

Odell, G. H. 1989. Experiments in lithic reduction. In D. S. Amick and R. P. Mauldin (eds), *Experiments in Lithic Technology,* pp. 163-198. BAR International Series 528, Oxford.

Odell, G. H. 1996. Economizing behaviour and the concept of "curation". In G. H. Odell (ed.), *Stone Tools: Theoretical Insights into Human Prehistory,* pp. 51-80. Plenum Press, New York.

Ozbun, T. L 1991. Boulders to bifaces: initial reduction of obsidian at Newburry Crater, Oregan. *Journal of California and Great Basin Anthropology* 13(2):147-159.

Parry, W. J. and R. L. Kelly 1987. Expedient core technology and sedentism. In J. K.Johnson and C. A. Morrow (eds), *The Organization of Core Technology,* pp. 285-304. Westview Press, Boulder.

Pavilides, C. 1999 The story of Imlo: the organisation of flaked stone technologies from the lowland tropical rainforest of West New Britain, Papua New Guinea. Unpublished Ph.D. thesis, La Trobe University, Melbourne.

Pelcin, A. 1997. The formation of flakes: the role of platform thickness and exterior platform angle in the production of flake initiations and terminations. *Journal of Archaeological Science* 24:1107-1113.

Perles, C. 1992. In search of lithic strategies: a cognitive approach to prehistoric chipped stone assemblages. In J. Gardin and C. Peebles (eds), *Representations in Archaeology,* pp. 223-247. Indiania University Press, Bloomington.

Petraglia, M. D. and D. T. Nash. 1987. The impact of fluvial processes on experimental sites. In D. T. Nash and M. D. Petraglia (eds), *Natural Formation Processes and the Archaeological Record,* pp. 108-130. BAR International Series 352, Oxford.

Pigdon, D. J. 1996. GIS and Terrain Modelling in Open Site Archaeology. Identifying Fluvial Processes in an Open Site from North-Western NSW. Unpublished M.A. Sc. thesis, Department of Geomatics, University of Melbourne.

Prentiss, W. C. and E. J. Romanski 1989. Experimental evaluation of Sullivan and Rozen's debitage typology. In D. S. Amick and R. P. Mauldin (eds), *Experiments in Lithic Technology,* pp. 89-99. BAR International Series 528 Oxford.

Purdy, B. A. 1984. Quarry studies: technological and chronological significance. In J. Ericson and B. Purdy (eds), *Prehistoric Quarries and Lithic Production,* pp. 119-127. Cambridge University Press, Cambridge.

Ramenofsky A. F. and A. Steffen 1998. *Unit Issues in Archaeology: Measuring Time, Space, and Material.* University of Utah Press, Salt Lake City.

Reardom, M. 1995. *The Corner Country: Where Outback Queensland, New South Wales and South Australia Meet.* Australian Geographic Society, Sydney.

Reeves, J. M. 1997. Hearths and Landscape: Creating an Archaeological Chronology for Stud Creek, NSW. Unpublished B.A. (Hons) thesis, Department of Archaeology, Latrobe University, Melbourne.

Reeves, J. M., S. Holdaway, D. Witter, P. Fanning and J. Shiner 1998. Open site radiocarbon chronologies in Sturt National Park, NSW. Poster presented at the Australian Archaeology Conference.

Reher, C. A. 1991. Large scale lithic quarries and regional transport systems on the high plains of East Wyoming. In A. Montet-white and S. Holen (eds), *Raw Material Economies Among Prehistoric Hunter Gathers,* pp. 251-284. Publications in Anthropology No. 19, University of Kansas, Lawrence.

Reid, I. and L. Frostick 1985 Arid zone slopes and their archaeological materials. In A. E. Pitty (ed.), *Themes in Geomorphology*, pp. 141-157. Croom Helm, Beckenham.

Renfrew, C. 1977. Alternative models for exchange and spatial distribution. In T. K. Earle and J. E. Ericson (eds), *Exchange Systems in Prehistory,* pp. 71-90. Academic Press, New York.

Rice, R. W., S. W. Freiman and P. F. Becher 1981. Grain size development of fracture energy in ceramics: I. A model for non-cubic materials, *Journal of the American Ceramic Society* 64:345-349.

Rice, R. W., S. W. Freiman and J. J. Jr. Mecholsky 1980. The development of strength-controlled fracture energy on the flaw-size to grain size ratio. *Journal of the American Ceramic Society* 63:129-136.

Rick, J. W. 1976. Downslope movement and archaeological intrasite spatial analysis. *American Antiquity* 41:86-107.

Ricklis, R. A. and K. A. Cox, 1993. Examining lithic technological organisation as a dynamic cultural subsystem: the advantages of an explicitly spatial approach. *American Antiquity* 58:444-461.

Robins, R. P. 1997. Patterns in the landscape: A case study in nonsite archaeology from southwest Queensland. *Memoirs of the Queensland Museum, Cultural Heritage Series* 1(1):23-56.

Root, M. 1992. The Knife River Flint Quarries: the organisation of stone tool production. Unpublished Ph.D. thesis, Washington State University, Washington.

Rose, G., P. O'Connell and I. A. Menzies, 1967. *Milparinka 1:250 000 Geological Sheet SH 54-7.* NSW Geological Survey, Sydney.

Rosen, S. 1997. *Lithics After the Stone Age: A Handbook of Stone Tools from the Levant.* Altamira Press, Walnut Creek, Canada.

Ross, A. 1981. An archaeological survey of Bunda Lake, Belarabon, western New South Wales. *Australian Archaeology* 12:17-26.

Ross, A. 1985. Archaeological evidence for population change in the middle to late Holocene in southeastern Australia. *Archaeology in Oceania* 20:81-89.

Ross, A., T. Donnelly and R. Wasson. 1992. The peopling of the arid zone: Human-environment interactions. In J. Dodson (ed.), *The Naïve Lands: Prehistory and Environmental Change in Australia and the South-west Pacific,* pp.76-114. Longman Cheshire, Melbourne.

Roth, W. E. 1897. *Ethnological Studies Among the North-West-Central Queensland Aboriginals.* Queensland Government Printer, Brisbane.

Sanders, W. T., J. R. Parsons, and R. S. Santley. 1979. *The Basin of Mexico Survey: Ecological Processes in the Evolution of a Civilisation.* Academic Press, New York.

Sappington, R. L. 1984. Procurement without quarry production: examples from southwest Idaho. In J. E. Ericson and B. A. Purdy (eds), *Prehistoric Quarries and Lithic Production*, pp. 23-34. Cambridge University Press, Cambridge.

Sassaman, K. E. 1994. Changing strategies of biface production in the South Carolina coastal plain. In P. C. Carr (ed.), *The Organization of North American Prehistoric Chipped Stone Tool Technologies*, pp. 99-117. Archaeology Series 7. International Monographs in Prehistory, Ann Arbor, Michigan.

Savage, S. H. 1990. GIS in archaeology. In K.S. Allen, S. W. Green and E. B.W. Zubrow (eds), *Interpreting Space: GIS and Archaeology*, pp. 22-32. Taylor and Francis, London.

Schick, K. D. 1986. *Stone Age Sites in the Making: Experiments in the Formation and Transformation of Archaeological Occurrences.* BAR International Series 319, Oxford.

Schick, K. D. 1987a. Modeling the formation of Early Stone Age artefact concentrations. *Journal of Human Evolution* 16:789-807.

Schick, K. D. 1987b. Experimentally-derived criteria for assessing hydrologic disturbance of archaeological sites. In D. T. Nash and M. D. Petraglia (eds), *Natural Formation Processes and the Archaeological Record,* pp.86-107. BAR International Series 352, Oxford.

Schiffer, M. B. 1972. Archaeological context and systemic context. *American Antiquity* 37:156-165.

Schiffer, M. B. 1975. The effects of occupation span on site content in the Cache River Archaeological Project. In M. B. Schiffer and J. H. House (eds) *The Cache River Archaeological Project, An Experiment in Contract Archaeology,* pp. 265-269. Research Series No. 8, Arkansas Archaeological Survey, Fayetteville.

Schiffer, M. B. 1976. *Behavioural Archaeology.* Academic Press, New York.

Schiffer, M. B. 1987. Formation processes of the Archaeological Record. University of New Mexico Press, Albuquerque.

Schiffer, M. B. 1996. Some relationships between behavioural and evolutionary archaeologists. *American Antiquity* 61:643-662.

Schlanger, N. 1994. Mindful technology: Unleashing the *chaine operatoire* for an archaeology of mind. In C. Renfrew and E. B. W. Zubrow (eds), *The Ancient Mind: Elements of Cognitive Archaeology,* pp.143-151. Cambridge University Press, Cambridge.

Seeman, M. F. 1994. Intercluster lithic patterning at Nobles Pond: a case study for "disembedded" procurement among early paleoindian societies. *American Antiquity* 59: 273-288.

Shackley, M. L. 1978. The behaviour of artefacts as sedimentary particles in a fluviatile environment. *Archaeometry* 20:55-61.

Sheets, P. D. 1975. Behavioural analysis and the structure of a prehistoric industry. *Current Anthropology* 16:369-391.

Sheets, P. D. and G. R. Muto 1972. Pressure blades and total cutting edge: an experiment in lithic technology. *Science* 175:632-634.

Shelley, P. H. 1990. Variation in lithic assemblages: an experiment. *Journal of Field Archaeology* 17:187-193.

Shermer, M. 1995. Exorcising Laplace's demon: Chaos and antichaos, history and metahistory. *History and Theory* 34(1):59-83.

Shott, M. J. 1986. Technological organization and settlement mobility: an ethnographic examination. *Journal of Anthropological Research* 42:15-51.

Shott, M. J. 1989. Technological organisation in Great Lakes Paleoindian assemblages. In C. J. Ellis and J. C. Lothrop (eds), *Eastern Paleoindian Lithic Resource Use*, pp. 221-238. Westview Press, Boulder.

Shott, M. J. 1992. Radiocarbon dating as a probabilistic technique: The Childers site and Late Woodland occupation in the Ohio valley. *American Antiquity* 57:202-230.

Shott, M. J. 1994. Size and form in the analysis of flake debris: review and recent approaches. *Journal of Archaeological Method and Theory* 1(1):69-110.

Shott, M. J. 1996. An exegesis of the curation concept. *Journal of Anthropological Research* 52:259-280.

Shott, M. J. 1997. Activity and formation as sources of variation in Great Lakes Paleoindian assemblages. *Midcontinenal Journal of Archaeology* 22(2):197-236.

Shott, M. J. 2003. Size as a factor in Middle Paleolithic assemblage variation in the Old World: A North American perspective. In N, Moloney and M. Shott (eds), *Lithic Analysis at the Millennium.* pp. 137-149. Archtype Press, London.

Sieveking, G. and M. Newcomer (eds) 1987 *The Human Uses of Flint and Chert: proceedings from the fourth International Flint Symposium.* Cambridge University Press, Cambridge.

Singer, C. A. 1984. The 63-kilometer fit. In J.E. Ericson and B.A. Purdy (eds), *Prehistoric Quarries and Lithic Production*, pp. 35-48. Cambridge University Press, Cambridge.

Singer, C. A. and J. E. Ericson, 1977. Quarry analysis at Bodie Hills, Mono County, New Mexico. In T. K. Earle and J. E. Ericson (eds), *Exchange Systems in Prehistory,* pp. 171-188. Academic Press, New York.

Simek, J. 1984. The k-Means Approach to the Analysis of Spatial Structure in Upper Palaeolithic Habitation Sites: Le Flageolit I and Pincevent Section 36. BAR International Series 205, Oxford.

Simek, J. F. 1994. The organisation of Lithic Technology and evolution: notes from the continent. In P. J. Carr (ed.), *The Organisation of North American Prehistoric Chipped Stone Tool Technologies,* pp. 118-122. International Monographs in Prehistory, Archaeological Series 7, Ann Arbor, Michigan

Smith, E. A. 1983. The application of optimal foraging theory to the analysis of hunter-gatherer group size. In B. Winderhalder and E. A. Smith (eds), *Hunter-Gatherer Foraging Strategies*, pp. 35-65. University of Chicago Press, Chicago.

Smith, M. A. 1986. The antiquity of seed grinding in Arid Australia. *Archaeology in Oceania* 21:29-39.

Smith, M. A. 1988. *The Pattern and Timing of Prehistoric Settlement in Central Australia.* Unpublished Ph.D. thesis, University of New England, Armidale.

Smith, M. A 1989. Seed gathering in inland Australia: Current evidence from seed-grinders on the antiquity of the ethnohistorical pattern of exploitation. In D. R. Harris and G. C. Hillman (eds), *Foraging and Farming. The Evolution of Plant Exploitation,* pp.305-317. Unwin Hyman, London.

Smith, M. A 1993. Biogeography, human ecology and prehistory in the sandridge deserts. *Australian Archaeology* 37:35-50.

Smith, M. A 1996. Prehistory and human ecology in Central Australia: an archaeological perspective. In S.R. Morton and D.J. Mulvaney (eds), *Exploring Central Australia: the Environment and the 1894 Horn Expedition* pp. 61-73. Surrey Beatty and Sons, Chipping Norton.

Smith, M. A. and B. J. Cundy 1985. Distribution maps for flaked stone points and backed blades in the Northern Territory. *Australian Aboriginal Studies* 2:32-37.

Snethkamp, F. E., C. A. Ebright and T. B. Serena. *The Blue Marsh Project: Archaeological Studies of the Late Archaic in the Pennsylvania Piedmont, Volume 1.* State University of New York, Binghamton and National Parks Service, Mid-Atlantic Office.

Soffer, O. 1991 Lithics and lifeways- the diversity in raw material procurement and settlement systems on the Upper Paleolithic East European Plain. In A. Montet-white and S. Holen (eds), *Raw Material Economies Among Prehistoric Hunter Gathers,* pp. 221-234. Publications in Anthropology No. 19, University of Kansas, Lawrence.

Soles, J. S. 1983. A Bronze Age quarry in Eastern Crete. *Journal of Field Archaeology* 10(1):33-46.

paulding, A, 1960. The dimensions of archaeology. In G. E. Dole and R. L. Carneiro (eds), *Essays in the Science of Culture in Honor of Leslie A. White*, pp.437-448. Crowell, New York.

Spence, M. W., J. Kimberlin and G. Habottle 1984. State-controlled procurement and the obsidian workshops of Teotihuacan, Mexico. In J.E. Ericson and B.A. Purdy (eds), *Prehistoric Quarries and Lithic Production*, pp. 97-105. Cambridge University Press, Cambridge.

Spencer, B. 1914. *The Native Tribes of the Northern Territory of Australia.* MacMillan, London.

Steffen, A., E. J. Skinner and P. W. Aimworth 1998. A view of a core, technological units and debitage analysis. In A. F. Ramenofsky and A. Steffen (eds), *Unit Issues in Archaeology, Measuring Time, Space and Material,* pp.131-146. Foundations of Archaeological Inquiry, University of Utah Press, Utah.

Stern, N. 1980. Taphonomy: some observations about its place in archaeology. Unpublished B.A. (Hons) thesis, University of Sydney, Sydney.

Stern, N. 1993. The structure of the Lower Pleistocene archaeological record. *Current Anthropology* 34:201-225.

Stern, N. 1994. The implications of time-averaging for reconstructing the land-use patterns of early tool-using hominids. *Journal of Human Evolution* 27:89-105.

Stern, N. 1995. The 'blue' tuff locality of Koobi Fora in northern Kenya: archaeological time and the record of past human behaviour. *The Artefact* 18:49-59.

Stevenson, C. M., L. C. Shaw and C. Cristino 1984. Obsidian procurement and consumption on Easter Island. *Archaeology in Oceania* 19:120-124.

Steward, J. 1937. Ecological aspects of southwestern society. *Anthropos* 32:87-104.

Steward, J. 1938. Basin-plateau aboriginal sociopolitical groups. *Bureau of American Ethnology Bulletin no. 120,* Washington, D.C.

Stockton, E. D. 1977. Pipeline survey of Aboriginal relics in northwestern New South Wales and southwestern Queensland. *Australian Arehaeology* 7:7-18.

Sullivan, A. P., III. and K. C. Rozen. 1985. Debitage analysis and archaeological interpretation. *American Antiquity* 50:755-779.

Sullivan, M. E. and S. Simmons, 1979. Silcrete: a classification for flaked stone artefact assemblages. *The Artefact,* 4:51-60.

Stoltman, J. B. 1978. Temporal models in prehistory: an example from Eastern North America. *Current Anthropology* 19(4):703-746.

Tait, M., 1998. Geology and landscape evolution of the Mt Wood Hills area, near Tibooburra, northwestern New South Wales. B.Sc. (Hons) thesis, Department of Earth Sciences, La Trobe University.

Tainter, J. A. 1998. Surface archaeology: perceptions, values, and potential. In A. P. Sullivan III (ed.), *Surface Archaeology*, pp. 168-179. University of New Mexico Press, Albuquerque.

Talent, J. A. 1970. *Minerals, Rocks and Gems: A Handbook for Australia.* Jacaranda Press, Melbourne.

Thomas, D. H. 1973. An empirical test for Steward's model of Great Basin settlement patterns. *American Antiquity* 38:155-176.

Thomas, D. H. 1975. Nonsite sampling in archaeology: Up the creek without a site? In J. W. Mueller (ed.), *Sampling in Archaeology,* pp. 61-81. University of Arizona Press, Tucson.

Thomson, D. 1983. Donald Thomson in Arnhem Land. Currey O'Neil, Melbourne.

Tindale, N. 1965. Stone implement making among the Nakako, Ngadadjara and Pijandjara of the Great Western Desert. *Records of the South Australian Museum* 15 (1):131-164.

Tindale, N. 1974. *Aboriginal Tribes in Australia; Their Terrain, Environmental Controls, Distribution Limits and Proper Names; an Appendix on Tasmanian*

Tribes. Australian National University Press, Canberra.

Torrence, R. 1982. The obsidian quarries and their use. In C. Renfrew and M.Wagstaff (eds), *An Island Polity: the archaeology of Exploitation in Melos*, pp.193-221. Cambridge University Press, Cambridge.

Torrence, R. 1983. Time budgeting and hunter-gather technology. In G. Bailey (ed.), *Hunter-Gatherer Economy in Prehistory: A European Perspective,* pp. 11-22. Cambridge University Press, Cambridge.

Torrence, R. 1984. Monopoly of direct access: industrial organization at the Melos quarries. In J. E. Ericson and B. A. Purdy (eds), *Prehistoric Quarries and Lithic Production*, pp. 49-64. Cambridge University Press, Cambridge.

Torrence, R. 1986. *Production and Exchange of Stone Tools*. Cambridge University Press, Cambridge.

Torrence, R. 1989a. Tools as optimal solutions. In R. Torrence (ed.), *Time, Energy and Stone Tools*, pp.1-6. Cambridge University Press, Cambridge.

Torrence, R. 1989b. Retooling: towards a behavioural theory of stone tools. In R. Torrence (ed.), *Time, Energy and Stone Tools*, pp. 57-65. Cambridge University Press, Cambridge.

Torrence, R. 1994. Strategies for moving on in lithic studies. In P. C. Carr (ed.), *The Organization of North American Prehistoric Chipped Stone Tool Technologies*, pp. 123-131. Archaeology Series 7. International Monographs in Prehistory, Ann Arbor, Michigan.

Turner, M. and D. Bonica 1994. Following the flake trail: adze production on the Coromandel East Coast. *New Zealand Journal of Archaeology* 16:5-32.

van der Leeuw, S. E. and J. McGlade (eds) 1997. *Time, Process and Structured Transformation in Archaeology.* Routledge, London.

Vehik, S. C. 1985a. *Lithic Resource Procurement: Proceedings from the Second Conference on Prehistoric Chert Exploitation.* Occassional Paper No. 4, Center for Archaeological Investigations, Southern Illinois University, Carbondale.

Vehik, S. C. 1985b. Late prehistoric settlement strategy and exploitation of Florence-A chert. In S. C. Vehik (ed.), *Lithic Resource Procurement: Proceedings from the Second Conference on Prehistoric Chert Exploitation*, pp. 81-98. Occasional Paper No. 4, Center for Archaeological Investigations, Southern Illinois University, Carbondale.

Veth. P. M. 1989. Islands in the Interior: a model for the colonisation of Australia's arid zone. *Archaeology in Oceania* 24:81-92.

Veth, P. M. 1993. *Islands in the Interior. The Dynamics of Prehistoric Adaptations within the Arid Zone of Australia.* International Monographs in Prehistory, Archaeological Series 3, Ann Arbor, Michigan.

Veth, P. M. 1995b. Marginal returns and fringe benefits: Characterising the prehistory of the lowland deserts of Australia (a reply to Smith) *Australian Archaeology* 40:32-38.

Wandsnider, L. 1992. The spatial dimension of time. In J. Rossignol and L. Wansnider (eds), *Space, Time and Archaeological Landscape*, pp. 257-282. Plenum Press, New York.

Wandsnider, L. 1996. Describing and comparing archaeological spatial structures. *Journal of Archaeological Method and Theory* 3(4):319-384.

Wandsnider, L. 1998. Regional scale processes and archaeological landscape units. In A. F. Ramenofsky and A. Steffen (eds), *Unit Issues in Archaeology: measuring time, space, and material* pp. 87-102. University of Utah Press, Salt Lake City.

Watchman, A. and R. S. Freedman 1978. Trace element analysis of Aboriginal greenstone artefacts to discriminate between the sources, Mount Camel and Mount William in Southeastern Australia. *Proceedings of the Royal Society of Australia* 90(2):27[illegible]-282.

Watts, S. H. 1974. The nature and occurrence of silcrete in a portion of inland Australia. Unpublished Ph.D. thesis, University of Sydney, Sydney.

Watts, S. H. 1978. The nature and occurrence of silcrete in the Tibooburra area of northern western New South Wales. In T. Langford-Smith (ed.), *Silcrete in Australia*, pp.167-186. Department of Geography, University of New England, Armidale.

Webb, C. 1993. The lithification of a sandy environment. *Archaeology in Oceania* 28:105-111.

Webb, C. 1996. The formation of a surface archaeological record in semi-arid Australia. Unpublished Ph.D. thesis, La Trobe University.

Whallon, R. 1973. Spatial analysis of occupation floors I: Application of dimensional analysis of variance. *American Antiquity* 38:266-278.

Whallon, R. 1978. Spatial analysis of Mesolithic occupation floors: A reappraisal. In P. Mellars (ed.), *The Early Postglacial Settlement of Northern Europe: An Ecological Perspective,* pp. 27-35. Duckworth, London.

White, J. P. and J. O'Connell 1982. *A Prehistory of Australia, New Guinea and Sahul.* Academic Press, Sydney.

Whittaker, B. N., R. N. Singh and G. Sun 1992. *Rock Fracture Mechanics. Principles, Design and Application.* Elsevier, Amsterdam.

Whittaker, J. C. 1987. Individual variation as an approach to economic organisation: projectile points at Grasshopper pueblo, Arizona. *Journal of Field Archaeology* 14(4):465-480.

Whittaker, J. C. 1994. *Flintknapping: Making and Understanding Stone Tools*. University of Texas Press, Austin.

Wilke, P. J. and A. B. Schroth, 1989. Lithic raw material prospects in the Mojave desert, California. *Journal of California and Great Basin Anthropology* 11(2):147-174.

Willey, G. R. 1953. *Prehistoric Settlement Patterns in the Viru Valley, Peru.* Smithsonian Institution Bureau of American Ethnology Bulletin 155, United States Government Printing Office, Washington.

Williams, E. 1987. Complex Hunter-Gatherers: a view from Australia. *Antiquity* 61:310-321.

Wilson, J. M. 1994. The use of Aboriginal axe quarries in the Tamworth region and their socio-economic context when seen in relationship to the Daruka "long transfer quarry". Unpublished B.A. (Hons) thesis, Department of Archaeology and Paleoanthropology, University of New England, Armidale.

Winterhalder, B 1981. Optimal foraging strategies and hunter-gather research in anthropology: theory and models. In B. Winderhalder and E. A. Smith (eds), *Hunter-Gatherer Foraging Strategies*, pp. 13-35. University of Chicago Press, Chicago.

Winterhalder, B. and E. A. Smith (eds) 1981. *Hunter-Gatherer Foraging Strategies*. University of Chicago Press, Chicago.

Witter, D. C 1992. Regions and Resources. Unpublished Ph.D. thesis, Department of Prehistory, Research School of Pacific Studies, Australian National University, Canberra.

Witter, D. C. 1995. A classification of Australian stone artefacts and principals of taxonomy. Unpublished manuscript presented at the Australian Lithics Conference, Australian Museum.

Wobst, H. M. 1978. The archaeo-ethnology of hunter-gatherers or the tyranny of the ethnographic record in archaeology. *American Antiquity* 43:303-309.

Wopfner, H. 1978. Silcretes of northern South Australia and adjacent regions. In T. Langford-Smith (ed.), *Silcrete in Australia,* pp. 93-141. Department of Geography, University of New England, Armidale.

Yerkes, R. W. and N. P. Kardulias 1993. Recent developments in the analysis of lithic artefacts. *Journal of Archaeological Research* 1 (2):89-119.

Yellen, J. E 1977. *Archaeological Approaches to the Present: Models for Reconstructing the Past.* Academic Press, New York.

Zvelebil, M. S. W. Green and M. G. Macklin. 1992. Archaeological landscapes, lithic scatters and human behavior. In J. Rossignol and L. Wandsnider (eds), *Space, Time, and Archaeological Landscapes*, pp.193-226. Plenum Press, New York.